**Benchmark Papers
in Geology / 26**

A BENCHMARK® Books Series

LOESS
Lithology and Genesis

Edited by

IAN J. SMALLEY
The University of Leeds

Dowden, Hutchinson & Ross, Inc.

STROUDSBURG, PENNSYLVANIA

Distributed by

HALSTED
PRESS

A division of
John Wiley & Sons, Inc.

Copyright © 1975 by **Dowden, Hutchinson & Ross, Inc.**
Benchmark Papers in Geology, Volume 26
Library of Congress Catalog Card Number: 75-30690
ISBN: 0-470-79901-3

77 76 75 1 2 3 4 5
Manufactured in the United States of America.

LIBRARY OF CONGRESS CATALOGING IN PUBLICATION DATA

Main entry under title:

Loess : lithology and genesis.

   (Benchmark papers in geology ; v. 26)
   Includes indexes.
   1. Loess--Addresses, essays, lectures. I. Smalley,
Ian J.
QE471.2.L63   551.3'7'5   75-30690
   ISBN 0 470-79901-3

Exclusive Distributor: **Halsted Press**
A Division of John Wiley & Sons, Inc.

ACKNOWLEDGMENTS
AND PERMISSIONS

ACKNOWLEDGMENTS

AMERICAN ASSOCIATION FOR THE ADVANCEMENT OF SCIENCE—*Science*
  Source and Deposition of Clay Minerals in Peorian Loess

AMERICAN JOURNAL OF SCIENCE (YALE UNIVERSITY)—*American Journal of Science*
  On the Delta and Alluvial Deposits of the Mississippi, and Other Points in the Geology
  of North America, Observed in the Years 1845, 1846

THE CLARENDON PRESS, OXFORD—*The Journal of Soil Science*
  Frost Soils on Mount Kenya, and the Relation of Frost Soils to Aeolian Deposits

GEOLOGICAL SOCIETY OF AMERICA—*Bulletin of the Geological Society of America*
  Lower Mississippi Valley Loess

GEOLOGICAL SOCIETY OF FINLAND—*Bulletin of the Geological Society of Finland*
  Stratigraphy and Material of the Loess Layers at Mende, Hungary

PERMISSIONS

The following papers have been reprinted or translated with the permission of the authors
and copyright holders.

ACADEMIC PRESS, INC.—*Quaternary Research*
  End of the Last Interglacial in the Loess Record

AKADEMISCHE VERLAGSGESELLSCHAFT GEEST & PORTIG KG—*Gerlands Beitraege zur
    Geophysik*
  The Origin of Loess

AMERICAN JOURNAL OF SCIENCE (YALE UNIVERSITY)—*American Journal of Science*
  Distribution Curves for Loess
  Glacial Versus Desert Origin of Loess
  Loess Types and Their Origin
  A Mechanical Analysis of Wind-Blown Dust Compared with Analyses of Loess

BRITISH LIBRARY, LENDING DIVISION—translation from *Pochvovedenie*
  V. V. Dokuchaev and the Loess Problem

ELSEVIER SCIENTIFIC PUBLISHING CO.—*Earth-Science Reviews*
  "In-Situ" Theories of Loess Formation and the Significance of the Calcium-Carbonate
  Content of Loess

Acknowledgments and Permissions

GEBRÜDER BORNTRAEGER VERLAGSBUCHHANDLUNG—*Zeitschrift für Geomorphologie*
  The Presence of Loess in Southeastern Spain (*summary*)

GEOLOGICAL SOCIETY OF POLAND—*Rocznik Polskiego Towarzystwa Geologicznego*
  Remarks on the Loess
  Wind Directions During the Accumulation of the Younger Loess in East-Central Europe
  (*summary*)

IOWA ACADEMY OF SCIENCE—*Proceedings of the Iowa Academy of Science*
  Cation Exchange Capacity of the Clay Fraction of Loess in Southwestern Iowa

NEW YORK ACADEMY OF SCIENCES—*Transactions of the New York Academy of Sciences*
  The Interaction of Great Rivers and Large Deposits of Primary Loess

POLISH ACADEMY OF SCIENCE—*Bulletin de L'Académie Polonaise des Sciences: Série
  des Sciences Géologiques et Géographiques*
  Influence of Capillary Ground Moisture on Eolian Accumulation of Loess

POLISH SCIENTIFIC PUBLISHERS
  Acta Universitatis Wratislaviensis Studia Geograficzne
  Loess Sedimentation in Poland (*summary*)
  Biuletyn Peryglacjalny
  The Problem of the Origin of Loess in Poland
  Report of the Sixth INQUA Congress in Warsaw, 1961
  Loess Genesis and Soil Formation

ROYAL GEOGRAPHICAL SOCIETY—*Geographical Journal*
  Loess of Central Asia
  Recent Observations on the Loess of North China

SOCIETY OF ECONOMIC PALEONTOLOGISTS AND MINERALOGISTS—*Journal of
  Sedimentary Petrology*
  The Formation of Fine Particles in Sandy Deserts and the Nature of "Desert" Loess
  Loess, an Eolian Product
  Loess, an Eolian Product: Discussion
  The Nature and Origin of Argentine Loess
  Origin of Quartz Silt
  Origin of Quartz Silt: Comments on a Note by Ph. H. Kuenen
  Petrographic Comparison of Some Loess Samples from Western Europe with Kansas
  Loess
  Some Remarks on the Mineral Epidote in Connection with the Loess Problem

SOIL SCIENCE SOCIETY OF AMERICA—*Soil Science Society of America Proceedings*
  Composition of Alluvial Deposits Viewed as Probable Source of Loess
  Loess Distribution from a Source

UNIVERSITY OF CHICAGO PRESS—*The Journal of Geology*
  An Observation on Wind-Blown Silt
  Wind—The Dominant Transportation Agent Within Extramarginal Zones to Continen-
  tal Glaciers

UNIVERSITY OF NEBRASKA PRESS—*Loess and Related Eolian Deposits of the World*
  The Loess Environment in Central Europe
  The Origin of Loesses and Their Relation to the Great Plains in North America

WEIZMANN SCIENCE PRESS OF ISRAEL—*Israel Journal of Earth Science*
  Petrography and Origin of the Loess in the Be'er Sheva Basin

THE WILLIAMS & WILKINS COMPANY—*Soil Science*
  Background of Model for Loess-Derived Soils in the Upper Mississippi River Basin
  Depth of Loess and Distance from Source

SERIES EDITOR'S PREFACE

The philosophy behind the "Benchmark Papers in Geology" is one of collection, sifting, and rediffusion. Scientific literature today is so vast, so dispersed, and, in the case of old papers, so inaccessible for readers not in the immediate neighborhood of major libraries that much valuable information has been ignored by default. It has become just so difficult, or so time consuming, to search out the key papers in any basic area of research that one can hardly blame a busy man for skimping on some of his "homework."

This series of volumes has been devised, therefore, to make a practical contribution to this critical problem. The geologist, perhaps even more than any other scientist, often suffers from twin difficulties—isolation from central library resources and immensely diffused sources of material. New colleges and industrial libraries simply cannot afford to purchase complete runs of all the world's earth science literature. Specialists simply cannot locate reprints or copies of all their principal reference materials. So it is that we are now making a concerted effort to gather into single volumes the critical material needed to reconstruct the background of any and every major topic of our discipline.

We are interpreting "geology" in its broadest sense: the fundamental science of the planet Earth, its materials, its history, and its dynamics. Because of training and experience in "earthy" materials, we also take in astrogeology, the corresponding aspect of the planetary sciences. Besides the classical core disciplines such as mineralogy, petrology, structure, geomorphology, paleontology, and stratigraphy, we embrace the newer fields of geophysics and geochemistry, applied also to oceanography, geochronology, and paleoecology. We recognize the work of the mining geologists, the petroleum geologists, the hydrologists, the engineering and environmental geologists. Each specialist needs his working library. We are endeavoring to make his task a little easier.

Each volume in the series contains an Introduction prepared by a specialist (the volume editor)—a "state of the art" opening or a summary of the object and content of the volume. The articles, usually some thirty to fifty reproduced either in their entirety or in significant extracts, are selected in an attempt to cover the field, from the key papers of the last century to fairly recent work. Where the original works are in foreign languages, we have endeavored to locate or commission translations. Geologists, because of their global subject, are often acutely aware of the oneness of our world. The selections cannot, therefore, be restricted to any one country, and whenever possible an attempt is made to scan the world literature.

To each article, or group of kindred articles, some sort of "highlight commentary" is usually supplied by the volume editor. This commentary should serve to bring that article into historical perspective and to emphasize its particular role in the growth of the field. References, or citations, wherever possible, will be reproduced in their entirety—for by this means the observant reader can assess the background material available to that particular author, or, if he wishes, he, too, can double check the earlier sources.

A "benchmark," in surveyor's terminology, is an established point on the ground, recorded on our maps. It is usually anything that is a vantage point, from a modest hill to a mountain peak. From the historical viewpoint, these benchmarks are the bricks of our scientific edifice.

RHODES W. FAIRBRIDGE

PREFACE

In the bibliography of his 1934 monograph, Scheidig listed over 600 papers but did not pretend that it was an exhaustive coverage of the literature. Publication has been going on at an accelerating rate since then, so there must be in existence at least several thousand units of loess literature. From this huge collection a choice of 50 items has been made. The restriction to lithology and genesis has reduced the field slightly, but it is obvious that much good material has had to be omitted. Since the choice has had to be so selective, a few words of explanation are in order. The choice is conventional rather than idiosyncratic, and it is hoped that each paper offers significant information about loess as a sediment. It must be realized, however, that this represents a subjective opinion by the editor, who is obviously influenced by his intellectual environment.

This essay has been written in English in Leeds, only some 600 km from the nearest major loess deposit in Europe but some 7000 km from the center of English-speaking loess research. The selection and the introduction would probably have been very different had they been produced in Wrocław or Prague or Moscow; perhaps not quite so different had they originated from Christchurch or Iowa City. The European loess literature of recent years appears to have been concentrated, by and large, on the problems of loess stratigraphy; and the center of gravity of loess-material investigations has been settled in the United States ever since the late 1940s and early 1950s, when Russell sparked off that great burst of activity.

The collection herein is designed to include something significant about each of the world's major loess deposits, although the unevenness of the choice will be apparent immediately. It is regretted that the Chinese literature is not better represented, because the great deposits of Kansu, Shensi, and Shansi are in one sense the cradle (along with the valley of the Rhine) of modern loess studies. Russian literature is also underrepresented; in the 1964 edition of Berg's book there are listed 452 Russian references and 103 in other languages. It is a regrettable but unavoidable fact that investigators are much more profoundly influenced by papers in their own language than by papers in other languages, and the truth of this is probably reflected in the selection of papers for this collection. Lugn (1962), in his *The Origin and Sources of Loess,* cites 74 references, every one in English; and Popov (1972), discussing the same topic, cites 17 references, all in Russian. Thus are we divided.

One might consider the present collection as presenting essentially a selection of American and European literature, with the papers from Poland, Germany, and so on, being considered part of the same tradition. The items are presented in roughly chronological order, with exceptions where obvious associations exist that make other juxtapositions desirable. Many of the papers have been shortened, often by the omission of illustrations.

Various friends and coworkers have contributed ideas and suggestions to this collection: I am particularly grateful to Jerzy Cegla, University of Wrocław; J. C. Frye, recently retired from the Illinois State Geological Survey; Vojen Lozek, Czechoslovak Academy of Sciences; Marton Pecsi, Hungarian Academy of Sciences; Roy Simonson, U.S. Department of Agriculture; and Dan Yaalon, Hebrew University of Jerusalem. The undertaking was carried out at the suggestion of Rhodes Fairbridge, Columbia University, whose assistance at all stages is acknowledged. The responsibility for the final selection is entirely mine. The last word, and some justification for the collection and the lithological approach to loess problems, is provided by Julian Tokarski: "There is no doubt that each fragment of any rock contains in its mineral and chemical composition and structure the history of its origin. The reading of this history is only a matter of using the correct methods of investigation."

I. J. SMALLEY

CONTENTS

Contents

Contents

CONTENTS BY AUTHOR

INTRODUCTION

In his great compilation of Quaternary data, Charlesworth (1957, p. 511) described loess as "by far the most important periglacial accumulation," and it has generated a commensurate amount of published material. Charlesworth himself appended 709 references to his chapter on loess. Flint (1957, p. 181) observed that loess is "one of the most remarkable of the Pleistocene deposits, around which an extensive literature accompanied by much controversy has accumulated...." This book is an attempt to select from this vast literature 50 significant papers and extracts. It was obviously impossible to adequately cover the entire enormous field of loess-related interests, and a fundamental distinction has been made. Investigations on loess fall basically into one of two categories: (1) the lithology and genesis of loess and (2) loess stratigraphy. This collection is concerned with category 1, although it will be appreciated that it is impossible to totally divorce the two fields; therefore, material of stratigraphic interest and relevance is included.

Figure 1 has been adapted from Pettijohn et al.'s (1965) pictorial description of the aspects of sand study and shows various possible approaches to loess. In this collection we attempt to concentrate on aspects 2 and 6 and to suppress aspects 1 and 7.

In 1934 Alfred Scheidig's book *Der Löss und Seine geotechnischen Eigenschaften* was published. This work, perhaps the most influential monograph published on loess, concentrates on the engineering geology of loess (aspect 7). Although difficult to locate, somewhat out of date, and difficult for nonfluent readers of German, it remains a useful basis

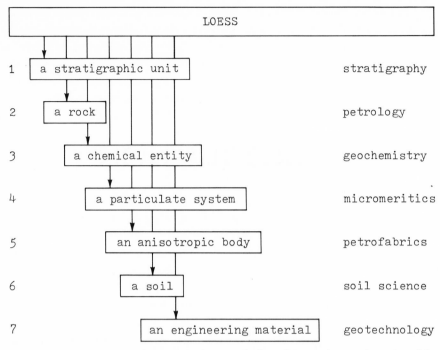

Figure 1. Approaches to loess, based on a similar diagram for sand produced by Pettijohn et al. (1965, fig. 1-3).

for the comprehensive study of loess sedimentology. It is hoped that this collection of papers will complement Scheidig's book and provide a fairly comprehensive review of developments from 1934 to the early 1970s. Only seven papers in this collection were published before 1934, and these were considered so significant that their omission was impossible. The remaining 43 items cover developments since 1934 and provide information about most of the significant loess developments and deposits in all parts of the world. Some apparent anomalies arise here because large deposits, such as that of the South American loess, have received comparatively little attention, whereas smaller deposits, thanks to accidents of development, have received considerable attention.

THE PROBLEM OF DEFINITION

To define loess is not easy, because there is still considerable discussion about what constitutes its truly characteristic features. The original German word *Löss* in its normal usage was simply a textural description, and a particular form of loose, crumbly earth was distinguished by this name. Subsequently, it was realized that this particular material was something special, in the geological sense, and exclusive additions have been made to the descriptive definition.

According to Flint (1957, p. 181), "Loess ... is a sediment, commonly nonstratified and commonly unconsolidated, composed dominantly of silt-size particles, ordinarily with accessory clay and sand, and deposited primarily by wind." This is a useful definition, but a compromise, because it includes elements from both "fundamental" types of definition. The Flint definition describes the material *and* suggests the way in which the deposit was formed. Russell (1944, Paper 9) argued earlier that the definition of loess should be rigid, descriptive, and "devoid of hypothesis." He deplored the tendency to incorporate a formation mechanism into the definition and cited, with disapproval, the definition by Smith and Norton (1935) who stated that loess is "the sediment which is believed to have been produced by the grinding action of glaciers, deposited in the bottom lands from streams carrying water from the melting glaciers and then picked up and redeposited on the uplands by the wind." This is, par excellence, a "mechanistic" definition of the sort favored by the writer of this introduction. The "descriptive" definition is, in this case, not scientifically satisfying.

One might describe a variety of volcanic rock (e.g., an ignimbrite) in careful descriptive terms, but to omit a mention of formation by volcanic action would seem absurd. A volcanic rock is essentially the product of a volcanic process, and the process is critical to its eventual state. Thus, it seems reasonable that process should feature in a satisfactory definition, or it would be reasonable if there were a consensus on the mode of loess formation. Russell (1944, Paper 9) propounding a fairly revolutionary theory, found the purely descriptive definition more useful. His is a long and careful definition:

> Loess is unstratified, homogeneous, porous, calcareous silt; it is characteristic that it is yellowish or buff, tends to split along vertical joints, maintains steep faces, and ordinarily contains concretions, and snail shells. From the quantitative standpoint at least 50 percent by weight must fall within the grain size fraction 0.01–0.05 mm, and it must effervesce freely with dilute hydrochloric acid.

He was at pains to emphasize the calcareous nature of the material, a factor not mentioned by Flint, and one which appears to be considered very important by proponents of *in situ* theories of loess formation.

In fact, carbonate content is a very poor loess indicator simply because it is so variable and so dependent on climatic conditions (this point has been made very clearly by Cegla, 1972; Paper 47). The constant material in loess is quartz, which is constant in both content and size. The process of loess formation has involved the production and supply of vast amounts of silt-sized quartz particles; this requires a geological event of considerable magnitude, which might usefully be included in the definition. Thus, quartz particles seem to define loess much more effectively than the carbonate content, and this fact was

3

used by Smalley and Vita-Finzi (1968, Paper 39) in their short definition: "Loess is a clastic deposit which consists predominantly of quartz particles 20–50 microns in diameter and which occurs as wind-laid sheets." This is a useful definition provided one believes in the eolian theory of loess deposit formation.

Useful working definitions have been provided by the Loess Commission of INQUA (the International Union for Quaternary Research). These were formulated to aid in the preparation of a loess map of Europe, and are summarized in Table 1 (see also Fink, 1974).

PROBLEMS OF LOESS FORMATION

The classical loess problem concerned the mode of formation of loess deposits, and many geologists would now contend that this has been solved. It will in fact be assumed that the critical event in the formation of a loess deposit is the transportation of silt-sized particles by the wind. This "loess problem" is actually only one of several concerned with the lithology and genesis of loess; the most critical questions appear to be the following.

1. How were the predominant silt-sized quartz particles formed?

2. How can the carbonate and clay mineral content of the loess be accounted for?

3. How were the deposits formed; that is, how were the particles transported and/or emplaced?

4. What significant events occur after the initial deposit has been formed?

The third of these problems has received by far the greatest amount of attention, because it is essentially a geological problem. The discovery is the deposit; the initial explanation sought was how the deposit was formed. Thus, most "theories of loess formation" (like the 20 assembled by Druif; see Scheidig, 1934, p. 42) are in fact explanations of the formation of loess deposits, not loess material. But, of course, the material is critical for the formation of the deposit; if there is no source of silt-sized particles, a loess deposit cannot form.

After the initial deposit has formed there may be changes of a pedological nature and the calcium (and magnesium) carbonate content may vary enormously. Also, the initial deposit may be eroded and related secondary deposits formed from the loess material [see Smalley, 1972 (Paper 46) for a discussion of this point].

The history of a loess deposit may be seen as a sequence of events;

Table 1. INQUA Loess Commission definitions

Name	Definition/description and synonyms
Loess	German synonyms: *Löss, typischer Löss* ("typical loess"). Characteristics: the definitely dominant fraction of the sediment is within 60–20 μ m (coarse silt, very fine sand), unstratified, primarily calcareous, quite porous capillary network; on the whole, dry material is yellow, buff, brownish yellow.
Sandy loess	German synonyms: *Sandlöss, Flottsand, lössiger Sand, sandiger Löss.* Characteristics: mixture of grains sized 60–20 μm and 500–200 μ m (fine sand, medium sand); often the distribution of particle sizes show a major peak within the silt range and a lesser peak within the medium sand range; sometimes there is an equal distribution among silt, (very) fine sand, and medium sand fractions; very often they are unstratified or in thin beds, usually noncalcareous, not so porous as loess, color similar to loess.
Clay-loess, clayey loess, argillaceous loess	German synonyms: *Tonlöss, toniger Löss, tonreicher Löss.* Characteristics: peak particle size of the sediment is within the range from 60–20 μm with 25–30% of particles being smaller than 2 μm (clay size); unstratified, low porosity; similar carbonate content and color to loess.
Loess-like sediments	German synonyms: *Lössderivate, lössartige Sedimente.* General characteristics: the term covers primarily eolian material that has been moved or redeposited in various (secondary) processes (allochthonous loess-like sediments) and/or modified *in situ* (autochthonous loess-like sediments); relevant processes are:

"Deluvial" (colluvial) processes and solifluction:
   hill-washed loess, solifluction loess, solifluxion loess (German terms: *Solifluktionslöss, Fliesslöss, Berglöss, Hanglöss*).
Fluvial (proluvial) processes:
   brickearth, brickearth (German: *Schwemmlöss, subaquatischer Löss*).
Modification caused by cryoturbation:
   cryoturbation loess (German: *Kryoturbationslöss*).
Eluvial and pedogenic processes:
   loess loam (German: *Lösslehm, Gleylöss, Staublehm*
   ("dust loam"), Decklehm ("covering loam").
Thorough, intense pedogenic modification and transformation (redeveloping):
   "semi-pedoliths" and pedoliths" (these terms are proposed by M. Pecsi for lithified soils formed from loess material).

Loess-like sediments may have originated from either loess, sandy loess, or clay loess; in any case their porosity is less than that of the original material; great variation of carbonate content, some may be essentially noncalcareous; colors may differ considerably in particular cases.

the task of the investigator is to isolate and understand their nature and to attempt to assess their significance. A satisfactory study of loess lithology and genesis will only be obtained in relation to this framework of events. A better understanding of the sequence of events should also lead to some reconciliation between the extreme views of loess formation. It seems that the "pedological" school of loess investigators sees the material as a result of considerably more events than do the "geological" investigators.

It would mean an injustice to many people but one could propose to divide the views on loess into "pedological" and "geological." These two distinct views of loess are current, and as it happens the strongholds of the respective views are the USSR and the United States. Thus, statements like "Western investigators accept the eolian theory of loess deposition" are encountered, but are not desirable. The dichotomy exists, but efforts should be made to close the gap and reconcile the views of the *in situ*/pedological theorists and the eolian/geological theorists. At the root of the division is the definition problem; if a certain calcium carbonate content is required before a material can be called loess, the *in situ* stance is almost comprehensible. The test of the more extreme theories should be based on the formation of the quartz material; if a theory makes no realistic allowance for the formation of the quartz particles, it cannot be valid. This is the weakness with Berg's view, the most extreme pedological theory [Berg, 1932 (Paper 8), 1964].

Figure 2 is an attempt to show, in outline, a possible sequence of the important events in the formation of a loess deposit; more systematic attempts have been made elsewhere [Smalley, 1971 (Paper 45), 1972 (Paper 46)]. It is generally agreed that silt-sized quartz particles predominate in loess, and the first significant event is the formation of these particles. Other constituents are carbonates and clay minerals, and their origin and variation should be explained. Various transportation modes may operate, although it seems that eolian transportation is characteristic. And several types of deposit may be formed. A reasonable ambition for the worldwide loess investigator would be to develop a history of each major deposit with the major variables and events identified and discussed.

Loess deposits have many things in common and it is this fact that makes the term useful in a geological sense. It allows a systematic approach to deposits as widespread as China, Argentina, and New Zealand. The similarity of deposits suggests the operation of common processes, and the recognition of significant natural processes is part of the business of science. But there are also significant and crucial differences among loess deposits, and in the past, because of the identification of investigators with particular deposits, characteristics that might be best

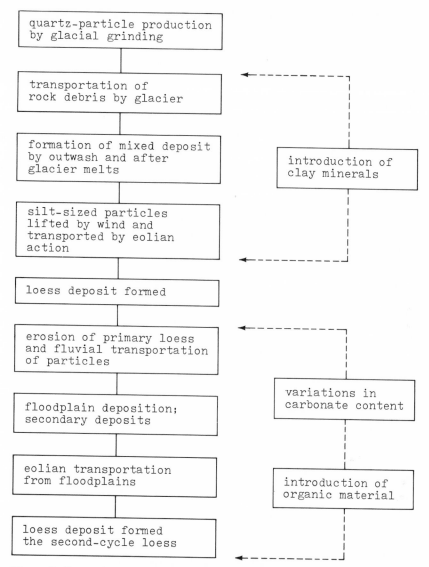

Figure 2. Events in the formation of loess deposits. Events in the right column cannot be assigned a definite place in the left sequence.

described as individual have been presented as general characteristics of loess deposits. To identify the special specific factors, at least the bare bones of a common framework for comparing loess deposits are necessary. For example, the Lower Mississippi Valley loess might be considered a special deposit because of its complex history, much more complex than the history of the Great Plains deposits from which it was derived. If the Mississippi loess were considered totally characteristic of the loess deposits of the world, a confusing picture could be produced.

GEOGRAPHICAL DISTRIBUTION OF LOESS DEPOSITS

There are large, readily identifiable loess deposits in Europe, the United States, South America, and China. There are smaller, sometimes disputable deposits in various other places. The question arises as to the significance, if any, of the distribution of loess material as loess deposits. Is loess, for example, a climatic indicator, so that the presence of a loess deposit records some particular paleoclimate, or is the climatic connection something of a delusion?

If there is a climatic connection, is it necessary to provide conditions for the production of loess material or for its transportation and emplacement, or both? It is apparent that there is a climatic connection but that it cannot be considered a direct effect. Loess deposits cannot form unless the characteristic material is available, and if this can only be produced in very large quantities by glacial grinding, there is a climatic connection. But the climate is primarily associated with the glaciers rather than with the loess, which appears more as a by-product. A cold, arid climate may be associated with glacial periods, and loess formation may be associated with glacial periods, but a direct connection between cold, arid climates and loess formation does not follow at all.

Many authorities [see Smalley and Vita-Finzi, 1968 (Paper 39), for a list] consider that there are two types of loess, hot and cold, or desert and periglacial, and that a distinction should be made between these two types when discussing loess distribution. Of the four major deposits listed, Charlesworth (1957, p. 541) recognized that three were periglacial and proposed that one, the Chinese, was of desert origin. If a deposit as large and significant as the Chinese loess were formed by a desert process, then that process would be of considerable importance and would merit equal status with the glacial–periglacial processes which gave rise to the three other major deposits. If, on the other hand, the Chinese loess is actually glacial material, any desert processes can only apply to relatively minor deposits and the desert process does not achieve equal rank with the glacial–periglacial process. Arguments against the existence of desert loess have been marshaled by Smalley and Vita-Finzi (1968, Paper 39) and its status has been defended by Yaalon (1969, Paper 31).

It should be noted that arguments about the possibility of forming silt-sized quartz grains may have to be modified in view of very significant observations on silt formation made by Moss et al. (1973). Their studies of quartz derived from granite revealed that many structural imperfections existed inside the granitic quartz crystals, and that these could be expected to facilitate breakage. Thus, although glacial grinding remains the most effective way of producing quartz silt, other methods deserve serious consideration. Brockie (1972) has reconsidered the possibility of frost shattering, and his preliminary results suggest that effective size reduction may be achieved by this agency.

REFERENCES

Berg, L. S. 1932. The origin of loess. Gerlands Beitr. Geophys. 35, 130–150.

——. 1964. Loess as a product of weathering and soil formation. Israel Program for Scientific Translations, Jerusalem, 205p.

Brockie, W. J. 1972. Experimental frost-shattering. Proc. 7th Geog. Conf. New Zealand Geographical Society, Hamilton, p. 177–186.

Cegla, Jerzy. 1972. Sedymentacja Lessow Polski. Acta Univ. Wratislaviensis No. 168 (Studia Geograficzne 17), Wroclaw, 71p.

Charlesworth, J. K. 1957. The Quaternary era, 2 vols. Arnold, London, 1,700p.

Fink, J. 1974. INQUA Loess Commission Circular Letter No. 11, Beilage 1 Quadrennial Report, p. 4–6.

Flint, R. F. 1957. Glacial and Pleistocene geology. Wiley, New York, 553 p.

Lugn, A. L. 1962. The origin and sources of loess. University of Nebraska Studies No. 26, Lincoln, Neb., 105p.

Moss, A. J., Walker, P. H., and Hutka, J. 1973. Fragmentation of granitic quartz in water, Sedimentology 20, 489–511.

Pettijohn, F. J., Potter, P. E., and Siever, R. 1965. Geology of sand and sandstone. Indiana Geological Survey and Indiana University, Bloomington, Ind., 205p.

Popov, A. I. 1972. Les loess et depots loessoides, produit des processus cryolithogenes. Biul. Peryglacjalny no. 21, p. 193–200.

Russell, R. J. 1944. Lower Mississippi Valley loess. Geol. Soc. America Bull. 55, 1–40.

Scheidig, Alfred. 1934. Der Löss und seine geotechnischen Eigenschaften. Steinkopf, Dresden, 233p.

Smalley, I. J. 1971. "In-situ" theories of loess formation and the significance of the calcium-carbonate content of loess. Earth Sci. Rev. 7, 67–85.

——. 1972. The interaction of great rivers and large deposits of primary loess. Trans. N.Y. Acad. Sci. 34, 534–542.

——, and Vita-Finzi, C. 1968. The formation of fine particles in sandy deserts and the nature of "Desert" loess. J. Sed. Petrol. 38, 766–774.

Smith, R. S., and Norton, E. A. 1935. Parent material of Illinois soils. *In* Parent materials, subsoil permeability and surface character of Illinois soils, p. 1–4. Illinois Agr. Exptl. Station, Urbana, Ill.

Yaalon, D. H. 1969. Origin of desert loess. Paper presented at 8th INQUA, Paris, 1969.

9

Editor's Comments
on Papers 1 Through 4

1   LYELL
    Excerpts from *Observations on the Loamy Deposit Called "Loess"*
    of the Basin of the Rhine

2   LYELL
    Excerpts from *On the Delta and Alluvial Deposits of the Missis-*
    sippi, and Other Points in the Geology of North America, Ob-
    served in the Years 1845, 1846

3   RICHTHOFEN
    On the Mode of Origin of the Loess

4   CHAMBERLIN
    Supplementary Hypothesis Respecting the Origin of the Loess of
    the Mississippi Valley

The scientific study of loess began in the nineteenth century, and by the end of the century a substantial loess literature existed. This great surge of publication was initiated by the studies made on the loess at Heidelberg by Karl Caesar von Leonhard (1824), an event that has been described by Kirchenheimer (1969). Loess became a material of prime geological importance owing to the observation and interest of Charles Lyell. The first phase of Lyell's geological development occupied the years up to the early 1830s (see Wilson, 1972), and *The Principles of Geology,* which established his reputation, was published in the period 1829–1833. Thus, by the time his first paper on loess was published in 1834 he exerted considerable influence and his work was widely read. The 1834 article (Paper 1) certainly introduced the material to many in the English-speaking world and brought the word "loess" into use.

Lyell visited America in 1841–1842 and again in 1845–1846. During this second visit he observed the Mississippi loess and recorded its likeness to the Rhine valley deposits. He enjoyed travel on the Mississippi steamers but was a little put out by the desire of his fellow passengers to look through his monocle. He was committed to the view that loess was a fluvial deposit, and in *The Antiquity of Man,* published in 1863, he stated:

> No doubt it is true that in every country, and at all geological periods, rivers have been depositing fine loam on their inundated plains . . . . But granting that loam presenting the same aspect has originated at different times and in distinct hydrographical basins, it is nevertheless true that during the glacial period the Alps were a great centre of dispersion, not only of erratics . . . but also of very fine mud . . . .

Lyell recognized the key role of glaciers in producing the material for loess deposits, even though he did not fully appreciate all the factors involved in loess transportation.

Mention of transportation, of course, immediately conjures up the image of Ferdinand von Richthofen, explorer extraordinary, and the man who "solved" the "loess problem." Controversy about loess in the nineteenth century concerned the "loess problem," the question of how loess material was emplaced in characteristic deposits. Richthofen established the eolian theory of loess deposition as the chief contender and, with a few notable exceptions, it is generally accepted. There is some dispute about who originated the eolian theory (see Paper 9), but there is no doubt that Richthofen established it as a major contribution to the study of loess. The theory appeared at a time of great geological activity; the volume of *Geological Magazine* that contained the first complete English statement of the eolian theory also contained the obituary for Charles Darwin (d. 1882), the papers by H. H. Howarth favoring a method of flood deposition, which provoked the contribution by Richthofen (Paper 3), a paper by S. V. Wood "On the Origin of the Loess," a rejoinder from Howarth, further remarks by Wood, a review of Richthofen's *China*, Volume 2, some suggestions by T. F. Jamieson, and so on.

A comprehensive theory of eolian deposition was first elaborated in Richthofen's major work *China: Ergebnisse eigner Reisen und darauf gegründeter Studien*, which was published in Berlin from 1877 to 1885. Soon after the *Geological Magazine* contribution appeared he was appointed professor of geography at the University of Leipzig, and in 1886 moved to a similar appointment in Berlin.

As the nineteenth century closed, the eolian hypothesis seemed to have accounted for the major problems of loess deposition; but various perceptive observers were not fully convinced. Paper 4 gives an end-of-the-century view of the loess problem by T. C. Chamberlin, with particular emphasis on the Mississippi valley loess. It appeared in the same volume of the *Journal of Geology* as Chamberlin's famous "Method of Multiple Working Hypotheses" and is, in fact, a good example of Chamberlin's proposed geological methodology. This allowed him to suggest that "The Mississippian loess, in its ultimate origin is glacial," a contention which was to be denied many times in the twentieth century.

REFERENCES

Kirchenheimer, Franz. 1969. Heidelberg und der Loss. Ruperto-Carola-Z. Ver. Freunde Studentschaft Universitat Heidelberg 46, 3–7.

Leonhard, K. C. von. 1823–1824. Charakteristik der Felsarten, 3 vols. Joseph Englemann Verlag, Heidelberg.

Wilson, L. G. 1972. Charles Lyell. The years to 1841: the revolution in geology. Yale University Press, New Haven, 553p.

1

Reprinted from *Edinburgh New Phil. Jour.*, **17**(33), 110–113, 118–120 (1834)

OBSERVATIONS ON THE LOAMY DEPOSIT CALLED "LOESS" OF THE BASIN OF THE RHINE

Charles Lyell

During the last summer, I had opportunities of examining the remarkable deposit called by the Germans " Loess," in several parts of the valley of the Rhine, between Cologne and Heidelberg, and also in some parts of the country of Baden, Darmstadt, Wurtemberg and Nassau. The observations made during this tour have caused me to modify some of the opinions which I formerly entertained and published respecting the probable origin and mode of deposition of this formation, and its relation to the newest volcanic products of the Lower Eifel. As much has been already written on this subject, I shall confine myself in this notice to what I saw during my late excursion, and shall give my observations nearly in the order in which I made them, pointing out afterwards the general conclusions to which they appear to me to lead.

13

It may be as well to state, that the Loess consists of a pulverulent loam of a yellowish grey colour, containing a certain quantity of carbonate of lime, according to Leonhard about a sixth part. When not associated with gravel it exhibits no signs of stratification. It contains almost everywhere imbedded terrestrial and aquatic shells of species still living in Europe, which have usually lost their colour, but are for the most part entire.

The Loess is found with its usual characters reposing here and there upon the gravel of the plains of the Rhine at Bonn, where I first examined it with attention, and patches of it are seen of much greater thickness on the flanks of the Siebengebirge, on the right bank, and at a corresponding height near the summit of the low hills which border the plain on the opposite bank. In all these localities terrestrial shells, chiefly Helix and Pupa, are by far the most abundant.

I employed a collector for a fortnight in obtaining shells from a deposit of Loess of considerable thickness, which is laid open on the right bank of the Rhine about a mile and a half below Bonn. The individual shells procured in an entire state amounted to 217 in number, not a seventh part of which were of aquatic species. The proportions were as follows:

Terrestrial—Helix 167, Pupa and Clausilia 18; 185 individuals; *Aquatic*—Lymnea 17, Paludina 10, Planorbis 5; 32 individuals;—217.

In order to compare these fossils with such shells as are now drifted down by the Rhine, I made a collection of the latter at low water from the mud and sand of the shore of the river for several miles above and below Bonn. Along the beach is a line of rubbish composed of small pieces of drift wood, leaves, weeds, sand, and other matter, cast up principally by the large waves raised by the steam-packets, as they cut through the water. Here the greater number of drift shells occur, and I collected 273 individuals which were in the following proportions.

Terrestrial—Helix 133, Pupa and Clausilia 12, Bulimus 2; 147 individuals; *Aquatic*—Paludina 48, Planorbis 34, Neritina 28, Lymnea and Succinea 5, Unio 6, Ancylus 3, Cyclas 2; 126 individuals;—273.

If I may be allowed to draw any general conclusion from this comparison, it would appear that, in the waters of the Rhine,

as in the loess, the drift-shells belong chiefly to terrestrial species, and in both the great mass of the shells are referable to the same genera, the principal difference consisting in the absence from the loess of species of the genera neritina, ancylus, and unio.   The only bivalve-shells I ever happened to meet with in the loess, were Cyclas fontinalis, *Drap* *.

It may be well to observe here, that, in some places where the bank of the river is wholly or partly composed of loess, the fossil shells are often washed out, and may be found entire on the shore; and they might, in such cases, unless great caution were used, be confounded with the more modern shells drifted down by the Rhine.   I was careful to guard against this source of error, by collecting chiefly from spots far from the loess, and by rejecting those which, by their want of colour, or by the circumstance of their being filled with loess, resembled the general characters of the fossils.   The colour of the more modern specimens affords in general a safe criterion for distinguishing them from the fossils; and I feel sure that there was scarcely any intermixture in the sets above compared, or only two or three lymnea, at least, were doubtful.

The greater part of the shells drifted by the Rhine agree specifically with those which are buried in the loess; and if I had enlarged my collection, the correspondence would no doubt have been much more perfect, for the shells of the loess vary in different localities, and those now brought down by the Rhine probably vary equally at different seasons.   As the drift shells of the Rhine agree with those of the loess, so the sediment of that river bears a very close resemblance to loess.   This was first pointed out to me by Professor Noegerath, and it has lately been confirmed by Mr Horner's experiments on the quantity and nature of the solid matter brought down in the waters of the Rhine at

---

* I found several specimens of this with both valves entire, together with Valvata piscinalis, in the interior of an individual of the Lymnea ovata, in loess at Odenau, near Bruchsal.   Hard calcareous concretions, in the same loess, contained shells of recent helix and clausilia, which were thus embedded in solid limestone.   In the third volume of my Principles of Geology, Appendix, p. 58, I included Cyclas palustris, and C. lacustris, Drap. in a list of loess shells; but I afterwards ascertained that they had been brought to the spot in mud used to fertilize the soil.   Probably they are to be found in loess.

Bonn *. The circumstance must, in part, be ascribed to the rapid degradation of loess, which is constantly going on throughout the valleys drained by the Rhine and its tributaries, but it also shews that the waste of other rocks in the same districts produces a sediment very similar in its nature to loess.

It is well known that the loess rests on the gravel of the plain of the Rhine. This superposition is well seen on the left bank of that river, about a mile above Bonn, where the loess fills up hollows in the gravel, and presents the appearance represented in the annexed sketch.

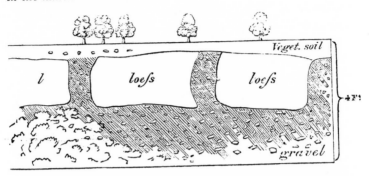

I conceive, that in this instance, small rills or torrents must first have furrowed the upper beds of gravel, leaving small trenches with vertical and occasionally overhanging walls, and then the waters holding loamy sediment in suspension must tranquilly have overflowed the spot and thrown down the loess until it first filled up the cavities, and then formed a continuous overlying mass.

[*Editor's Note:* Material has been omitted at this point.]

In the Spessart, and in the country immediately around Aschaffenburg, I observed no loess. The road which leads from Frankfort to the foot of the Taunus, passes first over the low flat plain of the Mayn, which is covered with yellow sand, for the most part very barren. (See section, No. 2.) At Höchst, on the Mayn, is a higher platform, composed of loess, and here the soil is extremely fertile. This platform afterwards rises to a still greater height between Höchst and Soden, which last town is situated in a valley cut through the loess, at the bottom of which the subjacent tertiary strata of the Mayence formation are laid open. On quitting Soden, I ascended the steep flanks of the Taunus mountains, and saw no loess. (See diagram, No. 2.)

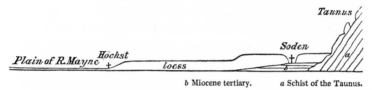

b Miocene tertiary.   a Schist of the Taunus.

I then crossed the highest part of the Taunus, where the greywacke passes into crystalline schists, and from thence descended towards Esch and Walsdorf, where the more ordinary greywacke of the Rhine, a yellow argillaceous and sandy rock, is very generally concealed under a deep covering of loam, which appears to have resulted from the decomposition of this greywacke, and not to have been transported from any distance. This loam has precisely the ordinary colour of the loess, and contains a great quantity of quartz pebbles.

The same alluvium is very general in the Westerwald, especially on the surface of that high table-land around Altenkirchen, Uckerath, and between that place and Siegburg, a district lying immediately behind the Siebengebirge.

No. 3.

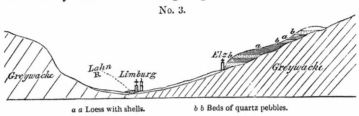

a a Loess with shells.   b b Beds of quartz pebbles.

The principal river which intersects the table-land of Nassau is the Lahn, which I crossed at Limburg, about twenty miles above its junction with the Rhine. The road from Limburg to Freilingen passes first by Elz. On the north of this village is a hill, which forms one boundary of the valley of the Lahn, and here loess is seen with all its usual characters, with many land and fresh-water shells; and alternating, as at Heidelberg, with gravel. I observed, in particular, a horizontal layer of white quartz peebles, a foot and a half in thickness, resting on a mass of loess fifteen feet thick, and covered by another bed of loess five feet in thickness; the loess, in both situations, including in it entire shells. Following the road, I found the slope of the hill above to consist of horizontal beds of quartz pebbles, which have a base of loess. Hence it appears that the valley of the Lahn, which is excavated through highly inclined greywacke, has, at some period since its excavation, been partially filled up with beds of gravel, alternating with loess, a great part of which has since been removed by denudation. (See Section, No. 3.) It appears that, during the accumulation of the mass, fine loam was sometimes thrown down, containing unbroken shells, then gravel, and then again the shelly loam.

On a review of the observations above mentioned, it appears

to me that the following conclusions may be legitimately deduced :—

1*st*, The loess is of the same mineral nature as the yellow calcareous sediment with which the waters of the Rhine are now commonly charged.

2*dly*, The fossil shells, contained in the loess, are all of recent species, consisting partly of land and partly of fresh-water shells.

3*dly*, The number of individuals belonging to land species usually predominates greatly over the aquatic, and this seems now to be the case with the modern shells drifted down by the Rhine.

4*thly*, Although the loess in general evinces no signs of stratification, we must yet suppose it to have been formed gradually, for the shells contained in it are very numerous, and almost all entire ; and sometimes beds of pure loess, fully charged with shells, alternate several times with strata of gravel, or of volcanic matter.

5*thly*, Although, in general, the loess overlies every formation, including the gravel of the plains of the Rhine, and the volcanic rocks, which have the most modern aspect, yet in some cases, as at Andernach, the volcanic matter is so interstratified as to indicate that some eruptions occurred during the deposition of loess.

These inferences seem to me sufficiently clear ; but if asked to account for the manner in which the loess, considering it as a fluviatile or lacustrine formation, was brought into the places which it now occupies, I must confess that the more I have studied the subject the more difficult I have found it to form a satisfactory theory.

[*Editor's Note:* Material has been omitted at this point.]

2

Reprinted from *Amer. Jour. Sci.*, **3**, Ser. 2, 34–35, 36–37 (1847)

ON THE DELTA AND ALLUVIAL DEPOSITS OF THE MISSISSIPPI, AND OTHER POINTS IN THE GEOLOGY OF NORTH AMERICA, OBSERVED IN THE YEARS 1845, 1846

Charles Lyell

THE delta of the Mississippi may be defined as that part of the great alluvial plain which lies below, or to the south of the branching off of the highest arm of the river, called the Atchafalaya. This delta is about 13,600 square miles in area, and elevated from a few inches to ten feet above the level of the sea. The greater part of it protrudes into the Gulf of Mexico beyond the general coast line. The level plain to the north, as far as Cape Girardeau in Missouri above the junction of the Ohio, is of the same character, including, according to Mr. Forshey, an area of about 16,000 square miles, and is, therefore, larger than the delta. It is very variable in width from east to west, being near its northern extremity, or at the mouth of the Ohio, 50 miles wide; at Memphis 30; at the mouth of the White River 80, and contracting again farther south, at Grand Gulf, to 33 miles. The delta and alluvial plain rise by so gradual a slope from the sea as to attain, at the junction of the Ohio, (a distance of 800 miles by the river,) an elevation of only two hundred

20

feet above the Gulf of Mexico. Mr. Lyell first described the low mud banks covered with reeds at the mouths of the Mississippi, and the pilot-station called the Balize ; then passed to the quantity of drift wood choking up the bayous, or channels, intersecting the banks; and, lastly, enlarged on the long narrow promontory formed by the great river and its banks between New Orleans and the Balize. The advance of this singular tongue of land has been generally supposed to have been very rapid, but Mr. Lyell and Dr. Carpenter, who accompanied him, arrived at an opposite conclusion. After comparing the present state of this region with the map published by Charlevoix, 120 years ago, they doubt whether the land has, on the whole, gained more than a mile in the course of a century.

[*Editor's Note:* Material has been omitted at this point.]

In attempting to compute the minimum of time required for the accumulation of the alluvial matter in the delta and valley of the Mississippi, Mr. Lyell referred to a series of experiments, made by Dr. Riddell, at New Orleans, showing that the mean annual proportion of sediment in the river was, to the water $\frac{1}{1245}$ in weight, or about $\frac{1}{1000}$ in volume. From the observations of the same gentleman, and those of Dr. Carpenter, and of Mr. Forshey, (an eminent engineer of Louisiana,) the average width, depth, and velocity of the Mississippi, and thence the mean annual discharge of water, are deduced. In assuming 528 feet (or the tenth of a mile) as the probable thickness of the deposit of mud and sand in the delta, Mr. Lyell founds his conjecture on the depth of the Gulf of Mexico, between the southern point of Florida and the Balize, which equals on an average 100 fathoms. The area of the delta being about 13,600 square statute miles, and the quantity of solid matter annually brought down by the river 3,702,758,400 cubic feet, it must have taken 67,000 years for the formation of the whole ; and if the alluvial matter of the plain above be 264 feet deep, or half that of the delta, it has required 33,500 more years for its accumulation,—even if its area be estimated as only equal to that of the delta, whereas it is, in fact, larger. If some deduction be made from the time here stated, in consequence of the effect of drift wood, which must have aided in filling up more rapidly the space above alluded to, a far more important allowance must be made, on the other hand, for the loss of matter, owing to the finer particles of mud not settling at the mouth of the river, but being swept out far to sea, and even conveyed into the Atlantic by the Gulf Stream. Yet the whole period during which the Mississippi has transported its earthy burthen to the ocean, though perhaps far exceeding 100,000 years, must be insignificant, in a geological point of view, since the bluffs or cliffs bounding the great valley, (and therefore older in date,) and which are from 50 to 250 feet in perpendicular height, consist in great part of loam, containing land, fluviatile, and lacustrine shells of species still inhabiting the same country. These fossil shells, occurring in a deposit resembling the *loess* of the

Rhine, are associated with the bones of the mastodon, elephant, tapir, mylodon, and other megatherioid animals; also a species of horse, ox, and other mammalia, most of them of extinct species. The loam rests at Vicksburg and other places on eocene or lower tertiary strata, which in their turn, repose on cretaceous rocks. A section from Vicksburg to Darien, through the States of Mississippi, Alabama and Georgia, exhibits this superposition, as well

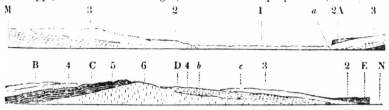

Section (M to N) about 750 miles in length from west to east, from Louisiana (on the west) through Jackson, Mississippi, to Tuscaloosa, in Alabama, and thence, by Montgomery, to the Atlantic, near Darien, in Georgia.*

as that of the cretaceous strata on carboniferous rocks at Tuscaloosa. Mr. Lyell ascertained that the huge fossil cetacean, named Zeuglodon, by Owen, is confined to the eocene deposits. In the cretaceous strata, the remains of the mosasaurus, and other reptiles, occur without any cetacea.

[*Editor's Note:* Material has been omitted at this point.]

* 1. The alluvium of the Mississippi.—2. Post-pliocene loam and sand, with recent shells and bones of extinct mammalia; the shells in this deposit, on the borders of the valley of the Mississippi, are of land and fresh-water species, those near Darien, of marine species.—3. Eocene formation.—4. Cretaceous strata.— 5. Carboniferous rocks.—6. Hypogene or granite, gneiss, mica schist, &c.

M to *a*. Louisiana.—A. Vicksburg, Mississippi.—B. Jackson.—C. Tuscaloosa, Alabama.—D. Montgomery.—E. Darien, Georgia.—N. Atlantic.

a. Mississippi river.—*b*. Chatahoochie river.—*c*. Flint river.

3

Reprinted from *Geol. Mag.*, **9**, Ser. 2, 293–305 (1882)

ON THE MODE OF ORIGIN OF THE LOESS

F. Richthofen

['T']HE following communication, although addressed to the Editor in the form of a letter, is of such importance that we need make no apology for treating it as an Original Article, feeling sure that our readers will be as much interested as ourselves in the observations of so eminent a geologist, who has spent many years in actual study of these vast deposits, as to the origin of which he is consequently able to speak with such profound knowledge and mature judgment. —EDIT. GEOL. MAG.]

SIR,—You will receive in a few days a copy of the second volume of my work on China, which I requested my publisher (Mr. Dietrich Reimer, of Berlin) to send to your address.[1] In this volume are embodied the results of my travels in *Northern* China, and it is chiefly devoted to the geology of that country. The third volume will comprise Southern China, while the fourth, which is to contain the description of a portion of the fossils collected by me, is now going through the press and will be ready within a few months. It will be accompanied by 52 plates. The palæontological memoirs

[1] Since received.—EDIT. GEOL. MAG.

in it have been prepared by Messrs. Dames, Kayser, Lindström, Schenk and Schwager. Unfortunately, the topographical and geological maps relating to Northern China (14 sheets), which will be published as the first portion of an Atlas of China, are not yet ready, and the second volume, to which this first series of maps belongs, must make its way for some time without their aid.

I hope that, if you should consider the book worthy of a special notice in the MAGAZINE, the reviewer will kindly take into consideration that, with the exception of the very able report of Pumpelly on a district of comparatively small extent in the vicinity of Peking, this is the first attempt towards the geological description of a vast region which, orographically as well as geologically, was entered by me as a *terra incognita*, and that as a solitary wanderer I did not enjoy the advantages offered to the geological member of a well-equipped expedition, who can devote all his energy to one single class of subjects, and is neither occupied with the construction of his own topographical maps, nor hampered by the daily-recurring care of pack-mules, carriage-bearers, etc.

The first volume of my work, which was published in 1877, has not been sent to the GEOLOGICAL MAGAZINE, because its contents were chiefly geographical and historical. One geological problem only was treated in it at considerable length; this is, the origin of the Loess and the mode of growth of the soil of steppes. It appears to me, therefore, quite natural that the book should have been taken notice of by only a few geologists. But I might have expected that a prominent scholar, who in his literary studies has moved over the same ground with me and, although with a far wider scope of learning than I acquired and at very much greater length than I was able to devote to the subject, has treated the history of Central Asia, should have at least glanced at the contents of the book. Such, however, has evidently not been done by Mr. H. H. Howorth, when he undertook to discuss the question of the origin of the Loess in two numbers of the MAGAZINE (January and February, 1882), and it appears that my publications on the subject have completely escaped his knowledge. All he knows respecting a theory on the mode of origin of the Chinese Loess, which I first advanced in 1870, is taken from the few lines of a foot-note of a paper by Mr. Kingsmill, who, in 1871, accepting the name of Loess, which I applied to the Chinese deposits, suggested a theory of a marine origin for them (Quart. Journ. Geol. Soc. vol. xxvii. pp. 376 to 383).[1] If the words in which Mr. Howorth mentions my theory did really render it, his

[1] The theory of the subaërial origin of the Loess, which, according to Mr. Howorth (p. 16), "has received the sanction of Richthofen and Pumpelly," but which, in fact, was started by me and endorsed by M. Raph. Pumpelly (*N. Y. Nation*, April 14, 1878), who had advocated before a fluviatile origin, was noticed first very briefly in my *Letter on the provinces of Honan and Shansi*, Shanghai, 1870, pp. 9-10, and at some greater length in my *Letter on the provinces of Chili, Shansi, Shensi*, etc., Shanghai, 1872, p. 13-18. The full discussion of the subject is given in *China*, vol. i. pp. 56 to 189, and a short abstract in *Verhandlungen der K. K. geologischen Reichsanstalt*, 1878, pp. 289 to 296. I could not avoid reverting to it repeatedly in *China*, vol. ii. (see for descriptions f. i. pp. 349-351, 422 427, 530-533, 550-551, and for discussion, pp. 741 to 766).

arguments against it would, at least in part, be well founded; but I consider it not improbable that he would not have started the controversy if he had taken the trouble to make himself acquainted with the subject against which it is directed.

Will you allow me, therefore, to offer to the readers of Mr. Howorth's article a short explanation of the views at which I arrived regarding the mode of origin of the Loess when I was gazing daily at its astounding deposits and grotesque features in the Chinese provinces of Honan and Shansi, views which I found not only corroborated during my further travels throughout all Northern China and in the Mongolian Steppes, but which, on the strength of comparative study, I was afterwards able to apply with equal force to Tibet, the region of Khotan and Yarkand, and great portions of south-western Asia, as well as to all Loess-covered regions of Europe, and of the continents of North and South America.

Any theory which undertakes to deal with the problem of the origin of the Loess must give a valid explanation of the following characteristic peculiarities of it, viz. :

1st. The petrographical, stratigraphical, and faunistic difference of the Loess from all accumulations of inorganic matter which have been deposited previously and subsequently to its formation, and are preserved to this day.

2nd. The nearly perfect homogeneousness of composition and structure, which the Loess preserves throughout all the regions in which it is found on the continents of Europe and Asia; it offers in this respect a remarkable contrast to all sediments proved to be deposited from water within the last geological epochs, excepting those of the deep sea, which are here out of the question.

3rd. The independence of the distribution of the Loess from the amount of altitude above sea-level. In China it ranges from a few feet to about 8000 feet above the sea,[1] and farther west it rises probably to much greater altitudes. In Europe it is known at all elevations up to about 5000 feet, at which it occurs in the Carpathians.

4th. The peculiar shape of every large body of Loess, as it is recognized where erosion has cut gorges through it down to the underlying ground without obliterating the original features of the deposit. These are different according to the hilly or level character of the subjacent ground. In hilly regions the Loess, if little developed, fills up depressions between every pair of lower ridges, and in each of them presents a concave surface; but where it attains greater thickness, it spreads over the lower hills, and conceals the inequalities of the ground. Its concave surface extends then over the entire area separating two higher ranges, in such a manner as to make the line of profile resemble the curve that would be produced by a rope stretched loosely between the two ranges.

---

[1] I met with it in China, in 1870, only at an altitude of 6000 feet, and this figure is given by Mr. Howorth (p. 76) erroneously as an observation of Mr. Kingsmill; in 1871 I found thick deposits of Loess at an elevation of 7000 feet in Southern Mongolia, and of 8000 feet on the Wu-tai-shan range in the province of Shansi.

This shape of surface is precisely similar to that which is character-istic of the salt steppes of Central Asia. It must, however, be re-marked that, just as in these, the development is frequently unequal on either side of a valley, and that the preponderance of the deposit on the same (f. i. the westerly) side can sometimes be observed in each basin throughout a larger region. The lowest portion of the surface of larger basins is frequently taken up by stratified soil consisting of the finest particles of Loess, and ex-hibiting a strong impregnation with alkaline salts. Over table lands and plains Loess is spread in the shape of most uniform sheets.

5th. The composition of pure Loess, which is the same from whatever region specimens may be taken, extremely fine particles of hydrated silicate of alumina being the largely prevailing in-gredient, while there is always present an admixture of small grains of quartz and fine laminæ of mica. It contains, besides, carbonate of lime, the segregation of which gives origin to the well-known concretions common to all deposits of Loess, and is always impregnated with alkaline salts. A yellow colouring matter caused by a ferruginous substance is never wanting.

6th. The almost exclusive occurrence of *angular* grains of quartz in the pure kinds of Loess.

7th. The complete absence of stratification. To this must be added the singular position of the laminæ of mica. When these are deposited by water, they are arranged horizontally and ac-cumulated in separate layers, while in Loess they are distributed without any order, and occur in every possible position.

8th. The capillary structure caused by the occurrence of innumer-able tubes, mostly incrustated with carbonate of lime, which have generally a vertical position, and ramify downwards like the roots of grass. Where Loess is covered by vegetation, the tubes may be seen taken up by rootlets to the depth of a foot or a few feet from the surface. In this internal structure, besides the mode of occurrence, is founded the chief difference of Loess from ordinary loam. The former may be designated as a kind of calcareous loam provided with internal structure.

9th. The tendency to vertical cleavage, which is the immediate consequence of the two last-named properties.

10th. The fact that land shells are imbedded in immense numbers throughout the Loess, and that the most delicate shells are perfectly preserved. Fresh-water shells are of extremely rare occurrence, as has been correctly pointed out by Mr. Howorth.

11th. The great quantity of bones of mammals found in the Loess, the genera and mostly the species, or the next relatives, of which are known to abound at present in steppes and on grassy plains. Herbivorous animals are represented as well as carnivorous preying on the former.

12th. The fact that wherever Loess fills a basin between hills, the inclined slopes of these are covered by angular fragments of the adjoining rock, on which the yellow soil rests. Layers of these fragments, beginning with a slight inclination and then passing into an horizontal position, extend from the hill-sides for some distance into the accumulation of the Loess itself, separating it in the neighbourhood of the encasing slopes into layers of varying thickness, while towards the central portion of each large basin this separation ceases almost completely, and the soil is very homogeneous from top to bottom, even in those instances where the vertical thickness is 1500 feet and more.

It is perfectly evident that no theory starting from the hypothesis of the deposition of Loess by water can explain all or any single one of these properties. Neither the sea nor lakes nor rivers could deposit it in altitudes of 8000 feet on hill-sides. Origin from water is perfectly unable to explain the lack of stratification, the profuse existence of capillary tubes, the vertical cleavage, the promiscuous occurrence of grains of quartz, the angular shape of these, the confused position of the laminæ of mica, the imbedding of land shells, and of bones of terrestrial mammals.

There is but one great class of agencies which can be called in aid for explaining the covering of hundreds of thousands of square miles, in little interrupted continuity, and almost irrespective of altitude, with a perfectly homogeneous soil. It is those which are founded in the energy of the motions of the atmospheric ocean which bathes alike plains and hill-tops. Too little weight has been granted hitherto by geologists to these agencies, and yet there is no other which has contributed in a greater measure to determine and to modify the character of the surface of any portion of the ground after its emergence from the sea, and to predestinate wide regions for the existence of certain kinds of plants and animals, and for the modes of nomadic or agricultural life of mankind.

Wherever dust is carried away by wind from a dry place, and deposited on a spot which is covered by vegetation, it finds a resting-place, and may be washed off and carried farther away by the next rain, if the ground is sloping, or it may be joined to the soil if the ground is flat or slightly inclined. If these depositions are repeated, the soil will gradually grow. At the same depth, therefore, to which the deepest rootlets of the grass of to-day are descending, the soil may have had its surface centuries ago. Remains of the past, such as buildings and entire cities, may in this way have been entombed by dust, provided that plants were growing on its deposits, and could secure a resting-place to all further supplies of atmospheric sediment.

In regions where the rains are equally distributed through the year, little dust is formed, and the rate of growth of the soil covered with vegetation will be exceedingly small. But where a dry season alternates with a rainy season, the amount of dust which is put in motion and distributed through atmospheric agency can reach enormous proportions, as is witnessed by the dust storms which in Central Asia and Northern China eclipse the sun for days in succession. A fine yellow sediment of measurable thickness is deposited after every storm over large extents of country. Where this dust falls on barren ground, it is carried away by the next wind; but where it falls on vegetation, its migration is stopped.

28

In rainless deserts the wind will gradually remove every particle of fine-grained matter from the soil, though a new supply of this may constantly be provided by the action of sandblast. The sediments of desiccated lakes, the soil which is laid bare by the retiring of the sea, the materials which are carried down by periodical torrents from glaciated regions to desert depressions, the particles which on every free surface of rock are loosened by constant decay—all these will be turned over again and again by the wind, and undergo an incessant sifting, until every earthy grain is blown off and nothing but moving sand and wind-worn pebbles remain.

The dust may travel great distances, and if the wind during the dry portion of the year blows constantly in one direction, that distance will increase, while the deposition of æolian sediments will be cumulative in places situated in the same direction. If the dust is deposited on mountain ranges endowed with considerable fall of rain and drainage towards the sea, it will be finally carried to this reservoir.

There are, however, chiefly two great classes of places where the dust of continents will rest permanently, and continue to accumulate through ages.

The first are what may be termed the central regions of continents, that is, those regions where, notwithstanding some rain which chiefly falls in one season of the year, the water has no drainage towards the sea, but is collected in inland basins from which it evaporates. This is the case in the Great Basin of North America, in Persia, and in Central Asia, from the Pamir to the Khingan range, and from the Himalaya to the Altai. The prevailing vegetation, independently of altitude, is that of the salt steppe. Grass and herbs take hold of the dust, whilst the debris that collects slowly on the hill-sides is, by very slow gradations, washed down the slopes by occasional rains, and will, if there happens to occur a period of heavier rainfall, be spread over a portion of the surface of the steppe, giving rise again to the growth of vegetation, which in its turn takes hold of the falling dust. In this way the dust will accumulate slowly but constantly through ages on those portions of the surface which are covered by vegetation. In the course of time it may reach a thickness of hundreds and perhaps of thousands of feet. The salts resulting from the decomposition of the rocks, and carried partly through the air together with the dust, and partly by water down the hill-sides, will remain in the soil, and collect chiefly in salt pools situated in the lowest portions of each basin, where, at the same time, stratified soil is deposited. As the surface which bears the vegetation is by slow degrees rising to a higher level, the tubes in the soil which contained the roots of former generations will retain their shape. The land-shells which feed on the steppe and withdraw to some depth underneath the surface in seasons of drought or cold will be entombed where they die, and the most delicate shells will be preserved. The same will be the case with the bones of mammals and birds living on the steppe, the dryness of the climate preventing the decay of any organic matter, as well as the formation of vegetable mould, which would be created in a moist climate through the decay of the organic matter.

In this way the deepest valleys, the wildest gorges, and the largest depressions in undrained regions may be gradually filled up with the deposits of dust, interchanging near the encasing slopes with the angular debris of rocks, but increasing in homogeneousness of composition and structure, and in freedom from any foreign ingredients towards the central portions of each basin. The inequalities of the ground will disappear, the lower hills will be buried, and the surface of the steppe will have a trough-like shape between every two protruding rocky ranges. If then, in consequence of a lasting change of climate, such a basin should gradually be filled with water, and an outward drainage be opened, erosion would soon furrow deep channels through the earthy deposit and expose its interior structure ; the fine tubes marking the site of the roots of countless generations of plants, the remains of the shells that had fed on the grass, and the bones of the mammals that have lived on the steppe would become visible; and the earth so exposed would be what is called Loess.

Such was the line of argument which I founded on the study of the Chinese Loess. I concluded that the same regions where the traveller of the present day moves between stupendous walls of yellow earth, and gazes with daily renewed wonder at the fantastic shape of rocks of earth produced by erosion ; where millions of people live in caves dug in the vertical faces of the Loess, while on its terraced surface they cultivate fields which are highly productive in wet summers, and terribly barren when moisture is not supplied in sufficient quantity—that these same regions, through which the Yellow River and its tributaries now take their courses, were once covered with dreary steppes only fit for nomadic life, and had no drainage towards the sea.

I had soon an opportunity of verifying the theory by a visit to Mongolia, where I saw precisely what my experience in the Loess regions had caused me to expect, namely, the very same shape of surface which I had observed in these, a steppe vegetation growing upon an impalpably fine earth mixed with grains of sand, and accumulations of the debris of rocks at the foot of the hill-sides. But in no place could I see the inner structure of the soil exposed to view. Proceeding, however, to the boundaries of the undrained region, where the drainage of some marginal basins had begun, but the channels of erosion were still shallow, the first sure signs of true Loess made their appearance on the side of every natural cut in the ground. From this first stage of the conversion of steppe basins into Loess basins, all grades of passage to the wildest and most grotesque landscapes, where the Loess was exposed to view in a thickness of a thousand feet, could be observed in rapid succession.

It appears to me that the theory answers all requirements as regards the Chinese Loess, in so far as it easily explains all its properties and every incident in the mode of its occurrence. It combines, moreover, into one class of natural processes two kinds

30

of phenomena, which, although they are almost the reverse of each other in regard to their outward appearance and the conditions they afford to human existence, are closely allied in nature in respect to their geographical distribution. It need only be noticed that the salt steppes of Central Asia are surrounded on all sides by Loess regions. The chief difficulty, when the theory was first advanced, was the want of a sufficient source whence the enormous amounts of dust required by it could have had their origin. But the problem has since been resolved in a most ingenious and, as I believe, satisfactory way by Mr. Raphael Pumpelly.[1]

There exist, besides the undrained salt steppe, regions of a somewhat different kind, which serve as permanent resting-places to the subaërially deposited dust. They are sufficiently distinct from the former class of places of deposition to be styled a second class of these, although, as a matter of course, there must be a series of gradations connecting both. To this second class belong those wide grass-covered plains which are known by the names of prairies, savannas, llanos, pampas, steppes of Southern Russia and Siberia, etc. They, too, are subjected to the alternation of a dry and a wet season. They are distinguished from the drainless salt steppe by their level surface and by the fact that they are crossed and partly drained by larger rivers, the origin of which lies almost exclusively beyond their boundaries. Some of these regions are very moist in the wet season, and bear a luxuriant vegetation of grass and flowering herbs, but dry up completely during the rest of the year. The rate at which the growth of soil takes place will depend upon the character of the adjoining regions from which the winds prevailing in the dry season remove the loose soil. It can no longer be doubted that the " black earth " of Southern Russia is growing in this way, and I am inclined to the same opinion with regard to the "Regur" of India. The black colour, which is proper to the uppermost layer only, appears to result solely from the formation of vegetable mould, the deeper portions showing the brown colour of the Loess, together with its structure, although this appears to be less perfect than in the former case. The bones of mammals will probably be badly preserved in this soil, because, in consequence of the ample rains and the slow rate at which the soil grows, they will partly decay before being perfectly covered up. This may not necessarily apply to those land shells, the animals of which die underground, at their places of refuge.

Another difference from the steppes of the first class must be produced by the circumstance that the salts will be removed, in part, by the water which percolates the soil and takes its way to the river channels.

When, after my return to Europe, I commenced to study more closely than I had done in former time the Loess of this continent, its perfect similarity in regard to composition, structure, and mode of occurrence with that of Asia, could not fail to strike me forcibly, and led me irresistibly to the conclusion that it must have been

---

[1] Amer. Journ. of Science and Arts, vol. xvii. 1879, p. 133.

formed by the same process of long-continued subaërial deposition of dust on steppes as that of the eastern continent, and the arguments which I had applied to this appeared to me to be no less valid for the Loess regions of North and South America.

The Loess-covered portions of Europe extend, as is well known, from the Pyrenees, the Alps, and the Balkan in the south to Belgium, the North German plains and Poland in the north, and from southern France in the west to beyond the limits of the continent in the east. Every portion of this entire region must have had the character of a steppe during a sufficient length of time to allow the deposit to be formed in at least such thickness as we observe at present. This thickness increases on an average as we proceed from north-west to south-east. It appears that, while east of the Alps the beginning of the steppe era may have been of earlier date, it commenced in Galicia, Germany and France during, or shortly after, the time of most extensive glaciation, and that one or the other kind of steppe was formed on the ground of the moraines as they were gradually laid bare by the retiring of the lowland glaciers. When Europe had its north-western limit beyond the present bathymetrical line of one hundred fathoms, and the summit line of the Alps was at greater elevation than at present, a continental climate must have prevailed such as is the prime condition for the formation of steppes, and it is probable that these had then their widest extent in a north-westerly direction. It would lead me too far now to explain why it appears that the conditions of climate, vegetation and animal life prevailing north of the Alps, after having gone through a stage resembling that of the tundras, must have been intermediate in character between those existing at the present time in Siberia and those prevailing in Southern Russia, while various evidence goes to show that farther south-east, in the Hungarian and Roumanian basins, there was no drainage to the sea, and the steppes of these countries resembled those of the drainless regions of Asia.

Gradually, when, with the renewed intrusion of the sea upon the land, the continental climate of Central Europe was converted into an oceanic climate, the change progressing slowly in the direction from north-west to south-east, the growth of the Loess ceased in the north-west, while it still continued in the south-east. Even now the soil is growing where it is covered by vegetation and sheltered from erosion. But the process is extremely slow and, with the exception of Southern Russia, is no longer regional, places of subaërial deposition being scattered among others of erosion.

At the same time when I published these arguments regarding the mode of origin of the Loess of Europe, Dr. Nehring, of Wolfenbüttel, came in the course of his admirable researches on the bones found in the Loess of Northern Germany to the well-known result, that the mammals which lived there at the time of the formation of that earth were identical with, or nearly related to, those which are living now on the steppes of Arctic regions, as well as in Siberia and Central Asia, and he concluded, that Germany must then have had the character of a steppe, and been subjected to a climate similar to that which prevails at present in western Siberia.

Thus, Dr. Nehring, who, at that time, had no knowledge of my researches, was led through the study of the fossil remains to precisely the same conclusion regarding a limited region in Europe, at which I had arrived with respect to a large portion of the continent by arguing on the structure and mode of occurrence of the Loess. Since then, the continued studies of the bones of mammals contained in the Loess, which have been made by Dr. Nehring and others, have yielded an overwhelming amount of evidence in the same direction, and have enabled us to extend the first conclusions to the whole of Germany, including the Rhine valley, Bohemia, and the vicinity of Vienna, and also to Hungary.

I believe I am correct in stating that, among those who have had extensive experience in Loess regions, all who have pronounced an opinion of late years are agreed that subaërial deposition is the only mode of origin by which all its peculiar features can be easily explained. Besides Mr. Raphael Pumpelly, who knows the Loess of Asia and North America, I mention chiefly Dr. Emil Tietze, of Vienna,[1] who studied it in Persia and Galicia, and the late Professor Karl Peters, of Gratz, who has probably examined a greater extent of European Loess regions than any other geologist, and, like Pumpelly, had, previous to 1877, advocated an aqueous origin as strongly as he afterwards did the subaërial. The celebrated M. von Middendorff has lately changed his views in a similar way.[2]

According to the subaërial theory as here pointed out, two different climatic stages are required for the formation of the typical Loess regions, the first of them marked by a continental and generally dry climate, during which the soil accumulated, the other distinguished by an increase in the fall of rain, in consequence of which the soil was furrowed by the erosive power of water and the steppe basins were converted into Loess basins. It is obvious that the conditions afforded for the existence of plants and animals and for the mode of life of mankind must have been almost the reverse of each other in either of the two stages, and that their change in time corresponded exactly to their change in space, as witnessed at present by the traveller when he descends from the Mongolian salt steppes with their uniform vegetation, their animals peculiarly adapted to a roving kind of life and their nomadic and unagricultural people, to the Loess basins of China, the characteristic feature of which consists in the labyrinthic ramification of very narrow gulches cut in the yellow soil to very

[1] Jahrbuch der K. K. geolog. Reichsanstalt in Wien, 1877, pp. 341-371; and more fully explained in the same journal for 1882, pp. 111-149. This last notice, which is of great importance, came to my knowledge after writing the present article.

[2] Mém. de l'Acad. Imp. des Sc. de St. Pétersbourg, t. xxix. 1881. It appears that Mr. W. T. Blanford has also adopted the theory of the subaërial origin of the deposits filling up undrained inland basins (see Proceed. R. Geogr. Soc. 1881, p. 79), and Mr. Clarence King informs me by letter that he ardently advocates the same mode of origin regarding the Loess regions of the Mississippi basin.

great depth and with perfectly vertical faces.   It is this character of surface, together with the fact of the existence of a drainage at low level and the removal, by it, of the soluble salts, which causes the remarkable contrast between the features of Northern China and those of Mongolia.   Vegetation offers there a far greater variety of forms ; those animals which are accustomed to roving on the boundless steppe cannot exist where the greatest possible unevenness of the soil is the distinctive mark ; and man is simply forced to adopt in the Chinese Loess regions a settled and agricultural mode of life, drained Loess being of all kinds of soil best adapted for the cultivation of cereals, while the innumerable recesses and naturally fortified positions afford him shelter and safety.

It seems hardly necessary to observe, how important, on account of this mode of origin, the occurrence of Loess is to the study of the causes of the present distribution of plants and animals.   The peculiar climatic conditions prevailing during the time of its accumulation, and the physical features of the regions covered by it must have influenced migration and variation in a considerable measure, and it is far from improbable that the habits of life and the migrations of primitive man over large portions of Europe and Asia have been directed by the same causes.   It must be added that a slight deterioration of the climate is sufficient to change the steppe, and chiefly the drainless salt steppe, into the most arid desert, and to cause the emigration of man and animals.

I have given this *exposé* a greater length than I intended.   It will refute, without any further discussion, the objections which Mr. Howorth has raised against the theory of the subaërial origin of the Loess, as, *e.g.* : that subaërial deposits such as this are nowhere being formed now (p. 16) ;—that it is incredible that subaërial deposits should have been deposited at a height of 6000 feet and to the depth of 1000 feet (p. 76) ;—that the subaërial theory treats the problem as a local Chinese problem, while it ignores that the Loess has to be accounted for in Europe as well as in China (p. 76) ;—that it cannot be understood how shells and animal debris could be carried by the wind (p. 76) ;—that the means would be inadequate to the end, as clay would not be acted upon by the wind (p. 77, after Kingsmill) ;—that the chemical composition of the Loess does not correspond with that of the inorganic elements of plants growing on its surface ;—that there is no known means by which these inorganic matters could have been supplied from the atmosphere ;—that, although silica might have been conveyed by the medium of dust storms, no way can be seen how the silicate of alumina could be conveyed ;—that there is no evidence of the ramifying tubes having their origin in the roots of plants ;— that the Loess is devoid of organic substances ;—that Loess, if subaërially deposited, should be different in composition according to the subjacent rock and could not be equal everywhere.

After having attempted to set aside, on these grounds, the theory of the subaërial origin of the Loess, which, evidently, was known to him by name only, but never by the arguments on which it rests, Mr. Howorth proceeds to expose his own hypothesis. Like that of his predecessor Dr. Hibbert Ware, of 1832,[1] it is boldly founded on the supposition of cataclysmic events of tremendous magnitude. But while Dr. Hibbert, who had with regard to the occurrence of the Loess the limited knowledge of his time, was satisfied with one event, namely a great flood, Mr. Howorth, to whom the whole amount of the present knowledge of the subject was accessible, pre-supposes two stages. The first was marked by a vast volcanic out-pouring of "subterranean mud," while in the second this mud was "largely steeped in floods of water," and "mixed with the ingredients of the superficial bed over which it poured." The mud was thereby spread over large extents of country, mammals and shells were imbedded in it, and the entire mass was deposited without stratification, although the flood must have reached up to at least 5000 feet in Central Europe, and to 8000 feet in China, overwhelming, as must be supposed, almost the whole extent of two continents. A similar event should have drowned simultaneously the two American continents and annihilated almost all beings living above the level of the sea. Granted, however, the flood, it is by no means easy to understand its character, as its author is opposed to the aqueous origin of the Loess, whether it be marine, or lacustrine, or fluviatile. It might be supposed that the flood consisted altogether of erupted mud, but we are not told why it did not leave a homogeneous sheet spread over the entire continent up to a few thousand feet, nor how it happened that the deposits we do find from this supposed flood occur in those regions only where the conditions of a continental climate have prevailed and do, in the greatest part, prevail to this day; nor do we learn the reasons why the Loess differs completely in composition and structure from all known kinds of volcanic mud ancient or modern, or how it is that the supposed fissures from which the enormous masses of muddy volcanic rock poured forth were rent at very great distance from the regions of the main distribution of the Loess, and at a time when volcanic energy, which had been most violent in the Tertiary age, was (at least in the regions in question) in its very last and dying stage.

I do not believe that any geologist will seriously take the trouble to argue against these fanciful views. It is strange that the same deposit which bears testimony in itself of having been formed in the slowest, the most quiet and undisturbed manner that can be imagined, should be considered the product of events grander and more violent in character than any heretofore devised in the long history of unfounded geological speculation. Whoever undertakes to advance a theory on a geological subject, should first observe, and observe again, and then compare his own results with what has been observed by others in other parts of the world. If the author of the volcanic theory of the Loess had devoted the same admirable industry, with which he has studied and written the history of the Mongols, to the personal observation of the soil on which the

[1] This theory was noticed in *China*, vol. i. p. 162.

wanderings and warfare of that people did very prevailingly take place, and which from their original seat in Central Asia and their imperial city of Khanbaligh spread to the west of Europe; if he had, besides, taken the trouble to make himself acquainted with the arguments of the existing theories respecting the mode of origin of this soil before undertaking to put them aside; and if he had for a moment considered from a geological point of view a few of the concomitant circumstances required by his own theory, he would hardly have ventured to adopt views which could be pronounced at an early and rather low stage of geological science, but are long since abandoned, and he would never have added to them suppositions which bear the character of the infancy of that science. If there is any subject to which the theory of Mr. Howorth may be applied, it is, in a figurative way, the history of the Mongols. The crowds in which they appear suddenly on the stage resembles the outpouring of volcanic matter from a hidden source, and the flood of them which soon inundated immense regions, "mixed with the ingredients over which it poured," may indeed be compared to the action of a sweeping wave. But the laws which govern the movements of mankind have but a very distant relation to those which can be discerned in the changes of the physical conditions of the surface of the globe.   F. Baron Richthofen.

Bonn, *May*, 1882.

4

Reprinted from *Jour. Geol.*, **5**, 795–802 (1897)

SUPPLEMENTARY HYPOTHESIS RESPECTING THE ORIGIN OF THE LOESS OF THE MISSISSIPPI VALLEY

T. C. Chamberlin

THE loess problem still remains obstinate. While it has yielded somewhat to progressive research, there is, I think, a nearly universal feeling of dissatisfaction with all theories thus far advanced. The eolian hypothesis appears to be the better supported so far as concerns the chief deposits of China and perhaps some of those of western America, while the aqueous hypothesis seems best supported so far as concerns the deposits of the Mississippi valley and western Europe. It is the judgment of some students that the ultimate solution will lie in the recognition of both hypotheses, but the means of discriminating between the two and of applying the criteria are as yet wanting. The present paper is intended to be a contribution in this direction. It is confined to the loess deposits of the Mississippi valley, but is probably applicable to the loess of western Europe.

The distribution of the loess in the Mississippi valley seems to be very significant in its peculiarities. These may be summed up in two great features.

1. The loess is distributed along the leading valleys. These embrace not only the great valleys, the Missouri and the Mississippi, but some of the subordinate valleys, as the Illinois, the Wabash, and others. The loess is found along the Missouri River from southern Dakota to its mouth; along the Mississippi River from Minnesota to southern Mississippi; along the Illinois and the Wabash from the points of their emergence from the territory of the later glacial sheets to their mouths. Along these valleys the loess is thickest, coarsest and most typical in the bluffs bordering the rivers and grades away into thinness, fineness and non-typical nature as the distance from the rivers

[1] Read before Section E, Am. Asso. Adv. Sci. Aug. 12, 1897.

37

increases.   In some instances the loess mantle rises to the divide and connects with the similar deposit of an adjacent valley, but the law of progressive fineness and thinness still holds.   This relationship is such as to create a very strong conviction that the deposit of the loess was in some vital way connected with the great streams of the region.

2. The second significant feature is the distribution of the loess along the border of the former ice-sheet at the stage now known as the Iowan.   (Strictly speaking there was more than one stage of loess formation, but for convenience only the main stage will be here discussed.)   The elaborate paper of McGee made us familiar some years ago with this relationship in eastern Iowa.   The studies of Calvin and his colleagues, Bain, Beyer, and Norton, of the Iowa Survey, of Winchell and Upham of the Minnesota Survey, of Todd of the South Dakota Survey, and of Salisbury, Leverett, Udden, Buell, Hershey, and the writer of the United States Survey, have greatly extended the evidence of this relationship.   It has recently been much advanced by the Iowa geologists and by Leverett and Hershey in northwestern Illinois. Next the border of the ice-sheet the loess is thick and typical, but graduates away with increasing distance from the ice border in a manner similar to the graduation away from the river valleys. On the border next the ice there are developed the formations designated by McGee paha, elongated domes of quasi-drumloidal contours which are mantled by loess.   This superficial loess graduates downwards into loess of coarser and coarser texture until it often passes into a nucleus of sand.   Below this there is often an embossment of till.   These pahas seem to be ice border phenomena.   Whatever their special mode of formation their distribution seems to connect them in some more or less direct genetic relationship with the ice.

It has been affirmed by several independent observers that the loess graduates into glacial clays and glacial till and this relationship further tends to confirm the association of the loess with glacial action.

It has been shown by the microscopical examinations of

Salisbury that the loess particles are composed in part of feld-spars, amphiboles, pyroxenes and other common constituents of the glacial clays. These silicates are decomposable under pro-longed weathering, and hence cannot well be supposed to come from residuary clays under the ordinary conditions of the Mississippi valley. The presence of the calcium and magnesium carbonates, independent of the presence of shells, points in the same direction. This inference is strengthened in a peculiar way by observations in the lower Mississippi valley. Above the Lafayette gravels and below the loess there is a stratum of silt which does not habitually contain the characteristic silicate particles of the loess. This stratum has been by most observers associated with the loess, but it is separated from it by a soil horizon as abundantly affirmed by the observations of Salisbury and the writer. On the other hand it graduates more or less freely into the Lafayette sands and gravels. The stratum is, as we interpret it, the last deposit of the Lafayette stage. It is a typical finishing deposit succeeding a fluvial sand and gravel. Now this has special significance in this relationship, in that it shows that in the stage closely preceding the loess deposit, the Mississippi did not lay down silts of the same constitution as the loess. The inference therefore is that the loess is not simply a fluvial silt brought down from the surface of the river basin, nor common wind drift borne into it, but that it had a special origin connected with glacial action which was competent to supply precisely the kind of silt of which the loess is made.

It is hard to resist the force of this argument from the constitution of the loess taken in connection with the two distributive relationships. Jointly they seem to force the conviction that the loess had its origin in some relationship to the ice of the Iowan stage and to the rivers that led away from the ice edge at that time.

But the hypothesis that the loess is simply an outwash of glacial grindings distributed along the river valleys by the glacio-fluvial waters is attended by grave difficulties. This remains true whether the deposition be supposed to have taken place either in a strictly fluvial fashion, or in a fluvio-lacustrine fashion, or in a true lacustrine fashion, or in an arm of the sea. In the first place, the vertical distribution of the loess cannot easily be explained. The extreme vertical range is not far from a thousand feet. The range within a score of miles is frequently from 500 to 700 feet. The loess sometimes seems to the field observer to have a special fondness for summit heights. It sometimes mantles topography of a pronouncedly rolling type. It does not then appear to be a deposit which once had a level or even a smooth surface out of which the rolling surface has been eroded, but to be a mantle laid down upon a previously undulatory surface. Such a mantle might perchance be laid down from water, but I am not aware that we have any demonstrative deposition of the kind which closely simulates the mantling of the loess in some of the upland territory. To suppose that the Mississippi, Missouri, Illinois, Wabash and lower Ohio rivers were so swollen that they united over their divides and threw down a mantle of fine silt over the southern and western half of Iowa and the southern parts of Illinois, Indiana and Ohio, is a somewhat severe tax upon belief. It is difficult to imagine the conditions which should have maintained such a body of water. This has been so much discussed that I need not dwell upon it. But even if such a body be supposed, it is difficult to imagine how the deposition could have been precisely what we find in the case of the loess. It is futhermore difficult to account for the presence of the land shells which abound in it; for if this great flood had the ice-sheet for its northern border, it is extremely difficult to imagine how it could have been peopled so widely with the terrestrial mollusks.

The limit of the loess does not appear to be a strictly topographic one. It is difficult to bring its border into strict accord with a horizontal plain as required by the lacustrine and marine phases of the hypothesis, or even into a consistent gradient as required by the fluvial phase, without an arbitrary warping of the surface. The spread of the loess in the lower Mississippi valley is more extensive and reaches greater heights on the east side than on the west side, so far as present knowledge goes. A similar fact seems to be true of the Missouri valley. I think this is generally true, but my observations are not sufficient to justify its unqualified affirmation as a generalization.

There are other difficulties attending the aqueous theory in its simple application, but I need not attempt an exhaustive recital here as they have received emphasis in the long battle between the eolian and aqueous hypotheses. The foregoing will I trust suffice to show that there is abundant occasion to still cast about for a more satisfactory explanation of the loess puzzle.

The supplementary hypothesis herewith proposed attempts to divide the honors between the aqueous and eolian agencies. It recognizes the tremendous force of the arguments from the distribution and the constitution of the loess in favor of the glacio-fluvial hypothesis, and it adopts that hypothesis as the fundamental explanation of the origin of the Mississippian loess. It assumes the presence of the Iowan ice at the chief stage of loess deposition. It assumes a very low slope of the land and a consequent wide wandering of the glacial waters. It assumes the development of extensive flats over which the silts derived from glacial grinding were spread. It assumes that the glacial waters were subject to great fluctuations; $1°$ as the result of periods of warm weather in the melting season, and $2°$ as the result of warm rains, which not only added directly to the volume of water, but forced the rapid melting of the ice. Gilbert has acutely observed that there is no way in which the atmosphere can convey its heat energy to a glacier so effectually as through warm rains.

Let it be imagined, therefore, that the silty waters from the margin of the ice-fields wandered over broad flats and constantly built them up by their sediments, and that at periodical flood stages they extended themselves widely over the plains, while between the flood stages they withdrew to more limited courses.

The territory covered by the maximum extension of the waters would be the zone of accumulation of fluvial loess. It is

41

not necessary to suppose that the periodic extensions of the floods were destructive of the vegetation over all the flat region. In some portions not only could vegetation persist, but the land mollusks and other animals dependent upon the vegetation could find a temporary retreat from the flood on the taller vegetation that may have prevailed.

After each of the periodical retreats of the water there would be left extensive silt-covered tracts facily exposed to the sweepings of the wind and from these, when dried, dust could be derived in great quantities to be borne away over the adjoining lands and lodged in their vegetation. The material thus derived would be essentially identical with the glacio-fluvial deposition, and thus the hypothesis seeks to account for the glacial element in the constitution of the eolian portion of the loess. The presence of land mollusks in the upland eolian loess finds in this way a ready explanation, while their presence in the lowland loess mingled with aqueous mollusks finds an almost equally obvious elucidation; for not only would the upland shells be washed into the lowlands, as we observe they are at the present time, but they would periodically invade the lowlands in the intervals between submergence and would be caught and buried there. Occasionally the shells of the lowland and aqueous mollusks would be borne to the uplands by organic agencies, and possibly in rare instances by the severest type of winds, and hence their occasional presence there is not remarkable.

To make this a good working hypothesis it would appear that there must be an accommodation between the breadth and fluctuations of the fluvial deposits and the extent and massiveness of the eolian deposits, for if we suppose the glacial floods to be confined within narrow channels, the sweeping ground of the winds would have been too scant to give origin to the great mantle of silt then attributable to them, for we must remember that in proportion as the river work is narrowed the wind work is expanded. It is obvious that the eolian factor will cut away its own ground if pushed too far.

There is little question that loess-like accumulations are now taking place on the bluffs adjacent to the Mississippi and Missouri valleys. Observation seems to clearly indicate this. But such accumulations are relatively scant in amount and limited in extent, and it is difficult, if not impossible, to believe that the great loess mantle had its origin from the wind drift of flood plains no more extensive than those of today. It must be constantly borne in mind that the eolian deposits are measured, not by the quantity of silt borne by the winds and lodged on the surface, but by the *difference* between such lodgment and the erosion of the surface. Under most conditions with which we are familiar the erosion is more than a match for the dust accumulations. The conditions must then have been extraordinary which would give a dust deposition sufficient to supply erosion and still leave so large a residuum as the loess mantle implies. The unleached and relatively unweathered nature of the body of the loess is specially in point here. These considerations warn us of the theoretical danger of too greatly circumscribing the fluvial action.

On the other hand, if we attempt to extend the fluvial hypothesis too greatly we fail to leave sufficient feeding ground for the molluscan life and we encounter the topographical and physical difficulties which have been previously urged against the pure aqueous theory. A Janus-faced hypothesis is here offered in the hope that by a judicious reference of a part of the loess to one class of action and a part to the other, a joint explanation may be found to afford a true elucidation of the perplexing formation. At any rate, it has seemed worth while to propose the hypothesis for trial. It will doubtless be extremely difficult to find a line of demarkation between the two classes of deposits. Such attempts as have been made in this line justify this apprehension. This supplementary theory has been in mind for several years and was briefly suggested in my paper on the *Genetic Classification of the Pleistocene Deposits*, presented at the Fifth Session of the International Congress of Geologists at Washington in 1891.[1] An effort has been made

[1] Compte-Rendu of the Fifth Session of the International Congress of Geologists, Washington, 1891, p. 192.

by some of my colleagues and by myself to find criteria of discrimination between aqueous and eolian loess. While individual types of both deposits are not difficult to find, a criterion or a series of criteria of general applicability which shall distinguish the two and assign to each its appropriate part is yet wanting.

Richtofen in his classic work on China urged as the explanation of the great Chinese loess an eolian hypothesis supplemented by a fluvio-lacustrine hypothesis. He insisted that the original and chief loess deposits were formed by dust blown from the great arid plateaus and lodged on the more fertile plains of China, and that from these primary deposits the streams gathered and subsequently redeposited in fluvial or lacustrine form a subordinate portion, thus giving origin to a secondary loess formation. Going beyond that field he and his supporters have apparently tried to apply this secondary factor to the explanation of difficulties in the European and American loess, to which its application is more than doubtful. It is interesting, however, to note that the loess puzzle of China, even in the mind of its chief exponent, finds a full solution only in a combination of eolian and aqueous hypotheses. The present writer herein urges the trial of a similar combination of hypotheses, but reverses the order of the terms in their Mississippian application. The aqueous loess is made primitive and the eolian loess secondary. The Richtofen loess may be said to be first eolian and secondarily aqueous; the Mississippian loess, first aqueous, and secondarily eolian. The Richtofen loess in its ultimate origin is residuary. The Mississippian loess, in its ultimate origin, is glacial. The Richtofen mode of origin may be said to be eolio-fluvial, the mode herein advocated, fluvio-eolian, in which terms the order of the words indicates the order of derivation and each word signifies a variety of loess.

Editor's Comments
on Papers 5 and 6

5   **KEILHACK**
Excerpts from *The Riddle of Loess Formation*

6   **PENCK**
Excerpts from *Loess of Central Asia*

A few closing remarks made by Keilhack at the end of his article (Paper 5) have been remembered, but the main message was ignored or forgotten. As a conclusion he suggested (possibly not at all seriously) that if no terrestrial source of loess material could be found the investigators should look to extraterrestrial sources. Thus, Keilhack's "cosmic loess" was born and it has obscured the serious content of the paper ever since. Keilhack was one of the first to perceive that an important aspect of the loess problem concerned the formation of the actual material of loess deposits.

That Keilhack could find no acceptable terrestrial source for loess material appears to lie in his overestimate of the amount of loess in existence. His distribution map certainly shows a highly optimistic view of loess distribution (optimistic from the point of view of the committed loess researcher); as regards the *nature* of the material, he was asking very pertinent questions (see Smalley and Perry, 1969, for more details).

The cosmic origin has contemporary supporters. Petersen (1973, p. 79) has stated that "The loess seems to provide an ideal example of a cometary dust accumulation. The deposits in Europe, Asia, and North America appear to have originated in a single encounter event because of close agreement in their form and composition."

Penck's major contribution to the study of loess was in his observations of the lack of fine material on the fringes of desert areas. He was publishing on loess in 1883 (the year he obtained his doctorate from the University of Munich), and as long ago as 1909 he noted the lack of a loess "girdle" around the Sahara desert. In 1930 (Paper 6), he wrote that "There is no loess in the steppes and deserts of Africa . . . . The deposit in Tripoli is not true loess." Also in the same paper he states: "Richthofen has the great merit of having recognized loess as an aeolian product. But it has not its origin in the decomposition of the rocks of deserts or steppes; it comes like the sand of many dunes from river deposits which are arranged by the wind." He thus recognized that, although Richthofen was right about eolian transportation, his theory had, by implication and with specific reference to the Chinese loess, come to suggest a desert source for the particles.

Penck's most significant contribution to Quaternary geology is contained in *Die Alpen im Eiszeitalter,* written in association with E. Bruckner and published in three volumes over the years 1901–1909. This was his greatest work and is the foundation of his lasting fame, but his contribution to the study of loess was of considerable consequence. He was born at Reudlitz, near Leipzig, in 1858, studied at the Universities of Leipzig and Munich, visited Yale and Columbia universities in 1908–1909, and died in Prague in 1945. His life and work are admirably described in a detailed obituary published in *Quartar*, no. 5 (1951).

REFERENCES

Petersen, R. G. 1973. A monograph on comets. Kroevel, Phoenix, Ariz., p. 124.

Smalley, I. J., and Perry, N. H. 1969. The Keilhack approach to the problem of loess formation. Proc. Leeds Phil. Lit. Soc. (Sci. Sect.) 10, 31–43.

5

THE RIDDLE OF LOESS FORMATION

Konrad Keilhack

*This translated excerpt was prepared expressly for this Benchmark
volume by W. Riha and I. J. Smalley, The University
of Leeds, from "Das Ratsel der Lossbildung," Zeit.
Deutsch. Geol. Ges., 72, 146–161 (1920)*

For many years I have been familiar with the deposits of loess in various parts of Europe. However, I have come to the conclusion that the certainty with which our textbooks discuss the question of the origin of loess and its formation is in inverse proportion to our knowledge. This may quite easily lead to the inference that the doctrine currently prevailing could consolidate to a dogma and thus become a hindrance to all further progress of our knowledge.

In the last few years many papers dealing with loess have been published; however, they deal with its exact dating and its relation to the glacial periods and the Paleolithic cultures without going into the problem of its origin (Wiegers; Soergel; Hilber; Geyer; Gutzwiller; Menzel).

Using a graphical representation of the north boundary of the loess, I quite recently *(Zeit. Deut. Geolog. Ges.,* 1919) made an attempt to destroy a legend about the deposition of loess which has prevailed since A. Penck's early days; he claimed that there was some causal relationship between this deposition and the various Ice Ages. I tried to replace this doctrine with the recognition of the complete independence of the distribution of loess from the glacial deposits. This map stimulated me to attempt a representation of the distribution of loess in the whole world. Such an attempt is feasible in the case of Europe; for Asia and America, however, there arise many difficulties, so that the representation on the accompanying map of the world in Figure 1 will probably have to be subjected to considerable amendments in the future. I therefore offer it as a first attempt which must be judged with leniency.

As a basis for this attempt I have used a copy of the geological world map in stereographic projection, in which the land zones are represented such that angles are conserved, whereas areas are only conserved in the middle part of the map. The loess domains on the Earth are indicated by dots; the two continental, the South American, and the Alpine domains of glaciation are hatched. For plane measurements a second copy of the map (in a projection-conserving atlas) was used.

The difficulties of the loess problem, one may even say of the loess riddle, have their roots in the following five facts:

1. Geographical distribution.
2. Very gigantic mass.
3. Restriction of its occurrence to such a small segment in the history of the Earth.
4. Regularity and the peculiarity of its composition.
5. Difficulty of determining its original and constituent substances.

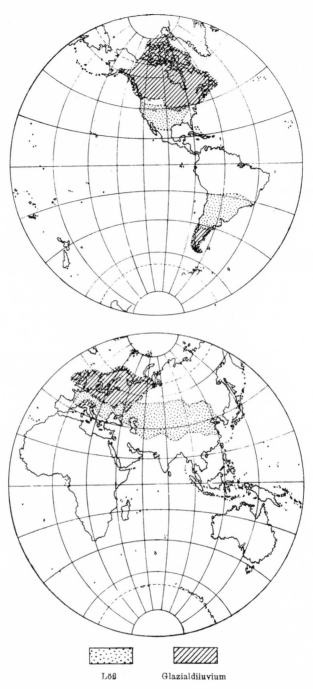

Figure 1. Range of distribution of loess and glacial diluvium. Löβ: loess; *Glazialdiluvium:* glacial diluvium.

THE FORMATION AND THE CONSTITUENT MATERIALS OF LOESS (SECTION 5 OF THE ORIGINAL PAPER)

The prevailing doctrine holds that loess is the result of a "blowing out" *(Ausblasung)* of minerals situated on the surface and affirms, in particular, that the glacial formations, especially the ground moraines and the *Geschiebemergel* (boulder clay), are the producers of loess deposits. Let us see which of these assertions can stand a factual test.

When we begin with the younger loess, which claims the lion's share as regards surface distribution, we first make the following observations: the widely varying opinions about the period of its formation—whether interglacial, glacial, or postglacial—are currently converging closer and closer to the view that the younger loess, especially on account of the vertebrate fossils that have been discovered, should be attributed to a glacial age, and its formation should be placed during the height of the last glacial period, not in its recession period. Assuming this, then only a fractional part of all the regions of glaciation on the Earth, namely the outer belt of the older glacial formations, would have been free and accessible to *Ausblasung* at the time of its formation. In what state then was the outer moraine belt? It had been exposed to disintegration for a long interglacial period and must have had the same surface structure as the glacial and fluvioglacial deposits of the same age, which were buried under the younger glacial formations. Consequently, it must have been free from lime *(entkalkt)* and mixed with clay *(verlehmt)* down to several meters deep. This conclusion cannot be dismissed. It follows then that the stratum generated by disintegration, and for several meters thick, must be first removed by *Ausblasung* before the wind could have a calciferous formation at its disposal. One might perhaps say that this first period of the *Ausblasung* of decalcified strata led to the formation of the decalcified *Lösslehme* in the form of deposits on which the wind might then have deposited normal calciferous loess. But this is not consistent with the proportion of their quantities, as calciferous loess is much more abundant than *Lösslehm;* the proportion should be exactly the reverse, or one would have to assume that the layers removed and transferred by wind denudation were so tremendously thick that even by the widest stretch of the imagination this assumption must be rejected.

But there is yet another, completely different way that leads to a rejection of the possibility of deriving loess from the moraines of the older glacial periods. If one tries to visualize the procedure that took place during the *Ausblasung* of the soil particles, when grains of 10–20 μm in diameter were selected, this can only happen as follows: the wind, relative to its capacity for transportation, lifts up particles according to their size, and carries them away, leaving behind anything more coarse grained. This causes the surface to become enriched with coarser particles, which eventually will form a protective cover over the subsoil, thus shielding the latter from the ravages of the wind. Only when fresh new soil, which has not been "blown out" *(ausgeblasen)*, is brought up to the surface will the wind have access to new material for blowing out. Such a conveyance of particles, from the bottom to the top, is effected almost exclusively by the activity of lower animals living in the soil. However, a redistribution of soil in this way rarely goes deeper than 1 mm. On planar surfaces, the wind thus cannot blow out more dust than is contained in a stratum of l-mm thickness. In our ground moraines the quantity of particles having a size suitable for the formation of loess amounts to about 30–40 percent, which means that an area of a certain ground moraine could produce material for an equally large loess stratum of 40- to 50-cm thickness. The sand and gravel formations predominating in the domain of the old moraines must be completely discarded. But the point is that the loess domains in Europe alone are several times larger than the open surfaces of the old moraines, and the thickness of the loess strata is again ten times the figure of 30–40 cm, derived above. Consequently, the old moraines cannot have produced more than 1 percent of the European loess. And how did the incomparably larger loess masses in Asia originate?

In North America the situation is a little, but not much more, favorable than in Eurasia. The deposits of the diluvial inland ice have to be discarded when we look for the original material of loess. But what else is then left? Granite and other eruptive rocks, crystalline schists, Paleozoic clay schists and graywacke, the medium- and coarse-grained sandstones; they are all completely unsuitable for consideration as the original material of loess. Thus there only remains the fine-grained sandstones and limestones from which loess may have originated. But where is there sandstone consisting of such fine quartz powder as we find in loess, and where is there a process that generates equally fine limestone powder when limestone disintegrates? I cannot think of an answer to either question and I am certain that here lies the greatest and most difficult problem of the loess mystery.

And something else: limestone and quartz, the two main components of loess, must have been picked up by the wind in entirely different parts of the world; for having had to exclude the glacial formations, we do not know of any minerals in which they both occur in the exact mixture and particle size as we find them in loess. But where, then, and in which way was the surprisingly homogeneous composition of two so completely different components effected? A composition not showing any deviations, not anywhere in the world! Can we explain this homogeneity in the composition of loess in Europe, Asia, and North and South America without having to assume that all the loess in these four immense domains obtained its material from the same mixing vessel? Must it not be true then that the composition occurred at considerable altitude? But this hypothesis is inconsistent with the size of the loess particles, which are much too big for them to float continuously in the high atmosphere—not only for days but for years, and hundreds of years. So many questions, so many completely unsolved problems!

From the assumptions of truth or the probability that the entire loess in all the world is a homogeneous mixture and must have originated from a common source which caused its deposition, it is only one step to the question of whether this completely precludes the possibility of an extraterrestrial or cosmic origin of loess. But this is for the astronomers to decide. However, I must point out that with such a seemingly bold assumption some of the questions I have raised could find a satisfactory answer: for example, the homogeneity of composition, the impossibility of a derivation from any terrestrial minerals, the restriction to the diluvium (i.e., the Quaternary), the zonal distribution over the whole world (just think of Saturn's rings!), and the causal relationship to the Glacial Periods.

6

LOESS OF CENTRAL ASIA

Albrecht Penck

[*Editor's Note:* In the original, material precedes this excerpt.]

Central Asia owes its daryas to the nival and subnival climate of its roof;
all derive their water from the melting of snow or glaciers. Their feeding by
rain or springs is insignificant. In spring or in summer they are fullest, during
the winter feeble. They bring water into the arid region and carry with themselves from the high regions enormous quantities of debris which they deposit
in the form of alluvial fans at the foot of the mountains. They have a freight of
sand and mud which accompanies them until they disappear. This freight is
then seized by the wind which blows the sand into dunes and carries away the
mud in the form of dust. This is the origin of the vast deserts of sand which
extend in the north of Central Asia, of the kums of Turan, and of the dunes in
the desert Taklamakan in the basin of the Tarim. Here and there the sands
are accompanied by loess, whilst along those rivers of Central Asia which
reach the sea there are neither sand-fields nor loess, with the exception of the
Hwang Ho.

Richthofen has the great merit of having recognized loess as an aeolian
deposit. But it has not its origin in the decomposition of the rocks of deserts
or steppes; it comes like the sand of many dunes from river deposits which are
rearranged by the wind. The mud is transported farther by the wind until it
was redeposited. This happens even on the barren ridge of the Kunlun up to
14,000 feet, as observed by Sir Aurel Stein; this is the rule in corners at the
foot of the mountains which in Central Asia are surrounded by a belt of loess.
In the Lop Nor region older river or lake deposits are strongly eroded by the
wind, and many loess deposits may come from here, but others are derived
directly from the mud which annual floods have left behind, and which was not
fixed by vegetation. Rivers which are very broad during the summer and
narrow in the winter are the birthplaces of many loess deposits. Thus it is
nowadays in Central Asia, thus it was in South-Eastern Europe during the

Ice Age. Then the Danube was swollen during the summer months by the waters coming from the alpine glaciation, and very low in the winter which was dry in the east of the ice caps of Northern Europe and of the alpine glaciation. The mud left by the Danube in the low plains of Hungary and near the Black Sea was whirled up by continental winds and spread over the neighbouring land.

This is also the origin of the loess of China. During the Ice Age the Hwang Ho brought thither in the summer months the waters and the mud derived from the Central Asiatic glaciation, which in winter was swept farther and deposited as loess in great thickness on the hills and mountains of Shansi, Shensi, Kansu, and Honan. From here it is eroded nowadays by rivers in innumerable gorges and brought to the great plains where it becomes again the prey of the wind so that the air is often full of dust. Thus it was during the Ice Age in the basin of the Mississippi. This mighty river was then during the summer full of the muddy melting water of the North American inland ice, whilst it was low in winter. Then the mud left behind by the flood was carried by the wind into the neighbourhood. Along the Mississippi are the loess deposits of North America, and not in the neighbourhood of the salt-steppes of the Great Basin. There is no loess in the steppes and deserts of Africa, and if storms bring dust from the interior of the Sahara, the deposit in Tripoli is not true loess. The origin of the loess is a very complex one; different causes—glaciers, rivers, and wind, nival and arid conditions—must work together. This happened during the Ice Age in the northern hemisphere, but at present it occurs only in Central Asia.

[*Editor's Note:* Material has been omitted at this point.]

Editor's Comments
on Paper 7

7 BARBOUR
Excerpts from *Recent Observations on the Loess of North China*

In the early 1920s Barbour started work in north China, and he later published several significant papers on the geology of this area and on the Chinese loess. One of his colleagues and coworkers was Teilhard de Chardin, and in 1965 he published a book that, although essentially about Teilhard, gives some details of the fieldwork and operations in Tientsin, Kalgan, and other regions. He describes the loess as follows:

> In Late Ice Age days, the same winds that swept coarse sediment into sand dunes on the pavement floor of the Gobi Desert, wafted the finer dust off the plateau lip to settle as loess, which makes the fertile soil in the valleys of the northern provinces. There it has remained as a national asset, unless washed away in the form of the yellow-grey mud that colors the Yellow River and the Yellow Sea (Barbour, 1965, p. 30).

Paper 7 offers a retrospective view of Barbour's loess investigations; it was published after he had spent more than 10 years considering the Chinese loess and two years before he joined the faculty of the University of Cincinnati. In 1930 he observed that

> In theory the natural procedure would be to restrict the term "loess" to deposits of a definite lithological type. In practice this proves extremely hard, because the intimate association and intermingling of the various types of sediment makes it impossible to be strictly consistent in delimiting any one facies. Moreover, taken as a whole, the loess *sensu stricto* represents a distinct set of climatic and physiographic conditions which held sway over a wide area, and the term has been an established formational name since the days of von Richthofen and Pumpelly (Barbour, 1930, p. 459).

Barbour was involved in the paleontological activity in North China around 1929, which was centered at Chou-kou-tien where Sinanthropus fossils had been discovered. Gunnar Andersson has described the events in his famous book *Children of the Yellow Earth,* which exhibits Barbour's superb draughtsmanship since many of the illustrative sketches are by him.

The paper presented here has been slightly shortened and the plates have been omitted.

REFERENCES

Andersson, J. G. 1934. Children of the yellow earth: studies in prehistoric China. Kegan Paul, Trench & Trubner, London, 345p.

Barbour, G. B. 1930. The loess problem of China. Geol. Mag. 67, 458–475.

———. 1965. In the field with Teilhard de Chardin. Herder & Herder, New York, 160p.

7

Reprinted from *Geogr. Jour.*, **86**(1), 54–56 (1935)

RECENT OBSERVATIONS ON THE LOESS OF NORTH CHINA

GEORGE B. BARBOUR

SINCE the days of von Richthofen the problems connected with the characteristic feature of the North China highlands have been keenly debated. The loess has been regarded in turn as an aeolian deposit, as a river-basin accumulation, as a lake silt, as a product of weathering, and as a complex of mixed origin involving the local redistribution by wind of material supplied by one of the other agencies—glacial and volcanic forces have never been seriously considered as far as this region is concerned. Recent detailed work along a variety of lines has thrown fresh light on the problem. A combined attack by the stratigrapher, the physiographer, the soil expert, the chemist, and the mineralogist is having its effect. Even the archaeologist has his word, for palaeolithic *foyers* occur at the base of the Upper Pleistocene loess, while more ancient deposits have yielded the implements associated with *Sinanthropus* (Peking Man).

Hitherto observers have not been in agreement as to what should be included in the term "loess." Hence a confusion has arisen like that surrounding the use of the word "laterite" in the tropics. The true loess is a yellow-gray, non-stratified, coherent, fine-grained loam deposit which shows the vertical cleavage characteristic of the loess-lands (Plate 1). The definition is thus in the first place a structural and textural one, predicating nothing as to its mode of origin or its composition. In point of fact however, as will appear below, a general correlation exists between all three aspects.

Three types of surface deposit

I. Loess of the type described accumulated in North China during the Malan (Upper Pleistocene) stage of aggradation. It is without bedding (see Plate 3) except in marginal areas where occasional layers of talus were washed down by thunderstorms, and along the axes of depressions in which, even during those more arid days, standing or flowing water was still present, at least intermittently.

Apart from shells of land-snails, occasional ostrich-egg fragments (*Struthiolithus*),[1] and still rarer hackberry seeds (*Celtis*),[2] the fine-grained facies is almost devoid of fossils. The characteristic Upper Pleistocene mammals are virtually restricted to the sandy lenses which mark the position of former water-holes and stream channels. The palaeolithic *foyers* found by Teilhard and Licent[3] near the base of the formation on the margin of the Ordos are similarly situated.

Reasons have been put forward elsewhere[4] for the belief that von Richthofen[5]

[1] J. G. Andersson, "Essays in the Cenozoic," *Geol. Surv. China, Mem. A.* 3, 1923, p. 53.
[2] G. B. Barbour, "Kalgan Area," *Geol. Surv. China, Mem. A.* 6, 1929, p. 68.
[3] P. Teilhard and E. Licent, "Discovery of a Palaeolithic Industry in N. W. China," *Bull. Geol. Soc. China*, vol. 3, 1924, p. 45.
[4] G. B. Barbour, *Geol. Mag.*, vol. 67, 1930, p. 467.
[5] F. von Richthofen, 'China,' vol. 1, Berlin 1877, p. 74.

[*Editor's Note:* Certain plates and figures have been omitted owing to limitations of space.]

was correct in regarding the bulk of the Malan loess as an aeolian deposit. The most convincing arguments lie (1) in its regional spread and variation as related both to the likely sources of detrital supply and to the trend of atmospheric pressure gradients, and (2) in its local topographic distribution, especially where it occurs far up the slopes of flanking mountains or overlooking the Huangho delta plain. Corroborative evidence is found in the character and distribution of the fossils; and in the thinning out and change of facies in the loess when traced down wind into the maritime belt and towards Manchuria, or up wind onto the Mongolian plateau where there is insufficient topographic relief to harbour, or vegetable cover to anchor, the finer wind-swept material. Aeolian activity of this kind has been described from Montana,[1] Patagonia,[2] the Sudan,[3] and elsewhere, and may still be seen operating on a feebler scale in North China to-day. The imported sediment which fell at Peking during one night of a heavy dustfall in March 1927 was estimated at 110 tons per square mile. Microscopic analysis showed it to be indistinguishable from typical loess of the Malan type.[4]

Failure to differentiate the formation from more ancient underlying ones has been responsible for exaggerated ideas as to its thickness. Travellers' references to "thousands of feet" of loess are also explained by the fact that a thin loess cover, veneering a dissected surface of 1000 feet relief, was assumed to imply a vertical thickness of that order of magnitude. In point of fact, there is no record of more than 600 feet thickness of undoubted loess, and throughout most of the region it does not exceed 100 feet (see Plate 2).

II. Over much of the North China highlands the Malan loess rests on an older deposit which shares some of its characteristics but has certain distinctive features. The colour is usually redder, the texture is slightly more clayey, and the vertical cleavage less perfect than is the case with the Malan loess (see Plate 8). Moreover the cliffs show a crude layer structure, marked by faint colour-banding and zones of concretions, the individual layers being from 2 to 30 or more feet in thickness (see Plate 7). The material in the centre of the thicker layers is less red in colour, loessic in character, and in the hand-specimen cannot be distinguished from the pure Malan loess. As many as twenty such layers can be counted in Plate 4. This approach to stratification frequently has a distinct dip conforming to the moulding of the depressions in which the material lies. In the centre of the basins the deposit often grades into stratified sandy beds or laminated lake silts (see Plate 5) with fresh-water mollusca, plant fragments, and occasionally gypsum.

There can be no doubt that the dark bands represent a succession of ancient "fossil" soil horizons from which the more soluble material was leached, the lime being redeposited in the concretionary layers which occur just below. Deposition must have been interrupted by long pauses during which the surface was exposed to weathering for sufficient time to develop mature soil profiles before being buried under a fresh accumulation. The origin of the

[1] N. S. Shaler, "Loess Deposits of Montana," *Bull. Geol. Soc. Am.*, vol. 10, 1899, p. 246.
[2] B. Willis, 'Northern Patagonia,' Scribners, 1914, p. 106.
[3] Mr. G. W. Grabham, personal communication.
[4] G. B. Barbour and C. J. Wu, *China Jour.*, vol. 7, 1927, p 305.

material will be considered below when comparing its mechanical and chemical composition with those of the Malan loess. But we may anticipate the conclusions by stating here that the banded loams are believed to be a series of loessic deposits which lost some of their original characteristics as the result of prolonged weathering.

The fluctuating nature of the climatic conditions responsible for the intermittent deposition awaits further investigation, but the process of accumulation undoubtedly covered a much longer span of time than that involved in the Malan stage. From a study of the mammal fossils Teilhard and Young have proved[1] the presence of two main faunal horizons—the Nihowan and Choukoutien associations, of Upper Pliocene and Lower Pleistocene age respectively. To the latter belongs *Sinanthropus*. At least one additional subhorizon is distinguishable (Stage *X* in Fig. C below).

Until this finer discrimination on a fossil basis was made, the entire series of banded loams was grouped with the Sanmen formation, but the palaeontologists have recently been inclined to restrict this useful inclusive term by making it equivalent to the Nihowan.

Over wide areas the loessic loams build physiographic features of major importance, occurring as dissected remnants of a prominent system of matched terraces flanking many of the rivers. But since they are usually unfossiliferous one cannot be sure to which substage they belong. However as the first aggradation was unquestionably of regional importance, it is safe to assume that Nihowan deposits are usually at the base of the terrace features. Physiographically therefore we may meanwhile continue for convenience to refer to the group collectively as Sanmen loams except where a slightly younger age can be demonstrated. Thus far this has only been possible in two or three isolated localities including the cave deposits made famous by the discovery of *Sinanthropus*.[2]

The question arises as to whether these deposits should be referred to as loess. Among engineers and in popular parlance they have gone by this name for so long that a situation has developed analogous to that regarding the use of the term laterite. If the word loess is to be allowed to include these loessic loams then care must be taken to distinguish the Malan loess from these older, partially altered Sanmen loess deposits. The alternative is to designate the latter as "banded loessic loams" or simply "loessic loams," adding a stratigraphic horizon where this is identified.

III. Still older than the loessic loams are a series of Red Clays first studied by Andersson[3] and referred to the Paote Stage (Pontian) of the early Pliocene. They are associated with a less rugged topography than that of to-day and have only escaped erosion on upland surfaces or in protected positions under the loams.

[Editor's Note: Material has been omitted at this point.]

[1] P. Teilhard and C. C. Young, "Preliminary Report on the Choukoutien Fossiliferous Deposits," *Bull. Geol. Soc. China*, vol. 8, 1930, pp. 173–202.

[2] For a discussion of the stratigraphic problems raised by recent palaeontological studies reference should be made to Teilhard and Pei, "New Discoveries at Choukoutien," *Bull. Geol. Soc. China*, vol. 13, 1934, pp. 385 ff.

[3] J. G. Andersson, "Essays in the Cenozoic," *Geol. Surv. China, Mem. A.* 3, 1923, p. 107.

Editor's Comments
on Papers 8 and 9

8   **BERG**
    Excerpts from *The Origin of Loess*

9   **RUSSELL**
    Excerpts from *Lower Mississippi Valley Loess*

Berg was the most persistent proponent of the *in situ* theory of loess formation. He held that the eolian theory of loess deposit formation was in error and that loess had essentially formed where it was found. His theory was published in many forms in various journals and books. Paper 8 has been chosen for inclusion as representing a fairly concise statement of Berg's position, and since it was originally published in English it presumably represents a text that met with Berg's approval. Another, more recent (1964) version of Berg's work in English exists but does not include any major developments of the basic theory. In fact, Berg claimed to have outlined his theory in 1916 and he continued to propagate it all his life.

Berg has rather neatly summarized the fascination of loess (1964, p. 2):

> The most problematic aspect of loess is the manner of its origin. Without solving this problem one cannot understand the history of regions which underwent glaciation. Indeed, while some authorities assert that loess was deposited on dry steppes—as wind-blown dust, others attribute an aquatic origin to the loess, considering it as a deposit due to glacial streams. The environments of physical geography presupposed by the one and by the other hypothesis are in direct opposition to each other . . . .
>
> Furthermore, loess quite often contains remains of Quaternary fauna and also human industries (or sometimes even skeletal remains of man). Therefore, as long as we do not solve the riddle of loess, we will fail to grasp the latest and most important stages of the history of both Earth and Mankind.

Paper 8 was cited with approval by Russell (Paper 9) and appears to have influenced him in the formulation of his famous theory of Mississippi loess formation, which provided so much controversy in the late 1940s and early 1950s. Berg has written "My own theory became known abroad by a brief review written by Mohr in Intern. Mitteil. für Bodenkunde 10, 1920; and later, by a more detailed exposition due to Anger and Wittschell (proponents of the eolian hypothesis) in Peterm. Mitteil, 1929 (p. 7-9); and lastly by my own article in Gerland's Beiträge zur Geophysik, 1932."

Paper 9 is difficult to introduce. It has caused the most controversy among loess investigators since the time of the acrimonious discussions on the "loess problem" at the end of the nineteenth century. The topic of discussion in each case was essentially the same. The first great discussion of the loess problem subsided into a fairly general acceptance that loess deposits were emplaced by eolian action (see Paper 3). This theory was challenged by Russell in his 1944 paper, and the response showed that geologists were still largely in favor of the eolian deposition theory. Berg had been offering an *in situ* theory for a number of years by the time Russell published his paper, but his views had not become widely known outside Russia and his major publication in English had not attracted much attention.

In retrospect, one can see the colossal oversimplification that distorted arguments about the nature and formation of loess. The total concentration on the "loess problem" probably prevented rather than encouraged the sensible investigation of the nature and formation of the material, the formation of the primary deposits, and subsequent important events.

There is a tendency to push Russell's paper into the background as an embarassingly eccentric view which it is polite not to notice. This would be a mistake; the paper contains several interesting and challenging observations. Russell's definition of loess is, par excellence, the observational definition; it contains no hint of formation mechanism, and arguments can be mustered that this is the best sort of definition for loess. The problem of definition—should loess be described via the mechanism of formation or via its observed nature—is basic and as yet unresolved. An entire scientific philosophy is involved in this problem and it cannot be adequately discussed in this short introduction. Russell also provided a first-class review of early loess investigations, and this alone would make the paper valuable to the contemporary loess investigator.

It can be argued that Russell's definition of loess led inevitably to his theory of loess formation. His definition stressed the carbonate content of the loess; thus, presumably, no material could be loess until it contained the requisite amount of carbonates. If the carbonate content is totally critical, the eolian deposition of fine quartz material may seem somewhat redundant. An attempt has been made to explain (justify?) the Russell approach to the "loess problem" and the interested reader is referred to Paper 45. For a very comprehensive study of the Mississippi loess, reference should be made to the work of Snowden and Priddy (1968) and Krinitzsky and Turnbull (1967).

The consensus, certainly in the United States and probably in most other places, is that Berg and Russell were wrong and their theories discredited. But when considering their work we should keep in mind the words of Oliver Wendell Holmes:

> The debris of broken systems and exploded dogmas form a great mound, a Monte Testaccio of the shards and remnants of old vessels which once held human beliefs. If you take the trouble to climb to the top of it, you will widen your horizon, and in these days of specialized knowledge your horizon is not likely to be any too wide.

REFERENCES

Berg, L. S. 1964. Loess as a product of weathering and soil formation. Israel Program for Scientific Translations, Jerusalem, 207p.

Krinitzsky, E. L., and Turnbull, W. J. 1967. Loess deposits of Mississippi. Geol. Soc. America Spec. Paper 94, 64p.

Snowden, J. O., Jr., and Priddy, R. R. 1968. Geology of Mississippi loess. Bull. Mississippi Geol. Econ. Topograph. Surv. 111, p. 1–203.

8

Reprinted from *Gerlands Beitr. Geophys.*, 35(2), 130–143, 149–150 (1932)

THE ORIGIN OF LOESS.

By

L. S. Berg, Leningrad.

In the following a short summary of my theory of the origin of loess is given.

I. Types of loess sediments.

Loess we call a sediment chiefly distinguished by peculiar texture: it is a loam, less frequently a sandy loam, mainly consisting of middle-sized particles of 0.05 to 0.01 mm. in diameter, which constitute half or even more of its mass; particles larger than 0.1 mm. are rare, and over 0.25 mm. all but deficient. Loess is an unstratified sediment (or at least, as a rule, not distinctly stratified), always porous, liable to split in vertical joints and being generally of a pale yellow colour. Loess is rich in carbonates containing 10—15 per cent, and occasionally more, of $CaCO_3$ and $MgCO_3$. Such a deposit is generally called a typical or aeolian loess, and is considered as arising from dust drifted by wind.

Deposits differing in one or several features from those just described are not called loess, but loess-like loam, loess-like clay, loess-like moraine, loess-like sand, loess-like sandy loam, surface clay loam, lake loess, etc. Loess-like deposits are now generally believed to be of aqueous origin.

Let us now say a few words on the chief hypotheses on the origin of loess.

II. Fluvio-glacial, deluvial, aeolian hypotheses.

1. Fluvio-glacial hypothesis.

Before RICHTHOFEN (1877) loess was generally believed to be an alluvial or, rather, fluvio-glacial deposit. Thus, LYELL (1834) considers the Rhine loess as an alluvium, deposited by the Rhine in the glacial period when great quantities of glacial mud were carried by rivers from the Alps.

The fluvio-glacial theory was and is still widely adopted in Russia. Of early authors we may here mention KROPOTKIN (1876) and DOKU-CHAEV (1892). Recently this theory begins anew to gain adherents.

K. GLINKA, who was formely (in 1908) a follower of the aeolian hypothesis, inclined 1923 (p. 126) to think that the loess of European Russia is an accumulation of muddy drift carried by glacial currents[1]). Soon after the retreat of the ice-sheet shallow valleys were incapable of holding the whole mass of the melted water of glaciers. This water periodically rose, overflowing the river banks, submerging the water-sheds and depositing a fine-grained material. "As high floods occurred but periodically, they could not altogether prevent the growth of vegetation, nor even the local formation of soils". This theory, according to GLINKA, helps to explain the presence in loesses, in loess-like loams and in similar deposits, of gravel and small boulders, which may have been transported on floating ice. TANFILIEV (1922, p. 109) suggests also that glacial waters participated in accumulation of at least some of the South Russian loesses.

The fluvio-glacial theory of the origin of White Russian loess was recently developed in detail by AFANASSIEV (1926). He points that glacial waters were evidently capable of depositing on preglacial areas not only sand and gravel, as is generally supposed, but a series of sediments from coarse to fine-grained ones. Such series of sediments can be actually observed in White Russia.

An objection is raised to the fluvio-glacial theory in the absence of any evidence that floods of glacial waters had ever submerged water-sheds now covered by loess. To this, on the other hand, it could be replied that neither has the submersion of water-sheds by fluvio-glacial waters been disproved. Certain data could be found in favour of the theory of their having been submerged. Thus, "pans", or small shallow depressions on water-shed steppe plateaus (Ukraine), are, evidently, remnants of such submersions. With reference to the inundation of water-sheds by fluvio-glacial waters K. GLINKA (1923) expresses these very weighty considerations: soon after the retreat of the ice-sheet river valleys, being filled in by moraine and fluvio-glacial deposits, could not be very deep. "Thus the Don-Voronezh water-shed within the Voronezh government proves to be composed throughout its entire width (4—5 km) of mighty fluvio-glacial sediments (several dozen meters). It is evident that under such conditions the shallow valleys of areas adjoining the ice-sheet were unable to hold all the quantity of water that was furnished by the melting of the stationary glacier. This water,

---

[1]) Latterly K. GLINKA (1927, p. 577) adhered to the soil theory of the origin of loess as advanced by myself.

especially in summer, overflowed the river banks, flooded the vast areas
of water-sheds and during its slow onward course deposited finely
grained material, which levelled out the inequalities of the surface and
eventually settled down in a sufficiently thick layer".

But to the fluvio-glacial hypothesis the following objection could
be urged: no recent fluvio-glacial or alluvial rocks can be recognized as
loess. In order to transform them into loess some loess-formative process
is necessary, but the fluvio-glacial theory does not even allude to it.

2. Deluvial hypothesis.

The second hypothesis, which numbers many followers in this
country, may be called deluvial. A. P. PAVLOV (1888) has proposed
to call deluvium (from deluo — I wash away) products of rock decay
drifted by rain currents and deposited on slopes and plains. PAVLOV
considers South Russian loess to be a deluvial deposit. This view, akin
to the hypothesis of ruissellement urged by LAPPARENT (1883), was
adopted by ARMASHEVSKY (1883, 1903) who had studied loess in the
Chernigov and the neighbouring governments, as also by others.

The deluvial theory explains very satisfactorily the presence of
loess on slopes; it must however be noted that the predominant type
of deluvium in loess regions is that which is derived from loess itself.
But as to loess of water-sheds (loess on plateaus) the deluvial hypothesis is unable to give an explanation of its origin, without presuming
that at some time there were heights, from which loess had been washed
off, and that later on these heights had been abolished by denudation.
This presumption, however, does not tally with well known facts
relating to the history of the landforms of loess regions.

Deluvial loesses do undoubtedly exist. They are the loesses of
slopes. But it must be borne in mind that at the present period the
formation of deluvial loesses has terminated, as their surface is covered
with soils.

3. Aeolian hypothesis.

The most widely adopted theory at present is the dust or aeolian
hypothesis. According to RICHTHOFEN (1877, 1886) loess is a deposit
formed of dust carried by the wind. Wind raises fine products of weathering in deserts and deposits them in steppes where the dust is retained
by the grassy vegetation and transformed into a porous, unstratified
rock — aeolian, or typical loess. Such an assumption RICHTHOFEN drew
from his travels in China, where he had repeated occasions to witness

this precipitation of dust. The celebrated traveller believed that Chinese loess is a recent formation being even now deposited in China by dust accumulation. When an attempt was made to apply this theory to European loess, it at once became evident that that loess is not a recent formation, but a geological one. Hence an explanation was sought for in former physico-geographical conditions: loess must be a product of former deserts. TUTKOVSKY (1899) advanced an opinion that South Russian loess was formed during the retreat of the great ice-sheet. The margin of the ice-sheet was the seat of anticyclonic winds of the character of foehns, i. e. relatively warm and dry. During the recession of the ice-sheet, its margin was bordered by a belt of deserts, where deflation of moraine deposits took place by glacial foehns, blowing from the north-east and the east. Throughout this desert belt barkhans (sand dunes) were formed, while the finest dust was transported further southward. Beyond this belt of deflation on the south, south-east and partly south-west extended "steppes with a continental climate, where fine moraine dust was deposited by the glacial foehns (zone of deposition)". This drift became a sediment of normal (typical) loess with all its varieties.

Objections to the aeolian hypothesis.

Leaving aside for the present a criticism of this manifestly untenable view, let us first adduce the objections to the wind-drift theory as it was formulated by RICHTHOFEN:

1) To begin with, the aeolian hypothesis is not derived from facts, observed to occur now, but takes for granted such physico-geographical phenomena that have never been observed. RICHTHOFEN (1877), B. WILLIS (1907), TUTKOVSKY (1899), V. OBRUCHEV (1911) and others thought that the loess of China and Turkestan is of modern growth and is still being formed by dust-drift. As was elsewhere shown by the author (1916), the processes of loess formation may still occur in a dry climate. But the main masses of Turkestan and China loess are not of recent formation, but are witnesses of former geological periods. Thus, in Ordos, the quaternary loesses are underlaid by mighty beds of similar rocks of Upper Tertiary age; the quaternary loess mantle is in its turn subdivided into many horizons[1]).

The dust observed by travellers in Turkestan, Central Asia, Mongolia, Manchuria and China is the result of man's activity, being a product

---

[1]) M. BOULE, H. BREUIL, E. LICENT et P. TEILHARD, Le paléolithique de la Chine. Arch. Inst. de paléont. humaine, Mém., IV, Paris 1928.

of deflation of upper horizons of soils developed upon, and from, loess. In Asia loess regions have for thousands of years been intensively cultivated, and the uppermost soil horizons can be easily carried away by the wind. "In Manchuria the winter is both very cold and dry; in ploughed fields the earth-clods very rapidly dry in the snow and are dispersed by the wind as dust over the surface of the snow. In winter on the way from Tsitsikar to Mukden may be seen snow drifts completely covered with black bands of dust. These drifts sometimes seem to consist not of snow, but of dust alone. Further on towards Peking the conditions remain the same, but there is still less snow, and the dust is no longer black, but yellowish"[1]).

Dust-storms — the result of excessive tilling — are frequently observed not only in Central Asia, but also in Ukrainian steppes[2]). Towards the end of April 1892 150 000 hectares of crops have been blown off in the Mariupol district; clouds of dust had completely obscured the sun on the 28th and 29th of April. It is also well known what masses of dust are raised on the roads of Turkestan by the slightest wind or by the motion of vehicles. Investigations of Ferghana dust, carried out in 1913 at a special dust-station at Osh, have proved that in Ferghana we have to do with a local Ferghana dust, derived from local soils tilled for ages. This dust, as well as dust of any other origin, is carried by winds until it reaches a lake, river or the sea. When the dust is beaten by the rain into the soil it becomes part of the latter as one of its constituents and does not differ from the usual soil particles; but in no way can anything like loess be obtained in this manner.

Since the studies of NEUSTRUEV (1910) it has become a matter of general knowledge that "aeolian" loess in Turkestan is now no longer formed. In these regions loess is generally overlaid with serozem — a normal zonal soil, to which a wind-drift origin could be attributed with as much reason as to soils of the brown, chestnut or chernozem zones.

2) The second objection, which has frequently been raised, is the following. It is absolutely incomprehensible, why the wind should drive sediment of only that texture which is characteristic of loess. The wind, according to its velocity, can carry either coarser or finer particles, but why it should give a preference to particles of 0.01 to 0.05 mm. in diameter, has never yet been explained by any follower

---

[1]) A. D. VOEYKOV, Climatic conditions of horticulture in Manchuria. Messenger of Manchuria, Harbin 1927, No. 2 (in Russian).

[2]) See my Elements of Climatology, 1927, pp. 141—142 (in Russian).

of the aeolian theory. Typical loess being characterized by the predominance of particles of the above mentioned diameter both in Europe, Asia and America, we should have been forced to conclude that the wind had everywere the same velocity. Moreover, the wind would have to blow in the same direction and with the same velocity during tens of thousands of years. Otherwise the sediment that had settled down could never be so uniform. As in actual fact winds blow with varying force and from varying directions, it is evident that had aeolian loess existed it would have been a mixture of particles of the greatest variety of coarseness, the texture of loess in neighbouring areas at the same time being very diversified. The adherents of the wind-drift hypothesis suppose that dust with a texture characteristic of loess had been drifted. But this is improbable: texture peculiar to loess must have been formed in situ.

3) The followers of the aeolian theory affirm that loess is formed not in deserts, whence loess dust is blown off by winds, but on the periphery of deserts, in steppes, where vegetation contributes to the accumulation of loess. Thus, according to this notion, the whole profile of loess had to pass through the stages of soil formation, namely of chernozem, or, at least, of the chestnut type of soils. But in that case a considerable quantity of humus should be present in loess, what, as is well known, does not take place: the content of humus in normal loess is manifested in tenths or hundredths per cent, and sometimes it runs down to naught. One might say that humus had been present once and had subsequently decomposed. But loess beds, as we know, are generally interstratified with one or occasionally several fossil humus horizons in which humus is unaltered; although little humus remains in these ancient soils, from 0.3 to 1.1 per cent on average, but a humus horizon is always distinct. Thus, if loess had been formed by steppe vegetation being buried under an accumulation of dust, the entire profile of loess should exhibit a semblance of a humus horizon. But such in fact is not the case. Therefore loess could not have been deposited in steppes.

4) From the point of view of the wind-drift theory the presence of such rocks as loess-like moraine loams is quite incomprehensible. The aeolians have not even attempted to explain this fact. I have observed loess-like boulder loams in the Gloukhov district of the Chernigov government. AGAFONOV has described a similar sediment containing a large number of boulders of crystalline rocks from the government

of Poltava (Khorol district). In the northern part of the government of Kiev (in the region of glaciation) peculiar boulder loams occur: "if the presence of boulders in these loams be ignored, in all other features they would represent typical loess" (FLOROV, 1916).

5) The occurrence of stratified loess has induced aeolians to advance quite untenable theories: they think that if dust reaches a lake it may give birth on its bottom to a peculiar stratified formation -- lake loess[1]).

But nothing can be more erroneous than this view: dust deposited on the bottom of a lake (supposing there is so much of it that it could form a considerable layer) will give rise to loam or sandy loam, but by no means to loess. Loess, when decaying, not only under water, but also on dry land, becomes transformed from a loose porous rock into a simple viscous loam enriched in fine-grained particles.

Not only ordinary dust deposited in a river or in a lake does not give birth to loess, but it is evident that even loess-dust, i. e. dust resulting from the deflation of loess, once fallen into water, must loose a greater part of the properties pertaining to loess and be transformed into the most ordinary loam or sandy loam, into the prosaic mixture of sand and clay, sometimes calcareous.

Similarly, chernozem can never been formed in a bog or under water. A bog soil is able to assume a chernozem aspect only when it dries up and ceases to be a bog. While a sediment remains under water, even if it be of the same mechanical composition as loess and contains carbonates of lime, it cannot assume the aspect of loess and should not be called either stratified, lake, or any other, loess, that sediment not having any relationship with the latter.

The notion of "lake" loess could only have arisen from observing terrestrial stratified rocks of a loess-like aspect. But this aspect was only obtained after the sediment had emerged from above the level of water. It is thus not a "lake" loess, i. e. a loess formed in the lake under water, but a loess from lake alluvium, which is far from being the same thing. Such stratified loess like sediments are of very common occurrence, for any fine-grained alluvium in a dry climate assumes the aspect of a loess-like sediment. Two instances can be adduced in illustration of the above.

---

[1]) Lake loess has been formed "by the settlement of aeolian dust transported from deserts of the deflation zone by glacial foehns" (TUTKOVSKY, 1912). Some authors (N. KRISHTAFOVICH, 1902), presume that dust deposited in a river can give birth to a special "alluvial loess".

If there is "lake" loess, why should there not be stratified "river" loess? And in fact there exists a recent loess-like river alluvium in the valley of the Lena near Yakutsk, where the climate is dry and continental[1]).

We shall now show an instance of stratified "marine" loess. On the banks of the Ural River near Uralsk, in the riverside cliffs may be seen a brown loess-like sandy loam, stratified below, with shells of the recent Caspian mollusc Adacna plicata (PRAVOSLAVLEV, 1913). On the lower Volga near Sarepta the Caspian sediments pass above into a formation very much resembling loess, but stratified[2]). A stratified marine deposit through the processes of weathering and soil-formation assumed on land a loess-like aspect.

Only in this manner can stratified loesses, river, lake, and marine, be obtained. They are stratified because they were subject to a process of loess-formation throughout a brief period and had not time to be transformed into true loess. They are loess-like because they emerged from water and became a terrestrial rock, undergoing the processes of weathering and soil-formation in a dry climate.

6) Let us now consider the "foehn" theory of TUTKOVSKY (1899). There can be no doubt that above the continental ice-sheet an anti-cyclonic system of isobars was prevalent. Yet, there can be as little doubt that in South Russia this circumstance could give rise to no foehns (except, possibly, to the so called foehns of the "free atmosphere" which do not interest us at present), as foehns are due to the falling of air from a more or less considerable height, this condition being wanting at the southern end of the ice-sheet. The "belt of sedimentation" in which loess accumulated was situated at too great a distance from the border of the ice-sheet to be affected by hypothetical glacial foehns.

According to the foehn hypothesis, these problematical foehns were supposed to blow from North-East and East: they formed "barkhans" in the belt of deflation and transported moraine dust to the South-West, West and South. But, as I have already shown in my recent work (1926, p. 80—82), the sand accumulations in the Polesie (Dniepr basin) described under the name of "barkhans" are in reality dunes formed by

---

[1]) G. N. OGNEV, Matériaux de la Commission Iakoute, No. 22, 1927, Acad. of Sciences.

[2]) A. ARKHANGELSKY, Mémoires du Comité géol., No. 155, 1928, p. 129. See also N. DIMO and B. KELLER, In the zone of semidesert, Saratov 1907, pp. 11—13.

the action of western and south-western and certainly not by north-eastern and eastern winds. During winter southward of the ice-sheet apparently blew mainly western winds (ENQVIST, 1916).

TUTKOVSKY supposes that during the advance of the ice-sheet and its stationary state the conditions for the deflation of moraines beyond the border of the ice-sheet and for the deposition of loess dust in steppes were unfavourable; but, according to the same author, during the retreat of the ice-sheet such conditions were present. The area liberated after the retreat of ice was an "absolute desert"; under the action of dry and warm foehns boulders split and gave rise to loose material drifted by the wind and deposited in the form of loess.

At present there can be no doubt whatever that what has just been described in no way corresponds with actual facts. Areas left behind the retreating glacier were not converted into a desert, but, as is self-evident, were first clothed with typical tundra vegetation, with such characteristic plants as Dryas octopetala, the dwarf birch, the polar willow. South of this belt spread out birch forests[1]. That such was the case we may judge from the occurrence of remains of plants in peat bogs, as also from the fact that under the upper horizon of the water-shed loess, on the moraine itself or on beds of fluvioglacial and alluvial deposits overlying the moraine, may be observed a horizon of fossil soil representing what may be called bog soil, swamped soil, podzol or (in the government of Kherson) even chernozem soil. During the period of formation of fossil bog, podzol and chernozem soils, there, evidently, could not have been deserts either there or, still less, further north.

In alluding partly to these facts, KROKOS, an adherent to the aeolian hypothesis, and AFANASSIEV, an adept of the fluvio-glacial hypothesis, both indicate that "no deflation of moraine deposits by the wind from retreating ice-sheets could have taken place" (KROKOS, 1924, p. 23, 28), that in this case "not the smallest trace of wind erosion" could ever be found (AFANASSIEV, 1924, p. 151).

It has been already pointed (A. P. PAVLOV, 1911), that a moraine (boulder loams and clays), owing to its solidity, constitutes, generally speaking, a material little adapted to deflation and to the production of dust. That moraines could not be subject to deflation may be inferred from the fact that in many parts of the Smolensk Gvt. were observed (NIKITIN, GLINKA, DOKTUROVSKY) fossil peats directly overlying mo-

---

[1] For details see my paper: Problem of the loess. II. "Priroda", 1929, pp. 319—323 (in Russian).

raines [in the Belsk district with remains of the hornbeam (Carpinus betulus) and Brasenia purpurea].

Thus, at the period of loess-formation there were neither foehns, nor eastern winds, nor deflation of moraines, remaining after the retreat of glaciers. Therefore, the foehn hypothesis must be wholly abandoned.

IV. The source of dust.

At present many followers of the aeolian hypothesis endeavour to regard as the source of dust not moraines, but fluvio-glacial and alluvial deposits (PENCK and BRÜCKNER, SÖRGEL, MIRCHINK and others).

If we agree with this supposition, we should be obliged to admit that vast accumulations of fluvio-glacial sands, sandy loams and loams were wafted by wind and partly transformed into sand dunes, partly drifted south and deposited in the form of loess, and partly (clay particles) driven into the sea. But the unlikelihood of such a supposition is obvious. Fluvio-glacial sand loams and loams, as any other deposit, would also be covered with vegetation. Only sands could have been subject to deflation during the dry post-glacial period. But how great must have been the quantity of sands if their deflation should give origin to the area of Ukrainian loess extending over more than one million sq. kilometres? Such sands must have covered an area about ten times as great. Let us take, for example, the analysis of sand from the Gorki district of the Mohilev government: according to AFANASSIEV (1924) it contains less than 7 per cent of particles 0.05—0.01 mm in diameter ("dust"), whereas loess of the Chernigov government possesses 50—80 per cent of such particles.

Material blown off fluvio-glacial deposits could not produce dust of a uniform texture, as it is impossible to imagine that upon river valleys both banks of which consist of loess, as for instance on the middle and lower Dniepr, should constantly blow winds in opposite directions. What far-fetched arguments are in this case resorted to, will be shown later.

V. Loess as a product of weathering and soil formation in situ.

Thus, neither the wind-drift, nor the alluvial or deluvial hypotheses can account for the origin of loess.

According to the hypothesis advanced by me in 1916 loess and loess-like sediments may be formed in situ from the most various fine-grained rocks rich in lime carbonates, through the agency of weathering and soil-formation in the conditions of a dry climate.

Origin of loess and origin of its parent rock.

Let us now examine the question, whether different kinds of loess as also loesses and loess-like sediments could not be of different origin. For example, could typical loess in one case be of aeolian and in another of alluvial origin; or should not rather all loesses be of aeolian and loess-like deposits of alluvial (fluvio-glacial) origin? It has been already mentioned that most scientists have adopted the last point of view. But this view is quite incorrect, as will be seen from the following.

What induced us to speak of loess-like moraines, loess-like sands and loess-like clay? A peculiar loess-like aspect of the above mentioned deposits. What is that aspect? Particles of loess-type, i. e. 0.05—0.01 mm. in diameter, begin to play an important part in this rock, the deposit becomes porous, calcareous, more or less yellowish, with a tendency to split in vertical joints.

Thus, we see that many rocks rich in carbonates can acquire a loess-like aspect. We must therefore distinguish the origin of the rock (let us call it the parent rock of loess) from the origin of its loess-like aspect. The origin of a parent rock may be very multifarious: it may be alluvial, deluvial, fluvio-glacial, glacial, aeolian, etc. But the loess-like aspect is, evidently, always acquired by the rock in some one way. The difference between loess and parent rock is the same as subsists between soil and rock: in order to transform rock into soil a soil-formative process is necessary; in order to convert rock into loess a loess-formative process is requisite. This process, which in detail allows many variations, is essentially the same: and from such a point of view we may speak of one family of loess-like sediments.

Process of loess formation.

What was that loess-forming process which transformed various superficial rocks into loesses and loess-like deposits?

The latest works of K. K. GEDROIZ (1912—1926) and GANSSEN (1922) have ascertained that process.

To acquire a loess-like aspect a sediment must necessarily contain: 1) a considerable quantity of the minutest particles of 0.0001 mm. or less in diameter, 2) a certain quantity of alumosilicates, notably those rich in SiO_2, 3) a considerable quantity of lime (and partly magnesium) carbonate; besides which, 4) the weathering process should occur in a dry climate peculiar to steppes or deserts.

The result of weathering in such conditions would be a sediment of "dusty" texture, i. e. with a considerable content of particles of 0.01 to 0.05 mm. in diameter, and of loose structure. The cause is that under the action of lime carbonate particles of silty fraction become as if cemented together forming coarser aggregates. These aggregates are rather stable and do not desintegrate even under the action of soil waters (no aggregates of any larger size can obviously be formed in that way). Owing to the formation of coarser particles from smaller ones pores are produced in rocks. This explains the porosity and looseness of loess.

Therefore loess may likewise be formed of material richer in fine particles than those of which it eventually consists. GANSSEN (1922, p. 41) has made the following remarkable experiment. A kaolin-sample was exposed to the action of alkaline silicates. Towards the end of the experiment it appeared that nearly half of the kaolin, which cannot be decomposed by muriatic acid, was transformed into zeolite silicates decomposable by muriatic acid. At the same time its texture had undergone the following peculiar alterations:

Particles of diameter	Before the experiment per cent	After the experiment per cent
finer than 0.01 mm.	93.5	45.3
from 0.01 to 0.05 mm.	3.3	43.2
coarser than 0.05 mm.	3.3	11.6

As may be seen, after the action of alkalies, the diameter of particles increased, and the texture approached that of loess. A similar process must have occurred in the parent rocks of loess also under the action of absorbed calcium and magnesium as also of carbonates of the same metals. The process of loess formation, says GANSSEN, consists in a nearly absolute hydratation of fine-grained alumosilicates, rich in silica, and in the envelopment of soil particles by calcium carbonate; owing to this process particles finer than 0.01 mm. in diameter are transformed into particles 0.01—0.05 mm. in diameter.

That is what occurs in a dry climate where, as we have seen, clays are transformed into loams. Whereas, if alumosilicates are weathering in a humid climate the quantity of (physical) clay increases, that is the number of particles under 0.01 mm. in diameter grows. When weathering loesses lose their carbonates and at the same time become more fine-grained and viscous.

Followers of the aeolian theory might say: Well, we admit that a rock acquires a loess-like aspect owing to loess-formative processes, but we are concerned in the origin of the parent rock of loess; and it is aeolian; the parent rock of loess is dust, an "aeolian" deposit.

But to this we can make the following reply. Therefore you admit, that by some process a rock may acquire a loess-like aspect. But why should you think that a parent rock may only be transported by wind? How can you account for that?

Recent formation of loess.

A question rises: is loess-formation a recent process or an ancient one?

To this the following answer could be given. Both recent and ancient loesses exist. But wherever loesses are met with they are always indicators of a dry climate. Transformation of alluvial and deluvial deposits into loess can be observed at present in Turkestan and Transcaucasia. NEUSTRUEV (1910) writes on the recent sediments in the Chimkent district, Turkestan: "On a closer examination of the deposits of Turkestan, we see that typical features of loess either all, or in certain combinations, are exhibited in all fine sediments. Alluvial deposits acquire porosity, a light coloured tint, a certain looseness and invariably effervesce on the application of an acid". "Nearly all alluvial and deluvial formations of the Chimkent district have the appearance of loess in some respect or other. They are all distinguished by possessing a high amount of $CaCO_3$, most of them are porous and of a very fine texture" (p. 20). In the Andizhan district loess-like layers occur very often in recent alluvial deposits of the rivers Kara-darya and Naryn (NEUSTRUEV, 1912). Sandy-clay river sediments of Eastern Turkestan (along the rivers Kashgar-darya, Yarkend-darya, Khotan, Keria and Nia) are called loess-like by BOGDANOVICH (1892, p. 193).

The soils of the Mugansteppe, Baku government, are derived from sediments carried by the Araxes and must be classed as alluvial deposits (ZAKHAROV, 1905). They are often represented by porous loams determined by ZAKHAROV as being "plain alluvial river-loess". As regards their texture, the Mugan alluvial soils are rather allied to Turkestan loesses (TULAIKOV, 1906). Alluding to this circumstance KOSSOVITCH (1911), a follower of the aeolian hypothesis, speaks that "the alluvial formation of rocks similar in texture to rocks of aeolian origin is possible". But alluvium can of course be transformed into loess by weathering only in a dry climate.

Hitherto we have given examples of recent loess formation in deserts. But the same process may be observed in the chernozem zone. As was indicated by Bogoslovsky (1899), in regions, where chernozem is developed upon moraine loam, the latter, in its upper horizons, acquires a loess-like character by becoming porous and highly calcareous and assuming a yellowish colour; the transformation extends down to a depth of $2^1/_2$—3 m from the surface. In the governments of Saratov and Simbirsk deluvium of tertiary silicious clays and sandstones, where they constitute the subsoil (parent rock) of chernozem, acquire an altogether loess-like aspect (l. c.). Bogoslovsky says that, under the influence of steppe weathering, soils of most diverse origin in many cases (owing to their texture) acquire a loess-like aspect. The same view is supported by Prasolov and Datsenko (in "Stavropol District", 1907, p. 62). Alluding to the loess-like aspect of the "terrace clays" of the Middle Volga, characterized by the content of carbonates and a columnar structure, these authors say that "any clay can acquire such features owing to the action of illuvial processes and the penetration of plant roots; in steppes all subsoils, except hard stony rocks, are loess-like clays". In reference to the subsoils of Tomsk and Omsk Neustruev (1925) says that their loess-like aspect may sometimes proceed from a steppe soil-formative process. As to deluvium, there can evidently be no doubt, that it is capable to acquire a loess-like aspect in the dry climate of steppes and deserts. With relation to slope loesses (Gehängelöße) even Wahnschaffe (1886) pointed that they "were formed, and possibly are still forming, on slopes by means of denudation of fine-grained material". As has been already mentioned, deluvial deposits of the Chimkent district as well as those of the Saratov and Simbirsk governments and many others have a loess-like aspect.

[*Editor's Note:* Material has been omitted at this point.]

XII. Conclusion.

Loess and loess-like rocks have one common origin — they are formed in situ from various rocks, always highly calcareous, being the result of weathering and soil-formation in the conditions of a dry climate. Some rocks of uniform texture are preeminently liable to give origin to loesses and loess-like rocks, such as certain alluvial and fluvioglacial deposits.

The not infrequent connection between glacial and loess regions is thus rendered intelligible.

The way and age of the deposition of the parent materials of loesses must be distinguished from the way and age of the transformation of these materials into loesses and loess-like loams.

The parent rocks of European loess have chiefly accumulated in the glacial period, when rivers discharged great quantities of muddy water which inundated the recent water-sheds. The transformation of these rocks into loess occurred in dry interglacial epochs and in the dry postglacial epoch.

It has been proved that many rocks of undoubtedly alluvial origin have a loess-like aspect: to these belong some loess-like deposits of the Northern Caucasus, Transcaucasia, Western Siberia and Turkestan.

It is likewise proved that highly calcareous clay rocks when weathering in a dry climate become more coarse-grained and acquire texture approaching that of loess.

A full transition of typical loesses into the most diversified loess-like sediments may be traced. To the latter an aeolian origin is not generally ascribed. Therefore, there is no reason to attribute it to typical loesses.

The aeolian hypothesis is not based upon facts observed at the present epoch. The dust to which the origin of loess of Turkestan and Central Asia has been ascribed is an artificial product — generally being the result of the deflation of loess itself. There is nothing to prove the formation of loess by means of dust-drift in the recent epoch.

Literature.

Full references are to be found in the following papers of mine:

BERG, L. S., On the origin of loess. "Izvestiya" (Bulletin) of the Russian Geographical Society, vol. 52, 1916. pp. 579—647 (the same in the book: Climate and Life. Moscow 1922, pp. 69—110) (in Russian).

BERG, L. S., The soil theory of the origin of loess. Bulletin de l'Institut Géographique, vol. VI. 1926, pp. 73—92, with a map (in Russian).

BERG, L. S., The problem of loess. "Priroda" (Nature), 1929, pp. 317—346 (in Russian).

Reprinted from *Bull. Geol. Soc. America*, **55**, 1–8, 23–30, 33–40 (Jan. 1944)

LOWER MISSISSIPPI VALLEY LOESS

BY RICHARD JOEL RUSSELL

CONTENTS

ILLUSTRATIONS

ABSTRACT

Lower Mississippi Valley loess resembles in all essential respects that of the Rhine Valley and other parts of Europe. Its field relationships preclude the possibility of eolian, lacustrine, fluvial, or other direct sedimentary origin. Typical loess grades upslope into parent material, from which it has been differentiated by a process here called loessification. Initial parent materials are terrace deposits physically similar to backswamp clays of the Recent Mississippi River. Parent materials weather into brown loam that creeps downslope, accumulating in greatest thickness in valleys and as mantles of bluffs. During loessification, carbonates accumulate, the size of particles becomes restricted mainly to 0.01–0.05 mm., snails are incorporated, and other loessial characteristics appear. More widespread development of loess east of the Mississippi results from wider areas of Pleistocene backswamp deposits. Loess occurs along major Mississippi tributaries, on such residual eminences as Crowleys Ridge, Sicily Island, and various hills west of the river in Arkansas and southeastern Missouri. Under some exposure conditions deloessification occurs, carbonates leach, coarse granules disintegrate, loessial characteristics disappear, and a type of brown loam is the end product. If a sharp distinction be made between loess and loesslike materials, it appears that the origin of loess in other regions is similar to that in the lower Mississippi Valley.

INTRODUCTION

When the writer began his lower Mississippi Valley field work, in 1928, he regarded the loess problem as settled and accepted the eolian hypothesis of origin. This belief was somewhat shaken when he found gravels and stratified beds in material he regarded as loess. During the next 10 years accumulating evidence favored fluvial origin, and investigations made during three trips to Europe tended to confirm that idea. At the annual meeting of the Association of American Geographers in 1936 he presented a short paper advocating fluvial origin, assigning the physical change from floodplain silt to typical loess to processes taking place on bluffs, where the deposits were being subjected to environmental conditions quite unlike those at sites of original deposition. Within more recent years evidence has accumulated that fluvial deposits do not change to loess *in situ*. They must first experience weathering and colluvial transportation.

Apparently a problem has arisen as a result of calling too many things loess. The necessity for a strict and wholly physical definition is inescapable before the question of origin may be considered seriously. After a definition of this kind is adopted it is possible to relegate many types of material to the general category, "loesslike deposits," and thereby remove many diverse, and often obvious, origins.

ACKNOWLEDGMENTS

To Professor H. N. Fisk the writer is indebted for many of the ideas advanced in this paper. Together we have examined in detail the deposits on Crowleys Ridge, many other localities west of the Mississippi, and outcrops along at least every main road and many of the byways in the loess belt of Kentucky, Tennessee, Mississippi, and Louisiana. To Professors W. Panzer, of Heidelberg, and Carl Troll, of Bonn, the writer is indebted for expert guidance to various loess localities in Germany, and to A. Penck for the term loessification.

A study of the literature has expanded gradually into a collection of abstracts and notes on more than 700 books and articles. The writer is indebted to Josephine G. Keller and Felix D. Richardson, of Louisiana State University, for their care in searching through complete sets of numerous American and foreign periodicals for titles of papers bearing on the problem, and to Marguerite M. Hanchey, Librarian in the School of Geology, for able assistance in the procurement of many items essential to the progress of the investigation.

For mechanical analyses the writer is indebted to Leo Hough, State of Louisiana, Division of Highways, and to Professor R. Dana Russell, Louisiana State University. Many helpful suggestions from Director H. V. Howe and Professor Chalmer J. Roy, School of Geology, Louisiana State University, have contributed to the progress of the study.

DEFINITION OF LOESS

The definition of loess should be rigid and descriptive,—devoid of hypothesis Materials resembling loess in one or more regards may be described as loesslike.

An examination of the literature reveals practical unanimity of opinion concerning the meaning of *Löss* for several decades after its introduction as a scientific term in about 1821 (Scheidig, 1934, p. 58), or its English equivalent, introduced by Lyell (1834, p. 111). After the middle of the century an increasing number of loesslike materials were being included in the term. The present century has witnessed a strong tendency to introduce hypothesis of origin into the definition. Smith and Norton (1935), for example, define loess as

"the sediment which is believed to have been produced by the grinding action of glaciers, deposited in the bottom lands from streams carrying water from the melting glaciers and then picked up and redeposited on the upland by the wind."

Todd (1918, p. 111) decried this tendency, but Tilton (1925) has insisted that eolian origin be made an essential part of the definition.

Observations appear irreconcilable to such a degree that over 20 different theories of origin have been proposed (Druif, 1927, p. 72–165). Interpretations are correct in many instances, but the diversity arises largely from the fact that many relate to material that is not loess at all.

No one seems to have formulated a precise, physical definition at an early date. A synthesis based on descriptions, however, indicates the intent to regard loess as homogeneous, unstratified, slightly indurated, porous, calcareous sedimentary rock consisting predominantly of particles of coarse silt size, ordinarily yellowish or buff in color, capable of splitting along vertical joints, and tending to stand in vertical faces. Everyone today seems to regard such material as "typical" loess, and it is doubtful that a better definition can be formulated. Quantitative limits might be added, but with caution. It is not safe to seek them in recent tabulations of mechanical or chemical analyses, such as appear in Scheidig (1934, p. 73–111), because they ordinarily include many samples of loesslike materials.

Most loesslike materials are excluded if a fair lime content be regarded as an essential characteristic. Scheidig regards as typical a calcium and magnesium carbonate content of 10–25%, though some lime-rich loess may contain 36%. Wailes (1854, p. 231) suggested 30% and over as typical in Mississippi. Keilhack (1920, p. 157) used 10–25% as typical. Calcium is the dominant carbonate and occurs as a cement throughout the entire mass. In addition it is present in concretionary nodules, root tubes, films concentrated along joints and faces, and in snail shells.

Sharp limitation in grain-size distribution is probably the most significant single diagnostic characteristic. Lyell (1834, p. 120) noted that it is far more homogeneous than alluvium. Leverett (1894, p. 111) observed that it contains far less fine material (< 0.005 mm.) than silt, and no coarse particles. Chamberlin and Salisbury (1885, p. 279) present detailed tabulations of grain sizes; their values are stated in millimeters but obviously refer to centimeters. Their counts show the predominance of particles in the fraction 0.01–0.05 mm., as do all subsequent studies. Keilhack (1920, p. 157) gives the following values as typical: 0.1–0.5 mm., 8–40%; 0.02–0.05 mm., 50–65%. In the writer's experience the single fraction 0.01–0.05 mm. ordinarily constitutes at least 50% (by weight) of a sample and in many cases amounts to 75%. Table 1 is representative of loess and some related materials.

TABLE 1.—*Grain-size distribution in samples of loess, leached loess, and a loesslike material*

(In per cent)

Sample (by weight) finer than	Loess			Leached loess	Loesslike
	1	2	3	4	5
Mm.					
0.5	99	98	100	100	100
0.4	99	98	100	100	100
0.3	99	98	100	100	99
0.2	98	98	100	100	96
0.1	98	97	99	100	84
0.09	97	97	98	100	83
0.08	96	96	98	100	82
0.07	95	95	98	100	81
0.06	93	94	97	99	79
0.05	90	91	96	96	77
0.04	82	86	90	90	75
0.03	64	76	78	81	71
0.02	33	50	59	54	60
0.01	15	19	30	23	38
0.009	14	16	28	20	34
0.008	12	14	25	18	32
0.007	11	11	22	15	29
0.006	10	9	19	13	25
0.005	9	8	16	11	21
0.004	8	6	14	9	18
0.003	7	5	11	8	16
0.002	6	3	8	5	11
0.001	4	1	6	2	8

Location of samples: 1. Haarlass, northside Neckar Valley, 1.5 km. above Alte Brücke, Heidelberg, Germany, 25 meters above river.

2. Five miles southeast of Natchez, Miss., U. S. 61, at junction between Woodville and Kingston roads, 4.5 miles east of Mississippi Valley bluffs, on divide between St. Catherine and Second creeks.

3. Center of Sicily Island, north-central Louisiana, west side of Mississippi Valley, on small stream leading to Ouachita River.

4. A few inches directly above sample 2.

5. Pleistocene terrace remnant 10 miles south-southwest of Delhi, La., 1.5 miles west of Lamar, west side of Mississippi Valley, northeastern Louisiana.

If the percentage of material finer than 0.01 mm. be subtracted from that finer than 0.05 mm., the remainders in samples 1, 2, and 3 are 75%, 72%, and 66%. This remarkable degree of sorting is also exhibited by leached loess sample 4, with a remainder of 73%. Although the outcrop represented by sample 5 exhibited so many loessial characteristics that many persons would identify it as loess, the degree of sorting is less perfect in several regards, and only 39% falls within the size limits appearing critical in the definition.

The definition should include the following essential characteristics: Loess is unstratified, homogeneous, porous, calcareous silt; it is characteristic that it is yellowish or buff, tends to split along vertical joints, maintains steep faces, and ordinarily contains concretions, and snail shells. From the quantitative standpoint at least 50%, by weight, must fall within the grain size fraction 0.01–0.05 mm., and it must effervesce freely with dilute hydrochloric acid.

In the discussion to follow the word loess is used only to describe materials that meet all parts of the definition. Leached loess is used only to describe material immediately overlying unaltered deposits, differing only in absence of calcareous content, which unquestionably was at one time loess. Other similar materials are in many cases described as loesslike.

LOESSLIKE SEDIMENTARY ROCKS

Many sedimentary rocks closely resembling loess in one or more regards should be called loesslike. In literature they are ordinarily designated by such terms as river-loess, lake-loess, marine-loess, loess-loam, noncalcareous loess, or stratified loess. The German language is particularly rich in such expressions. With regard to position it offers: *Deckenlöss, Höhenlöss, Plateaulöss, Flankenlöss, Beckenlöss, Hanglöss, Gehänge-löss, Flusswinkellöss, Bogenlöss, Terrassenlöss, Steppenlöss, Luvlöss, Leelöss*, and others, and with regard to origin: *Primärlöss, Sekundärlöss, Dejektiver- or Schwemmlöss*, and *Inundationslöss*. Age distinctions are recognized, such as: *jünger Löss, älterer Löss, fossiler Löss*, and *rezenter Löss*. There are also such terms as *Lösslehm, entkalker Löss*, and *vertoner Löss*. French is also rich in loess terms, and many other languages exhibit the same tendency, as may be appreciated by the fact that almost every expression listed has its direct equivalent in English. Some of these words refer to loess, but most have originated as the result of uncritical definitions, extended beyond all reasonable limits.

To counterbalance the tendency toward inclusion of altogether too many loesslike materials in the definition of loess many writers use such terms as typical, true, genuine, pure, and, in all, over 50 similar adjectives in the more common European languages. This confusion disappears when it is realized that so many references to loess actually mean loesslike materials.

The residual weathering of many types of silt and clay yields loesslike products. Jongmans and Van Rummelen (1938, p. 120) report Cretaceous bryozoans and echinoids in loess in southeastern Netherlands. The writer has visited the field locality and confirms all conclusions but one—the material is actually loesslike. Quaternary alluvial deposits are especially prone to assume loessial characteristics during weathering on outcrops providing good drainage. The writer recently regarded such materials as immature loess (1938, p. 75), failing to realize that only under special conditions, where colluvial transportation occurs, can any of them become loess.

The physical characteristics most commonly responsible for confusion are color and tendency to fracture vertically. Obscure stratification, superficial textural similarity, and impalpability also lead to confusion.

Terrace materials, to a greater degree than other types of deposits, assume many loessial characteristics. The writer estimates that over half of the American literature on loess actually refers to loesslike terrace silts. This confusion exists elsewhere. Many of the photographs by Willis (1907) and the excellent sketches and photographs by Barbour (1929) clearly show flat-surfaced alluvial terraces in China.

The areal extent of loess deposits has been grossly exaggerated by widespread inclusion of loesslike materials. Tillo (1893) considered 4% of the land surface of the earth covered by loess. For the entire latitudinal zone between 20°S. and 40°S. his estimate was 13%, for South America, 10%, and for North America, 5%. Keilhack (1920) raised the estimate to 9.3% for the earth's land area. Detailed distribution maps (Scheidig, 1934, p. 7–25; Jenny, 1941, p. 58) include nearly all areas where alluvial terraces are extensively developed. In the lower Mississippi Valley the areal extent of loess is an extremely small fraction of the belt ordinarily so designated. The writer has found the same to be true in the Rhine, Danube, Rhone, and other valleys in Europe.

Lower Mississippi Valley literature is difficult to interpret because loesslike deposits may be called loess by one writer and by some other name, such as brown or yellow loam, by another. In addition brown loam has been used for dozens of distinctly different materials. The weathering of many Tertiary formations yields a number of residual products that are ordinarily called brown loam. These may be individually distinct, and all are quite different from brown loam that has weathered from Pleistocene terrace deposits. The weathering of loess yields still another type of brown loam. The term is also used for various soils. As a rule all such materials are either noncalcareous or only faintly so. They fail to show diagnostic sorting characteristics of loess. Some are decidedly loesslike at the outcrop. Others are so different that they have never caused confusion.

The least justified terminological practice is that of calling a material loess because of known or suspected eolian origin. Deposits of accumulating dust ordinarily bear no physical resemblance to loess, and few deserve to be called loesslike. Davis (1905, p. 62) and Huntington (1907, p. 103) applied the name to dunelike accumulations of powdery white dust in Turkestan. R. W. Pumpelly (1908, p. 271) experienced dust storms that led him to the conclusion that he was in a region of "still living loess." Hobbs (1931, p. 383) speaks of the difficulties in walking in deep loess freshly deposited on tundras. Such statements are quite in contrast to those of Schokalsky (1932, p. 85), who clearly distinguishes between loess and the dust covering it in the Salt Range of India. Berg (1932, p. 133) has observed that much of the dust so widespread in Asia is "the result of man's activities, being a product of deflation of the upper horizons of soils developed upon, and from, loess . . . but in no way can anything like loess be obtained as a result of its subsequent deposition." Clapp (1920) and Cressey (1934, p. 75) clearly relate dust in the air of northern China to the cultivation of fields or to the presence of extensive loess deposits.

Redeposited loess is a confusing term. In the lower Mississippi Valley it ordinarily means silty terrace deposit and implies evident stratification. This expression should be restricted to materials of known genesis and used only in a strict, wholly literal sense. About the only examples are such things as miniature fans associated with gully dissection.

THEORIES ON THE ORIGIN OF LOESS

Theories on origin run almost the entire gamut of geological possibilities. Direct volcanic deposition was postulated by Howorth (1882) and others. I. C. Russell (1897, p. 59–63; 1901, p. 81–83) regarded loess as a residual deposit from weathered basalt in eastern Washington. An examination of his localities reveals the presence of loesslike material that is indeed residual, with remnants of veins extending through it, toward the surface. Keilhack (1920, p. 161) concluded that loess is of cosmic origin, being derived from a dust ring resembling those of Saturn. A strange cause of Pleistocene glaciation is involved in his hypothesis. At the other extreme is the conclusion of Campbell (1899) that loess was derived from animal and vegetable matter. Orton and others proposed the idea that earthworms and burrowing animals created it by bringing fine materials to the surface (G. F. Wright, 1890, p. 104; Savage, 1915, p. 106). Druif (1927) and Scheidig (1934) give more complete statements of various theories, but they overlook recent Russian literature.

Ehrlich (1848) originated the eolian hypothesis according to some writers, and it may be that his study of a dust storm that covered much of central Europe, an area in which loess is relatively abundant, was the basis for this belief, but an examination of several of his writings revealed no direct statement of the hypothesis.

Virlet d'Aoust (1857) deserves credit for the eolian hypothesis, having definitely prescribed it for deposits in Mexico. Freudenberg (1909, p. 274) visited the localities in question and confirmed the identification of "typical yellow loess."

Credit for the eolian hypothesis is ordinarily given to Richthofen, who popularized it and is responsible for its vogue today. The original reference is a report on the Provinces of Honan and Shensi, written in Shanghai in 1870.

American geologists did not receive the eolian hypothesis well. Todd (1878), Child (1881), Winchell (1884, p. 263), Chamberlin and Salisbury (1885, p. 287), and Dana (1895) attacked it vigorously. Fluvial or lacustrine origin were the theories then in vogue. Among notable adherents to the former were Lyell (1834; 1838), Binney (1846), Dumont (1852), D. D. Owen (1852), Newberry (1889), Upham (1892), Todd (1897, p. 50), and, more recently, Fuller and Clapp (1903), L. A. Owen (1905), and G. F. Wright (1921), Lacustrine origin was favored by R. Pumpelly (1866), Hayden (1872), Call (1882), Mcfarlane (1884), Warren (1878), and Witter (1892b). Chamberlin and Salisbury (1885) thought both origins probable. Salisbury (1892, p. 317) restricted deposition to waters of glacial origin. Near coasts estuarine origin was favored by Gordon (1892), Hopkins (1872, p. 137), and Hilgard (1874). Lyell favored fluvial origin on the whole, but where valleys widened he pictured deposition in bodies of relatively still water. Thus, in the upper Rhine Valley loess was deposited in a "pre-existing basin or strath, bounded by lofty mountains" (1863, p. 330). Marine origin, advocated by Kingsmill (1871) and Skertchly and Kingsmill (1895), attracted few adherents. Some kind of aqueous origin explained loess according to most geologists for nearly 3 decades after Richthofen's eolian origin was advanced.

In the Central States there had long been a feeling that the lacustrine theory was unsatisfactory. Deposits in Nebraska were some 2000 feet higher than those along the Mississippi. Satisfactory boundaries of lakes could not be demonstrated. Shimek's (1897) patient research emphasized the terrestrial character of the fauna. Localities were too closely spaced and too numerous to permit lacustrine origin (Shimek, 1903; Chamberlin and Salisbury, 1907, p. 410; Leverett, 1932).

R. Pumpelly was so impressed by Richthofen's thesis that he retracted a whole series of observations and conclusions, advanced in support of lacustrine origin in

1866 as a result of his own explorations in China (1879; and many restatements, *e.g.*, 1908; 1918, p. 611ff.).

European sentiment also swung toward the eolian hypothesis. Land mammals were being found in material regarded as loess, and many exhibited dry-climate affinities, suggesting that deposition of dust from Asiatic or African deserts had occurred during inter-glacial stages.

It was easier for Americans to abandon the lacustrine than to accept the eolian hypothesis. Some favored fluvial origin, but their evidence, on the whole, suggests stratified, loesslike deposits. Barbour (1929, p. 70) is correct in stating that such materials are "one cause for the slow death of the aqueous theory."

Chamberlin and Salisbury (1885, p. 287) proposed a combination theory invoking both fluvial and eolian agencies, but after many pages of discussion rejected it in favor of aqueous origin. Chamberlin (1897) later regarded the combination theory more favorably, and it was eloquently restated by Keyes (1898). Sardeson (1899) advocated eolian origin. Reports from China and Turkestan (Obrutchev, 1895; Huntington, 1907) to the effect that loess was being formed by deposition of dust appeared to clinch all arguments in favor of eolian origin.

Today most Americans appear to favor the idea that rivers transported fine glacial debris to broad flood plains, from whence it was picked up by winds and deposited on, or near, adjacent bluffs, particularly along the eastern sides of valleys. This is essentially the combination theory of Chamberlin and Salisbury (1885) and has been well stated many times (Fairchild, 1898; Willis, 1907, p. 249; W. B. Wright, 1914; Emerson, 1918; Fenneman, 1938, p. 97; Lobeck, 1939, p. 377). Todd (1897) and later L. A. Owen (1904; 1926) attacked it vigorously. Recently Flint (1941, p. 27) stated that 3 decades have elapsed since "the last serious opposition to the eolian view was voiced."

Minor theories deserve only brief notice here. Some run along such obsolescent lines as a marine catastrophe involving waters at least 700 feet deep (Hibbert, 1834). Others invoke snow drifts (Davison, 1894), alternate freeze and thaw (Kerr, 1881; Wood, 1882), or residual weathering (Peters, 1859; Linstow, 1910; Rummelen, 1923).

The close relationship between loess and Quaternary terrace deposits was noticed by Fallou (1867), who differentiated it from brown loam on the basis of carcareous content. Harris and Veatch (1899, p. 177) regarded it as "a local development of the loam." Winchell (1903, p. 141) regarded it as a direct and contemporary variation of till. Oefelein (1934) noted that weathering of "gray loess" produces a buff color. Observations along these lines have been receiving increasing attention in Russia, where they have been summarized by Berg (1932) in a paper that should not be overlooked.

[*Editor's Note:* Material has been omitted at this point.]

ORIGIN OF LOWER MISSISSIPPI VALLEY LOESS
GENERAL STATEMENT

No theory in vogue today appears to account for the physical characteristics, distribution, or stratigraphic position of lower Mississippi Valley loess. The closest approach has been advanced by Berg (1932), but insistence that it be formed under arid climatic conditions is locally untenable.

The essential part of loessification, or the process by which loess is differentiated from its parent material, is concerned with producing from backswamp deposits a sedimentary material having the typical grain size, sorting, porosity, and calcareous content of loess. No widely held theory seems capable of doing this.

OBJECTIONS TO EARLIER LOESS THEORIES

The objection may be raised against fluvial or lacustrine origin on the basis that loess is not a stratified deposit *in situ*. The Pleistocene deposits of the lower Mississippi Valley are fluvial terrace formations, and their stratigraphy is thoroughly understood. Loess forms no part in the stratigraphic sequence. It extends down slopes across truncated edges of both terrace deposits and Tertiary strata. In vertical range it lies between the higher, finer deposits of the oldest, Williana, terrace and the active flood plain. In southwestern Mississippi at many places this range amounts to over 300 feet within 1 or 2 miles. Similar observations have been made elsewhere.

Lyell (1834, p. 120) found it overlying all other formations in its vicinity, including the latest gravels. In Indiana, Shaw (1915, p. 107) gives the vertical range of "bluff loess" as from 700–800 feet at its top to 375–400 feet at its base. Worthen (1866, p. 58–83) found that in Illinois it occurs as coatings on bluffs, failing to reach their tops but extending to their bases, as did Fowke (1908, p. 37) near St. Louis, Missouri. Richthofen (1877, p. 156) reported deposits near Vienna that extend upslope for 300 feet, nowhere more than 50 to 60 feet thick at any given place. Barbour (1929, p. 64) stated that in China: "Though loess can be traced uphill uninterruptedly for distances of more than 500 feet, the greatest vertical thickness actually measured in unquestionable primary loess is only 35 feet." Such observations preclude a lacustrine, fluvial, or marine origin.

Against eolian origin it may be urged that no actual or hypothetical directions of winds could account for its distribution. It covers slopes leading in all possible directions and is ordinarily as strikingly developed on one side of a ridge as on the other. In pseudoanticlinal exposures, whatever their orientation may be, one limb ordinarily resembles closely the other. It occurs on both sides of the Mississippi and other large rivers.

The shape of a deposit is that ordinarily assumed by colluvial materials. Richthofen (1877, p. 154) described it as "leaning" against mountains and (1882) likened its upper surface to a rope stretched loosely between two ranges, a fortunate simile that has been quoted many times. Such shapes are typical of lower Mississippi Valley deposits. Nowhere are such forms as dunes or deposits heaped with steep faces toward the valley and gentle slopes leading eastward to a feather edge as many supporters of the eolian hypothesis believe.

The sorting appears to be too uniform to be the result of direct deposition from a current. It seems improbable that either wind or water could move with the uniform velocity required to permit the accumulation of material so homogeneous that at least a half and ordinarily about three fourths of its particles (by weight) fall within the limited diameter range 0.01–0.05 mm.

The ratio between calcium and aluminum is one not likely to be encountered in any variety of primary sediments (Polynov, 1937, p. 175). Detailed studies by Krokos (1926) show that Ukranian loess could not be a product of mechanical disintegration of rock, such as rock flour or dust.

In the lower Mississippi Valley loess development correlates mainly with two main factors: (1) the presence of backswamp deposits in terrace formations, and (2) deep dissection. To be acceptable a theory of origin must harmonize with these facts.

LOESSIFICATION

The process of loessification starts in parent material that originally was deposited as alluvium on flood plains during the Pleistocene. It affects the finer parts of such deposits, especially those that have accumulated in backswamps and are present only in minor amounts along Pleistocene meander belts. It is restricted to parts of terrace formations that now stand considerably above flood plains. The deposits must consist mainly of silt and clay. They are somewhat calcareous and contain carbonaceous matter derived from plant remains.

The initial stage of the process is weathering and differentiation of soil profiles. While pedogenic processes are active much of the original calcareous content, including any fossil shells that may be present, is lost to ground waters. The resulting product is a brown loam that thickens residually on flats but is relatively mobile on slopes. In deeply dissected territory it creeps into valleys, where it accumulates to considerable thicknesses.

The colluvial phase of the brown loam is derived from the upper parts of the profile of weathering and soil development and hence is characterized by coarser particles than the average present elsewhere. The loss of finer materials goes on at all stages of colluviation and is intensified by churning movements. Surface washing probably contributes to some degree. With increasing distance downslope comes closer approach to the remarkable sorting and uniform texture of loess.

Toward the lower parts of colluvial slopes is a zone of carbonate enrichment, the carbonates having been derived from terrace materials and brown loams on surfaces upslope. Snail shells introduced during colluviation are preserved only where carbonate enrichment takes place and hence characterize materials advanced far in loessification. The introduction of carbonates effects a measure of structural competence, retards creep, makes fracturing possible, and renders faces relatively stable. By the time significant enrichment has occurred loessification is practically complete.

There are several distinct processes involved in loessification. We may turn to their details.

CALCAREOUS CONTENT: Mature loess contains 10 to 25% calcareous content (Keilhack, 1920, p. 157), varies from 3.25 to 25.60% within vertical sections of Ukranian loess (Krokos, 1926, p. 117), or, on the basis of weight lost under hydrochloric acid treatment, is somewhat less than 36% for lower Mississippi Valley deposits. Concentration of salts is one of the most essential parts of loessification.

Bluffs facing flood plains are especially favorably located from the standpoint of salt accumulation. Low positions in valleys of deeply dissected territory are also favored. Substances leached from topographically higher, calcareous backswamp deposits are carried by ground water to lower levels, and many are concentrated toward exposed surfaces.

The calcareous content of loess is secondary according to most observers. Chamberlin and Salisbury (1885, p. 304) emphasize the point that the amount present in snail shells is disproportionately small and that a magnesium carbonate element could not have come from shells. Willis (1907, p. 249) regarded water as the depositing agent. Hay (1914, p. 41), Krokos (1926), and others have noted increasing calcareous content toward lower positions in exposures.

Polynov (1937, p. 175) notes that alkalies derived from the leaching of such materials as backswamp clays move in ground water mainly as bicarbonates. During capillary movements in which the solutions approach the surface an intensive separation of CO_2 takes place, and the remaining salt is precipitated as the normal carbonate.

"We can thus assume that the calcareous south Russian loess is a region of concentration of that lime (and partly magnesia) which was lost in solution some time by the moranic drifts adjoining the loess on the north, and now to a considerable extent depleted of their alkali and alkaline earth bases, and in particular calcium. Analogous relations may be observed in over more limited areas in Northern Mongolia, where the leached crust of weathering covering summits and upper slopes of the Kentei and Khangai massifs passes at lower levels into a mantle of calcareous loess."

The source of the carbonates in lower Mississippi Valley deposits is evident for they occur below the most calcareous parts of Pleistocene flood plains. Pleistocene meander-belt deposits are not only unfavorable to loessification from the physical standpoint but also because they are less calcareous. The transitional increase in calcareous content of materials involved in downslope gradations between terrace deposits and loess and the striking concentration of salts at low levels is proof that accumulation is of secondary origin.

Materials other than loess collect lime under similar conditions of exposure in the lower Mississippi Valley. Bluffs of sand and gravel are more calcareous than their stratigraphic equivalents in flatter territory. Huge volumes of concretionary material are common in such deposits, and along some layers tabular masses of secondary limestone have accumulated. Excellent examples occur south of the bridge at Natchez.

Lime cements many of the grains in loess into coarse aggregates. Surplus lime forms concretions. A surplus is ordinarily available, so that even rearranged deposits develop concretions (Todd, 1899, p. 99). Exposures favorable to the escape of moisture develop crusts, as was noted by Hilgard (1860, p. 314), Witter (1892b, p. 330), and many others.

The mineralogical investigation of carbonates such as occur in loess is practically a virgin field for research according to Polynov (1937, p. 31–32).

GRAIN-SIZE CHARACTERISTICS: In typical backswamp deposits there is less sorting and, on the whole, much finer material. Loessification witnesses a change in which sorting becomes emphasized and the size of the average particle is increased.

The idea that loess grades into alluvial clays is not new. Hilgard (1879, p. 107) noted loess passing laterally into clayey loam. Leverett (1897, p. 157) records loess passing into clayey loam and then into loamy clay. Mabry (1898) emphasized the close relationship to brown loam. Harris and Veatch (1899, p. 117) regarded loess as a special development of loam. Todd (1894) noted that loess contains more clay away from streams until it so resembles adjoining clays that it is difficult to separate the two. Shaw (1915, p. 107) also found increasing clay content with greater distance from streams. This observation has been used to support eolian origin,—it being supposed that coarser particles lodge lower on bluffs. Some believe the entire deposit was loess originally, the clays and loams being an expression of weathering (Mabry, 1898; Sardeson, 1899).

Earlier works support the idea that loess is colluvial. Hilgard (1860, p. 195) noted that the deposits thicken downslope, and Moyer (1932) found the same true in China, as did Worthen (1868, p. 123) and Winchell (1879, p. 168) in the central United States. Chamberlin (1890, p. 471) and Todd (1918, p. 116) found upper and lower parts of exposures connected as the result of creep.

A relationship between colluviation and uniformity in grain size was suggested by Worthen (1866, p. 329), who visualized a "partial sifting process" as taking place on slopes. Wahnschaffe (1886, p. 360) thought that some German loess is still being formed as the result of removal of finer materials while it occupies positions on slopes. That loess has been derived from deposits including a larger proportion of finer materials is demonstrated by mechanical analyses, such as those of Table 1, for these invariably show a tail of fine grain sizes extending far beyond the coarse silts that constitute an overwhelming proportion of each sample.

There are more factors in the "partial sifting process" than Worthen, Wahnschaffe, or others contemplated. It is initiated during weathering, intensified by pedogenic processes, stimulated during colluvial transportation, and appears to end only after the deposit has become thoroughly cemented. Selective truncation and concentration of upper, coarser parts of the profile is characteristic during each stage of loessification. Downward translocation of finer particles occurs during each stage. Other losses, such as material entrained by various currents at the surface, favor removal of finer and concentration of coarser particles in the residual material.

The transition between parent material and loess is so gradual that field geologists in the lower Mississippi Valley encounter difficulties in mapping contacts between lithologic units. Many Tertiary and Quaternary deposits yield a brown loam that closely resembles loess. As a group the brown loams are relatively noncalcareous and do not effervesce with dilute acid. Their loss in weight when treated with hydrochloric acid in the laboratory is ordinarily about 5 per cent, and this includes removed free iron oxides and soluble colloids as well as carbonates. Mechanical analysis ordinarily indicates degrees of sorting and materials of sizes quite unlike those of loess. Brown loams taking part in loessification attain such close resemblance that mechanical analyses of samples collected near loess are almost identical, and calcareous content is the best guide as to whether it is loess or not. The samples in Table 2 indicate the last step in loessification from the grain-size standpoint.

The comparison presented in Table 2 is typical of that between loess and material immediately upslope that has undergone all stages but the final one, lime enrichment, in loessification. At the outcrop the material of sample 6 effervesced, contained snails, and was covered by moss requiring a calcareous environment, whereas all these features were absent in the other case. The loess contained somewhat more

TABLE 2.—*Comparison between loess and nearly loessified brown loam*

(In per cent)

Per cent of sample (by weight) finer than	Loess		Loesslike brown loam	
	6	6D	7	7D
Mm.				
0.06	94	97	99	99
0.05	86	90	97	98
0.04	76	82	94	96
0.03	63	71	84	90
0.02	44	54	52	66
0.01	18	24	20	32
0.009	14	21	18	28
0.008	11	19	16	26
0.007	11	18	15	24
0.006	10	17	14	22
0.005	9	17	13	21
0.004	9	16	12	19
0.003	5	13	10	17
0.002	0	11	6	15
0.001	0	8	4	13

Description of samples: 6. Loess from cut shown in Pl. 1, fig. 1, 1.6 miles east on U. S. 84 from junction with 61 at Washington, Miss. Sample taken at central part of exposure. Of the original material 19.7% was soluble in dilute hydrochloric acid. Analysis run without dispersing agent, with slight flocculation toward end of analysis.

6D. Same sample, same type of analysis, but with sodium oxalate dispersion.

7. Brown loam from next cut east, 0.4 mile away, at slightly higher elevation. At exposure was loesslike in all regards except that it would not effervesce with acid. In laboratory 5.2% was soluble in dilute acid. Analysis run without dispersing agent and without flocculation.

7D. Same sample, same type of analysis, but with sodium oxalate dispersion.

coarse material, 37% of sample 6 and but 16% of sample 7 being material larger than 0.03 mm., before dispersion. After treatment with sodium oxalate the proportions were, respectively, 29% and 10%.

In all cases in Table 2 dispersion reduced the amount of coarse material present, indicating the presence of aggregates that were to some extent broken down into constituent particles.

Berg (1927; 1932) regarded aggregation of clay particles into granules of silt size as the dominant factor in producing the sorting characteristics of loess. His evidence rests heavily on an experiment by Ganssen (1922, p. 41), who subjected a sample of kaolin to the action of alkaline silicates (Table 3). Loesslike texture was produced in kaolin. Ganssen gives few details concerning his methods. The end product was certainly not loess. The experiment was conducted as an attempt to explain how wind-transported dust might be changed to resemble loess. Ganssen apparently failed to realize that his experiment might be used to discredit the eolian hypothesis, and Berg applied his results to that end.

TABLE 3.—*Textural change in a kaolin sample*
Subjected to exposure in a solution of alkaline silicates

Grain size of particles	Original material	After treatment
above 0.05 mm.	3.3%	11.6% (weight)
0.05–0.01	3.3	43.2
below 0.01	93.5	45.3

The lower Mississippi Valley affords ideal conditions for testing Berg's hypothesis. Pleistocene backswamp deposits, to a peculiar degree, contain all the elements favorable to consummation of the process of building aggregates out of finer particles.

Clay suspensions are flocculated by calcium salts (Baver, 1940) such as exist in backswamp deposits. Flocculation itself, however, does not insure the stability of aggregates. Removal of the flocculating agent may result in dispersion to original physical state. Where aggregates are cemented they may become granules that retain identities even after being exposed to an environment unfavorable to aggregation. Backswamp deposits not only contain the necessary alkaline materials to favor aggregation but also several agents favorable to preservation of granules.

The relatively high organic content of backswamp deposits favors stability of aggregates. Improvement in physical properties of clay soils treated with lime and humus is chiefly the result of aggregation of smaller particles into larger, relatively stable units. Lime promotes aggregation and with humus produces relative stability. Some contend that humus alone accomplishes both ends (Myers, 1937). That carbon deposits are disseminated in loess has been noted by Hershey (1896, p. 296), Udden (1898), and Shaw (1914, p. 298). They are common in residues of coarse fractions from mechanical analyses of lower Mississippi Valley loess.

Aggregate stability is also favored by the irreversibility of colloidal iron hydroxide with respect to dehydration effects. In many soils this factor is important in producing stable aggregates. Roberts (1933) has noted aggregation of 95 per cent of all original clay and silt particles into units larger than silt in the iron-rich Nipe clay of Puerto Rico. Lower Mississippi Valley backswamp deposits and loess are both relatively rich in iron.

Cemented aggregates are characteristic of loess. Hilgard (1860, p. 110) noted this and stated that they are permanently fixed by calcareous incrustations. Todd (1899, p. 95) found coatings of both calcium carbonate and iron oxides. Various European investigators (Scheidig, 1934, p. 61) regard *Krümelstruktur* (crumb structure) as characteristic of loess.

The degree of aggregation in a sample depends largely on the treatment it has received prior to examination. In thin sections the whole body is cemented into essentially a single mass. Samples kept in bags form lumps that are aggregates in much the same sense that clods are in soils. Powders on slides exhibit clusters that are essentially aggregates of dimensions that depend chiefly on the amount of crushing they have experienced. Sand and silt fractions retained after vigorous treatments in mechanical analyses still display aggregates. The fundamental size of particles in such aggregates is that of silt, rather than the clay of Berg's hypothesis. Individual minerals ordinarily appear fresh and angular.

Tables or curves of grain sizes in loess attempt to refer to individual mineral particles rather than to aggregates. Their validity depends to a great extent on the treatment of the sample in analysis. Samples washed gently on 200-mesh sieves display exaggerated sand fractions. Violent agitation, long soaking, and effective dispersion must occur in order that most aggregates be broken down. In samples 6D and 7D of Table 2, resolution to fundamental mineral sizes is about as complete as possible. Few residual aggregates were found in petrographic examination of the various fractions, even in the case of the finest, when examined for Brownian movement. Such aggregates as were found consisted of material mainly larger than clay-particle size.

The most rigorous method followed in the present study involved treatment with dilute hydrochloric acid at various temperatures below boiling for prolonged periods, until every sign of effervescence stopped. Samples were then washed, the water being drained through a Pasteur-Chamberland filter. The loss in weight varied from between 19.7 and 35.8 per cent in loess and between 5.2 and 5.8 per cent in loesslike materials. Mechanical analyses were by pipette method, with sodium oxalate as the dispersing agent. No sample thus treated showed appreciably more clay than samples subjected to ordinary mechanical analyses. Petrographic examination of dried residues of various size fractions showed that loess is composed mainly of primary mineral and rock particles in such proportions as are cited in Tables 1 and 2. The silt particles are not aggregates of clay-size material, as postulated by Berg. Parent materials, well advanced in loessification, differ mainly with regard to carbonate enrichment, not with regard to constituent particles. Those less advanced differ mainly in having greater proportions of fine material. In all cases quartz grains are predominant. Calcite grains are conspicuous. Minor amounts of rock fragments, various heavy minerals, and other minerals characteristic of floodplain deposits are present in varying proportions.

POROSITY: This is an essential feature of loess. Most writers note conspicuous openings and many comment on the ease with which water travels through them. These descriptions are concerned mainly with visual porosity. Examination under a hand lens reveals open ends of root tubes, irregular fractures, crumb structures, and a general appearance not unlike moist sand. Specific porosity is comparatively low, much lower than that in backswamp deposits, but individual void sizes are large, so that loess is highly permeable. The high coefficient of permeability promotes effective drainage, encourages leaching at higher levels and soluble salt enrichment below, lowers water tables, and favors deep penetration of plant roots. All such features give loess a characteristic appearance that differentiates it from brown loam or other loesslike materials.

Todd (1897), Leverett (1897, p. 156), and others have noted that permeability decreases away from bluffs where the material exhibits less complete loessification, finer grain sizes, less structural competence, and fewer root tubes either on slopes above typical deposits or at depth behind exposed faces.

Berg (1932, p. 141) regards porosity as the result of aggregation of clay particles into coarser units, so that voids are left as a result. There is not enough fine material present during the final stage of loessification to make this possible, and its absence from loess aggregates disproves the idea. Aggregates affecting visual porosity are common but are composed of coarse materials.

RELATIVE COMPETENCE: This character of loess as contrasted with loesslike materials is the result of cementation. The colluvial transportation involved in loessification is inhibited or prevented so that loess undergoes only such mass movements as are characteristic of rocks of some strength and durability. Rock fall of a variety that may be called loess fall and slump are typical. Open fractures may be maintained. Such competence disappears gradually when loess is leached, and it is ordinarily wanting in parent materials or other loesslike deposits.

SUMMARY: Loessification is the process through which loess is formed from its parent materials. In the lower Mississippi Valley the initial parent material is Pleistocene backswamp sediment. Weathering and pedogenic processes convert its upper part into brown loam that is relatively noncalcareous, is devoid of snails, and tends to display pinnacly surfaces on steep exposures. It creeps freely on slopes and undergoes textural changes, mainly in the direction of losing its finer particles, in the process. Residual concentration of coarser materials, which are mainly of silt size in the initial parent material, starts with the weathering and pedogenic phase of loessification and continues until the final stage, when mass movement is checked by cementation. Vertical translocation of smaller solids is essentially equivalent to that taking place in the differentiation of the A-horizon of a soil. Surface losses of fine material occur for various reasons. Material involved in creep is that present at the surface and therefore the coarsest in typical profiles. There is thus a progressive loss in proportion of finer particles downslope and a consequent approach to grain sizes and sorting characteristics of loess in that direction. Carbonates leached from upper surfaces are translocated in solution and concentrated toward lower parts of slopes, where loessification becomes complete. Snails living on slopes are probably incorporated in moving materials at various places, but their shells are preserved only where carbonate content is increasing and for that reason occur only in loess.

[*Editor's Note:* Material has been omitted at this point.]

APPLICATION OF CONCLUSIONS TO LOESS IN OTHER REGIONS

It is admittedly hazardous to apply conclusions reached in the lower Mississippi Valley to other regions. If the conclusions are sound, however, to deny their universal application is possibly unwarranted conservatism. Fact, opinion, and hypothesis are so interwoven in commonly held concepts of loess that they have become indistinguishable. The problem is presented with the hope that facts can be separated from theories and that the application of conclusions to other regions will be considered on the basis of the evidence presented.

The writer is convinced that the definition of loess used in this paper applies as well along the Rhine as in the lower Mississippi Valley. He is convinced that Lyell (1847), Hilgard (1879), and others are correct in regarding the materials as being essentially identical. His own samples, including sample 1 of Table 1, display physical identity. His field work indicates identity in origin. Deep cuts in brickyards near Aachenheim, in Alsace, reveal loess on slopes between alluvial terraces. Along the Neckar Valley, east of Heidelberg, loess also covers slopes associated with Pleistocene terraces that exhibit basal gravels, with finer materials above from which it has been derived.   At Poppelsdorf, on the outskirts of Bonn, the creep of loess over truncated edges of Pleistocene gravels is clearly exhibited, and for many miles along the foot-hills loess thickens downslope into valleys, where it is intensively utilized for truck crops.   The same relationships occur in southeastern Netherlands, in the early Pleistocene delta region of the Rhine.   At Kapel St. Rosa, 1 km. southeast of Sittard, are excellent outcrops, and many exposures reveal advanced stages of deloessification. It seems improbable that a material with such striking physical and field relation-ships has more than one origin.   There are suggestions that it may be formed through loessification of "red clay" in China (Barbour, 1927, p. 289; Andersson, 1923, p. 107).

That there is a loess problem seems to result from two main causes: (1) the identi-fication of many loesslike materials as loess, and (2) the insistence that the origin be eolian.

To eliminate confusion resulting from the first cause many deposits should be re-examined.   The adobe of the Southwest should be dropped from consideration. Terrace deposits should be excluded.   Only material such as occurs at Council Bluffs or Muscatine, Iowa, should be accepted as loess.   Widespread horizontal deposits of any kind and beds containing numerous vertebrates, such as occur in the vicinity of Lincoln, Nebraska, may be suspected as not being loess.   In a general way all materials not containing terrestrial snail shells should be regarded with suspicion.   Acceptance of a rigorous physical definition will exclude so many loesslike deposits that the problem is reduced to relatively simple and soluble terms.

The insistence that loess be eolian has not only kept the problem alive but has also led to some amazing conclusions.   Deposits of dust have been called loess in spite of almost total physical dissimilarity.   The even topography of the Great Plains has been ascribed to deposition by wind, and hence surface materials have been called loess (Matthew, 1899).   Elaborate hypotheses have been evoked to show that deserts to the west are sources of material entrapped by vegetation to the east (Keyes, 1912). On the other hand, loess described as occurring along the eastern sides of streams in the Central States is supposed to form "billowy ridges," more or less dune-shaped (Savage, 1915, p. 101).   In the lower Mississippi Valley a topography has been created in the minds of many that does not exist in fact.   It has taken the form, not of the dust dunes of Turkestan, the flats of the Great Plains, nor the billowy ridges of Illinois, but of a gigantic dunelike ridge extending from the Ohio River to Louisiana, with a high scarp facing the Mississippi and a gentle eastward slope leading to a feather edge some 10 to 15 miles away.

Barbour's (1929, p. 65) statement that loess in China "must have extended as a meniscus surface far up the mountain slopes" is an excellent topographic characteri-zation.   To explain a mantle of that sort an appeal to colluvial agencies appears more rational than any suggestions thus far provided by the group some of our European colleagues term "the aeolians."

REFERENCES CITED

Andersson, J. G. (1923) *Essays on the Cenozoic of China*, Geol. Survey China, Mem. ser. A, no. 3, vi + 153 pages.
Aughey, Samuel (1880) *Sketches of the physical geography and geology of Nebraska*, Omaha, 265 pages.

Baltz, V. A., and Polynov, B. B. (1930) *On the soils of Manchuria*, Dokuchaiev Inst. Soil. Sci., Contrib. knowledge soils Asia, pt. 1, p. 31–44.

Barbour, E. H. (1903) *Report of the state geologist*, Nebr. Geol. Survey, vol. 1, Lincoln.

Barbour, G. B. (1927) *The loess of China*, Smithson. Inst., Ann. Rept. 1926, p. 279–296.

—— (1928) *A re-excavated Cretaceous valley on the Mongolian border*, Geol. Soc. London, Quart. Jour., vol. 84, p. 719–727.

—— (1929) *The geology of the Kalgan area*, Geol. Survey China, Mem. ser. A, no. 6, 148 pages.

—— (1936) *The loess of China*, XVI Intern. Geol. Cong. (Washington, 1933), Rept. p. 777–778.

Baver, L. D. (1940) *Soil physics*, John Wiley and Sons, New York.

Berg, L. S. (1927) *Loess as a product of weathering and soil formation*, Pedology (Moscow), year 22, no. 2, p. 21–37.

—— (1932) *The origin of loess*, Gerl. Beitr. Geophys., vol. 35, p. 130–150.

Binney, Amos (1846) *The bluff formation at Natchez, Mississippi*, Boston Soc. Nat. Hist., Pr., vol. 2, p. 126–130.

Braun. Alexander (1842) *Vergleichende zusammenstellung der lebenden und diluvialen Molluskenfauna des Rheintales mit der tertiären des Mainzer Beckens*, Ber. Vers. deutsch. Naturf., vol. 20, p. 142–152.

—— (1847) *Brief, Okt. 28, 1846*, Neues Jahrb. Mineral., Geognosie, Petrefakten-Kunde 1847, p. 49–53.

Broadhead, G. C. (1879) *Origin of the loess*, Am. Jour. Sci., 3d ser., vol. 18, p. 427–428.

Brown, C. A. (1938) *The flora of Pleistocene deposits in the western Florida Parishes, West Feliciana Parish, and East Baton Rouge Parish, Louisiana*, La. Dept. Conserv., Geol. Bull. 12, p. 59–96.

Call, R. E. (1882) *The loess of North America*, Am. Nat., vol. 16, p. 369–381, 542–549.

—— (1889) *The geology of Crowleys Ridge*, Ark. Geol. Survey, Ann. Rept. 1889, vol. 2, p. 1–249.

Campbell, J. T. (1889) *Origin of the loess*, Am. Nat., vol. 23, p. 785–792.

Carman, J. E. (1929) *Further studies of the Pleistocene geology of northwestern Iowa*, Iowa Geol. Survey Ann. Rept. 1929, vol. 35, p. 15–195.

Carmony, F. A. (1903) *Jefferson County*, Nebr. Geol. Survey, Rept. 1903, p. 235–242.

Chamberlin, T. C. (1890) *Some additional evidence bearing on the interval between the glacial epochs*, Geol. Soc. Am., Bull., vol. 1, p. 469–480.

—— (1897) *Supplementary hypothesis respecting the origin of the loess of the Mississippi Valley*, Jour. Geol., vol. 5, p. 795–802.

——, and Salisbury, R. D. (1885) *Preliminary paper on the Driftless Area of the upper Mississippi Valley*, U. S. Geol. Survey, 6th Ann. Rept., p. 205–322.

——, —— —— (1907) *Geology*, 2d ed., rev., vol. 3, Henry Holt, New York.

Child, A. L. (1881) *The loess of the western plains—subaerial or subaqueous?* Kansas City Rev. Sci. Ind., vol. 4, p. 293–294.

Clapp, F. G. (1920) *Along and across the Great Wall of China*, Geog. Rev., vol. 9, p. 221–249.

Clendenin, W. W. (1896) *A preliminary report upon the Florida Parishes of eastern Louisiana and the bluff, prairie, and hill lands of southwest Louisiana*, La. State Exp. Sta., Geol. Agric., pt. 3, p. 159–247.

—— (1897) *A preliminary report upon the bluff and Mississippi alluvial land of Louisiana*, La. State Exp. Sta., Geol. Agric., pt. 4, p. 257–290.

Cressey, G. B. (1934) *China's geographic foundations*, McGraw-Hill, New York.

Dana, J. D. (1895) *Manual of geology*, American Book Co., New York.

Davis, W. M. (1905) *A journey across Turkestan*, Carnegie Inst. Washington, Pub. 26, p. 23–118

Davison, Charles (1894) *On the deposits from snowdrift, with especial reference to the origin of the loess and the preservation of mammoth remains*, Geol. Soc. London, Quart. Jour., vol. 50, p. 472–487.

Druif, J. H. (1927) *Over het ontstaan der Limburgsche löss in verband met haar mineralogische samenstelling*, Proefschrift, Bosch and Zoon, Utrecht, 331 pages.

Dumont, A. H. (1852) *Loess or lehm*, Geol. Soc. London, Quart. Jour., vol. 8, p. 278–281.

Ehrlich, C. (1848) *Berichtet über die Abstammung des am 1 Februar d. J. in Wien beobachteten Meteorstaubfalles*, Ber. Mitt. Freunden Naturwiss. Wien (W. Haidinger), vol. 4, p. 304–308.

Emerson, F. V. (1918) *Loess-depositing winds in the Louisiana region*, Jour. Geol., vol. 26, p. 532–541; (Abstract) Geol. Soc. Am., Bull., vol. 29, p. 79.

Fairchild, H. L. (1898) *Glacial geology in America*, Am. Geol., vol. 22, p. 154–189.

Fallou, F. A. (1867) *Über den Löss, besonders in Bezug auf sein Vorkommen, im Königreiche Sachsen*, Neues Jahrb. Mineral., 1867, p. 143–158.

Fenneman, N. M. (1938) *Physiography of eastern United States*, McGraw-Hill, New York.

Fisk, H. N. (1938) *Geology of Grant and La Salle Parishes*, La. Dept. Conserv., Geol. Bull. 10, 246 pages.

—— (1940) *Geology of Avoyelles and Rapides Parishes*, La. Dept. Conserv., Geol. Bull. 18, 240 pages.

Flint, R. F. (1941) *Glacial geology*, in Geology, *1888–1938*, Geol. Soc. Am., 50th Ann. vol., p. 19–41.

Fowke, Gerard (1908) *Surface deposits along the Mississippi between the Missouri and Ohio rivers*, Mo. Hist. Soc., Coll., vol. 3, p. 31–52.

Freudenberg, Wilhelm (1909) *Geologische beobachtungen im Gebiete der Sierra Nevada von Mexico*, Deutsche. geol. Gesell., Monatsber., vol. 61, p. 254–274.

Fuller, M. L. (1922) *Some unusual erosion features in the loess of China*, Geog. Rev., vol. 12, p. 570–584.

——, and Clapp, F. G. (1903) *Marl-loess of the lower Wabash valley*, Geol. Soc. Am., Bull., vol. 14, p. 153–176.

Ganssen (Gans) R. (1922) *Die Entstehung und Herkunft des Löss*, Preuss. Geol. Landes., Mitt., vol. 4, p. 37–46.

Glenn, L. C. (1906) *Underground waters of Tennessee and Kentucky west of Tennessee River and of adjacent area in Illinois*, U. S. Geol. Survey, W. S. Paper 164, 173 pages.

Gordon, C. H. (1892) *Quaternary geology of Keokuk, Iowa, with notes on the underlying rock structure*, Am. Geol., vol. 9, p. 183–190.

Gow, J. E. (1913) *Preliminary note on the so-called "loess" of southwestern Iowa*, Iowa Acad. Sci., Pr., vol. 20, p. 221–230.

Harper, L. L. (1857) *Preliminary report on the geology and agriculture of the state of Mississippi*, Jackson, 350 pages.

Harris, G. D. (1902) *The Tertiary geology of the Mississippi embayment*, La. State Exp. Sta., Geol. Agric., pt. 6, p. 5–39.

——, and Veatch, A. C. (1899) *A preliminary report on the geology of Louisiana*, La. State Exp., Sta., Geol. Agric., pt. 5, 345 pages.

Hay, O. P. (1914) *The Pleistocene mammals of Iowa*, Iowa Geol. Survey, Ann. Rept. 1912, p. 7–662.

Hayden, F. V. (1872) *Preliminary report of the U. S. geological survey of Wyoming and portions of contiguous territories*, Washington, 511 pages.

Hershey, O. H. (1895) *River valleys of the Ozark Plateau*, Am. Geol., vol. 16, p. 338–357.

—— (1896) *Early Pleistocene deposits of northern Illinois*, Am. Geol., vol. 17, p. 287–303.

Hesse, Richard, Allee, W. C., and Schmidt, K. P. (1937) *Ecological animal geography*, John Wiley and Sons, New York.

Hibbert, S. (1834) *Geschichte der erloschenen Vulkane im Becken von Neuwied*, Neues Jahrb. Mineral., 1834, p. 657–688.

Hilgard, E. W. (1860) *Report on the geology and agriculture of the State of Mississippi*, Jackson, xxiv + 391 pages.

—— (1874) *Silt analyses of Mississippi soils and subsoils*, Am. Jour. Sci., 3d. ser., vol. 7, p. 9–16.

—— (1879) *The loess of the Mississippi Valley and the eolian hypothesis*, Am. Jour. Sci., 3d. ser., vol. 18, p. 106–112.

—— (1884) *The steep slopes of the western loess*, Science, n. ser., vol. 4, p. 302.

Hobbs, W. H. (1931) *Loess, pebble beds, and boulders from glacial outwash of the Greenland continental glacier*, Jour. Geol., vol. 39, p. 381–385.

Hopkins, F. V. (1872) *Annual Report of Prof. D. F. Boyd, Superintendent of the Louisiana State University, for the year 1871, to the Governor of the State of Louisiana*, Am. Jour. Sci., 3d. ser., vol. 4, p. 136–138.

Howorth, H. H. (1882) *The loess—a rejoinder*, Geol. Mag., 2d ser., vol. 9, p. 343–356.

Huntington, Ellsworth (1905) *A geologic and physiographic reconnaissance in central Turkestan*, Carnegie Inst. Washington, Pub. 26, p. 159–216.

—— (1907) *The pulse of Asia*, Houghton Mifflin, Boston.

Jenny, Hans (1941) *Factors in soil formation*, McGraw-Hill, New York.

Jongmans, W. J., and Rummelen, F. H. van (1938) *Esquisse géologique de Nederlandsche Limburg avec, considération spéciale de la partie méridionale*, XV Intnat. Geog. Cong., (Amsterdam, 1938), Guide to excursion Bl, 148 pages.

Keilhack, K. (1920) *Das Rätsel der Lössbildung*, Deutsche geol. Gessell. Zeitschr. Monatsber., no. 6/7, p. 146–161.

Kerr, W. C. (1881) *On the action of frost in the arrangement of superficial earthy material*, Am. Jour. Sci., 3d. ser., vol. 21, p. 345–358.

Keyes, C. R. (1898) *Eolian origin of loess*, Am. Jour. Sci., 4th ser., vol. 6, p. 299–304.

—— (1912) *Relations of Missouri River loess mantle and Kansan drift-sheet*, Am. Jour. Sci., 4th ser., vol. 33, p. 32–34.

Kingsmill, T. W. (1871) *The probable origin of deposits of "loess" in north China and eastern Asia*, Geol. Soc. London, Quart. Jour., vol. 27, p. 376–384; (Abstract) Geol. Mag., vol. 8, p. 284–285.

Krokos, W. I. (1926) *Chemische Characteristik des Lösses im früheren Chersoner-Gouvernement, Lieferung II*, Wissenschaftl. Forschungs Inst. in Odessa, Ber., vol. 2, no. 4, p. 100–123.

Landes, K. K. (1933) *Caverns in loess*, Am. Jour. Sci., 5th ser., vol. 25, p. 137–139.

Le Conte, Joseph (1897) *Elements of geology*, 4th ed., D. Appleton & Co., New York.

Leverett, Frank (1894) *Notes on Pleistocene geology*, Am. Geol., vol. 13, p. 109–116.

—— (1897) *The Illinois glacial lobe*, U. S. Geol. Survey, Mon. 38, 608 pages.

—— (1932) *Quaternary geology of Minnesota and parts of adjacent states*, U. S. Geol. Survey, Prof. Paper 161, 149 pages.

Linstow, O. von (1910) *Das Alter des Lösses am Niederrhein und von Köthen-Magdeburg*, Preuss. Geol. Landes., Jahrb. 1910, pt. I.

Lobeck, A. K. (1939) *Geomorphology*, McGraw-Hill, New York.

Lyell, Charles (1834) *Observations on the loamy deposit called "loess" of the basin of the Rhine*, Edinburg New Philos. Jour., vol. 17, p. 110–122.

—— (1838a) *Observations on the loamy deposit called loess in the valley of the Rhine*, Geol. Soc. London, Pr., vol. 2, p. 83–85.

—— (1838b) *On the occurrence of fossil vertebrae of fish of the shark family in the loess of the Rhine, near Basle*, Geol. Soc. London, Pr., vol. 2, p. 221–222.

—— (1838c) *Presidential address of 19 February, 1836*, Geol. Soc. London, Pr., vol. 2, p. 357ff., references on p. 387–388; Am. Jour. Sci., vol. 33, p. 76–117.

—— (1847) *On the delta and alluvial deposits of the Mississippi River, and other points in the geology of North America, observed in the years 1845, 1846*, Am. Jour. Sci., 2d ser., vol. 3, p. 34–39, 267–269.

—— (1849) *A second visit to the United States*, Harper, New York.

—— (1863) *The geological evidences of the antiquity of man*, 2d. Am. ed., Childs, Philadelphia.

Macfarlane, James (1884) *The formation of cañons and precipices*, Science, n. ser., vol. 4, p. 99–101.

Mabry, T. O. (1898) *The brown or yellow loam of north Mississippi and its relation to the northern drift*, Jour. Geol., vol. 6, p. 273–302.

Marbut, C. F. (1935) *Soils of the United States*, U. S. Dept. Agric., Atlas Am. Agric., pt. 3, 98 pages.

Matthew, W. D. (1899) *Is the White River Tertiary an aeolian formation?* Am. Nat., vol. 33, p. 403–408.

McGee, WJ (1891) *The Pleistocene history of northeastern Iowa*, U. S. Geol. Survey, 11th Ann. Rept., pt. 1, p. 199–586.

Mill, H. R. (1895) *The glacial land-forms of the margin of the Alps*, Am. Jour. Sci., 3d ser., vol. 49 p. 121–126.

Moyer, R. T. (1932) *Introduction to a study of the soils of Shansi Province, China*, Dokuchaiev Inst. Soil Sci., Contrib. knowledge soils of Asia, pt. 2, p. 9–16.

Myers, H. E. (1937) *Physico-chemical reactions between organic and inorganic soil colloids as related to aggregate formation*, Soil Sci., vol. 44, p. 331–359.

Newberry, J. S. (1889) *The origin of the loess*, School Mines Quart., vol. 10, p. 66–69.

Obrutchev, W. (1895) *Geographische skizze von Centralasien und seiner südlichen umrandung*, Geog. Zeitschr., vol. 1, p. 257–285.

Oefelein, ĸ. T. (1934) *A mineralogical study of loess near St. Charles, Missouri*, Jour. Sedim. Petrol., vol. 4, p. 36–44.

Owen, D. D. (1852) *Report of a geological survey of Wisconsin, Iowa, and Minnesota*, Philadelphia, 638 pages.

Owen, L. A. (1904) *The loess at St. Joseph*, Am. Geol., vol. 33, p. 223–228.

—— (1905) *Evidence on the deposition of loess*, Am. Geol., vol. 35, p. 291–300.

—— (1926) *Later studies of loess*, Pan-Am. Geol., vol. 45, p. 173–174.

Peters, K. (1859) *Geologische studien aus Ungarn*, Geol. Reichsanst. Wien., Jahrb., vol. 10.

Polynov, B. B. (1937) *The cycle of weathering*, Trans. A. Muir, (Murby, London).

Pumpelly, Raphael (1866) *Geological researches in China, Mongolia, and Japan during the years 1862 to 1865*, Smithson. Inst., Contrib. knowledge, no. 202, 162 pages.

—— (1879) *The relation of secular rock disintegration to loess, glacial drift and rock basins*, Am. Jour. Sci., 3d ser., vol. 17, p. 133–144.

—— (1908) *Ancient Anau and the oasis-world*, Carnegie Inst. Washington, Pub. 73, vol. 1, p. 3–80.

—— (1918) *My reminiscences*, 2 vols., Henry Holt, New York.

Pumpelly, R. W. (1908) *Physiography of central Asian deserts and oases*, Carnegie Inst. Washington, Pub. 73, vol. 2, p. 270–282.

Richthofen, F. F. von, (1877) *China*, vol. 1, Berlin (Dietrich Reimer), many reviews exist, *see* Stuntz, S. C. and Free, E. E., for list, also Am. Jour. Sci., 3d ser., vol. 14, p. 487–491.

—— (1882) *On the mode of origin of the loess*, Geol. Mag., n. ser., vol. 9, p. 293–305.

Roberts, R. C. (1933) *Structural relationships in a lateritic profile*, Am. Soil Survey Assoc., Bul., vol. 14, p. 88–90.

Robinson, G. W. (1936) *Soils: their origin, constitution, and classification*, 2d ed., Murby, London.

Rummelen, F. van (1923) *Zie Handelingen 19*, Natuur en Geneesk. Cong., (Maastricht, 1923).

Rungaldier, R. (1933) *Bermerkungen zur Lössfrage, besonders in Ungarn*, Zeitsch. Geomorph., vol. 8, p. 1–40.

Russell, I. C. (1897) *A reconnaissance in southeastern Washington*, U. S. Geol. Survey, W. S. Paper, vol. 4, 96 pages.

—— (1901) *Geology and water resources of Nez Perce County, Idaho*, U. S. Geol. Survey, W. S. Paper, vol. 53, 141 pages.

Russell, R. J. (1938) *Physiography of Iberville and Ascension Parishes*, La. Dept. Conserv., Geol. Bull. 13, p. 3–86.

—— (1940) *Quaternary history of Louisiana*, Geol. Soc. Am., Bull., vol. 51, p. 1199–1234.

Safford, J. M. (1869) *Geology of Tennessee*, Nashville, 550 pages.

Salisbury, R. D. (1892) *The drift of the north German lowland*, Am. Geol., vol. 9, p. 294–319.

Sardeson, F. W. (1899) *What is the loess?* Am. Jour. Sci., 4th ser., vol. 7, p. 58–60.

Savage, T. E. (1915) *The loess in Illinois: its origin and age*, Ill. Acad. Sci., Tr., vol. 8, p. 100–117.

Scheidig, Alfred (1934) *Der Löss und seine geotechnischen eigenschaften*, Th. Steinkopff, Dresden and Leipzig.

Schmitthenner, Heinrich (1919) *Der chinesische Lösslandschaft*, Geog. Zeitschr., vol. 25, p. 308–322,

Schokalsky, Z. J. (1932) *The natural conditions of soil formation in India*, Dokuchaiev Inst. Soil Sci., Contrib. knowledge soils Asia, pt. 2, p. 53–155.

Shaw, E. W. (1914) *So-called waterlaid loess of the central United States*, Washington Acad. Sci. Jour., vol. 4, p. 298.

—— (1915) *On the origin of the loess of southwestern Indiana*, Science, n. ser., vol. 41, p. 104–108.

Shimek, Bohumil (1888) *Notes on the fossils of the loess at Iowa City, Iowa*, Am. Geol., vol. 1, p. 149–152.

—— (1896) *A theory of the loess*, Iowa Acad. Sci., Pr., vol. 3, p. 82–89.

—— (1897) *Is the loess of aqueous origin?* Iowa Acad. Sci., Pr., vol. 5, p. 32–45.

—— (1902) *The loess of Natchez, Mississippi*, Am. Geol., vol. 30, p. 279–300.

—— (1903) *The loess and the Lansing man*, Am. Geol., vol. 32, p. 353–369.

—— (1916) *The loess of Crowley's ridge, Arkansas*, Iowa Acad. Sci., Pr., vol. 23, p. 147–157.

Skertchly, S. B. J., and Kingsmill, T. W. (1895) *On the loess and other superficial deposits of Shantung, (North China)*, Geol. Soc. London, Quart. Jour., vol. 51, p. 238–254.

Smith, R. S., and Norton, E. A. (1935) *Parent Materials, Subsoil Permeability and Surface Character of Illinois Soils*, in *Parent material of Illinois soils*, Univ. Ill., Agric. Exp. Sta., Urbana, p. 1–4.

Stephenson, L. W., Logan, W. N., and Waring, G. A. (1928) *The ground-water resources of Mississippi*, U. S. Geol. Survey, W. S. Paper, vol. 576, 515 pages.

Stuntz, S. C., and Free, E. E. (1911) *Bibliography of eolian geology*, U. S. Dept. Agric., Bur. Soils, Bull., vol. 68, p. 174–263.

Swallow, G. C. (1855) *The first and second annual reports of the geological survey of Missouri*, Jefferson City.

Tarr, R. S., and Martin, Lawrence (1914) *College physiography*, Macmillan, New York.

Tillo, Alexis von (1893) *Die geographische Verteilung von Grund und Boden*, Pet. Mitt., vol. 39, p. 17–19.

Tilton, J. L. (1925) *The definition of loess*, Science, n. ser., vol. 62, p. 83.

Todd, J. E. (1878) *Richthofen's theory of the loess, in the light of the deposits of the Missouri*, Am. Assoc. Adv. Sci., Pr., vol. 27, p. 231–239; reviewed in Am. Jour. Sci., 3d ser., vol. 18, p. 148 (1879).

—— (1894) *Pleistocene problems in Missouri*, Geol. Soc. Am., Bull., vol. 5, p. 531–548.

—— (1897) *Degradation of loess*, Iowa Acad. Sci., Pr., vol. 5, p. 46–51.

—— (1899) *The moraines of southeastern South Dakota and their attendant deposits*, U. S. Geol. Survey, Bull., vol. 158, 171 pages.

—— (1918a) *Aqueous loess*, Kans. Acad. Sci., Tr., vol. 29, p. 115–116.

—— (1918b) *Eolian loess*, Kans. Acad. Sci., Tr., vol. 29, p. 200–203.

Udden, J. A. (1898) *Some preglacial soils*, Iowa Acad. Sci., Pr., vol. 5, p. 102–104.

—— (1902) *Loess with horizontal shearing planes*, Jour. Geol., vol. 10, p. 245–251.

Upham, Warren (1891) *A review of the Quaternary era, with special reference to the deposits of flooded rivers*, Am. Jour. Sci., 3d ser., vol. 41, p. 33–52.

—— (1892) *Inequality of distribution of the englacial drift*, Geol. Soc. Am., Bull., vol. 3, p. 134–148.

—— (1895) *Climatic conditions shown by North American interglacial deposits*, Am. Geol., vol. 15, p. 273–295.

Veatch, A. C. (1905) *The underground waters of northern Louisiana and southern Arkansas*, La. State Exp. Sta. Geol., Bull., vol. 1, p. 82–91; U. S. Geol. Survey, W. S. Paper, vol. 114, p. 179–187.

—— (1907) *Geology and underground water resources of northern Louisiana with notes on adjoining districts*, La. State Exp. Sta., Geol., Bull., vol. 4, p. 261–457.

Virlet d'Aoust, P. Th. (1857) *Observations sur un terrain d'origine météorique ou de transport aérien qui existe au Mexique, et sur le phénomène des trombes de poussière auquel il doit principalement son origine*, Geol. Soc. France, Bull., 2d ser., vol. 2, p. 129–139.

Wahnschaffe, Felix (1886) *Die lössartigen Bildungen am Rande des norddeutschen Flachlandes*, Deutsche Geol. Gesell., Zeitschr., vol. 38, p. 353–369.

Wailes, B. L. C. (1854) *Report on the agriculture and geology of Mississippi*, Jackson, 371 pages.

Warren, G. K. (1878) *Valley of the Minnesota and of the Mississippi River to the junction of the Ohio; its origin considered*, Am. Jour. Sci., 3d ser., vol. 16, p. 417–431.

Willis, Bailey (1907) *Research in China*, Carnegie Inst. Washington, Pub. 54, vol. 1, references p. 246–251.

Winchell, N. H. (1879) *The loess of Minnesota*, Am. Jour. Sci., 3d ser., vol. 17, p. 168–170.

—— (1884) *Geology of Minnesota*, Minn. Geol. Survey, Final Rept., vol. 1, 673 pages.

—— (1903) *Was man in America in the glacial period?* Geol. Soc. Am., Bull., vol. 14, p. 133–152.

Witter, F. M. (1892a) *Arrow points from the loess at Muscatine, Iowa*, Am. Geol., vol. 9, p. 276–277.

—— (1892b) *Geology*, in *History of Muscatine County (Iowa)*, p. 323–338 (copy examined lacked title page).

Wood, S. V. (1882) *On the origin of the loess*, Geol. Mag., 2d ser., vol. 9, p. 339–343.

Worthen, A. H. (1866) *The geology of Hancock County*, Ill. State Geol. Survey, vol. 1, p. 327–349.

—— (1868) *Geology and Paleontology* in *Geology of Jackson, Jersey, and Green Counties*, Ill. State Geol. Survey, vol. 3, p. 58–133.

Wright, G. F. (1890) *The glacial boundary in western Pennsylvania, Ohio, Kentucky, Indiana, and Illinois*, U. S. Geol. Survey, Bull., vol. 58, 112 pages.

———— (1901) *Recent geological changes in northern and central Asia*, Geol. Soc. London, Quart. Jour., vol. 57, p. 245–250.

———— (1904) *Evidence of the agency of water in the distribution of the loess in the Missouri Valley*, Am. Geol., vol. 33, p. 205–222.

———— (1911) *The ice age in North America*, 5th ed., Oberlin (Bibliotheca Sacra).

———— (1921) *Origin and distribution of the loess*, Geol. Soc. Am., Bull., vol. 32, p. 48–49.

Wright, W. B. (1914) *The Quaternary ice age*, Macmillan, London.

LOUISIANA STATE UNIVERSITY, BATON ROUGE, LOUISIANA.
MANUSCRIPT RECEIVED BY THE SECRETARY OF THE SOCIETY, APRIL 13, 1943.

Editor's Comments
on Paper 10

10   HOBBS

Wind—The Dominant Transportation Agent Within Extramarginal Zones to Continental Glaciers

Hobbs was a prolific writer on loess and his most significant contribution was probably in emphasizing the importance of the wind systems controlled by glaciers in distributing loess material. He also (1931) pointed out the existence of loess deposits in Greenland.

Hobbs' approach is interesting: having established his model of anticyclonic wind systems associated with glaciers he can then suggest where loess should be found—with success. His most elaborate paper on the subject was published in 1943 by the American Philosophical Society (Hobbs, 1943a). It is a very long paper and we cannot include it here. A year earlier he published his general views on the role of wind as a transportation agent in the marginal zones of continental glaciers and that paper is included in the present collection. Apart from the more specific application of his ideas to the North American continent (Hobbs, 1943a), he also considered the glacial anticyclone associated with the European continental glaciers (Hobbs, 1943b).

The Hobbs anticyclonic idea fits the European and North American glaciations very neatly and is possibly applicable to the Chinese loess. If Sun and Yang (1961) have actually established the existence of large Quaternary continental glaciers in China, an associated loess deposit might be expected, and the anticyclonic winds could deposit debris from southern glaciers in the northern marginal regions. The resultant loess deposit would presumably thin in the southwest–northeast direction, which does not appear to accord with published observations.

REFERENCES

Hobbs, W. H. 1931. Loess, pebble bands and boulders from glacial outwash of the Greenland continental glacier. J. Geol. 39, 381–385.

——. 1943a. The glacial anticyclone and the continental glaciers of North America. Proc. Amer. Phil. Soc. 86, 368–402.

——. 1943b. The glacial anticyclones and the European continental glaciers. Amer. J. Sci. 241, 333–336.

Sun Tien-ching and Yang Huan-jen. 1961. The great Ice Age glaciation in China. Acta Geol. Sinica 41, 234–244 (in Chinese).

10

Reprinted from *Jour. Geol.*, **50**(5), 556–559 (1942)

WIND—THE DOMINANT TRANSPORTATION AGENT WITHIN EXTRAMARGINAL ZONES TO CONTINENTAL GLACIERS

WILLIAM HERBERT HOBBS

University of Michigan

ABSTRACT

The dominant transportation agent within the extramarginal zones to continental glaciers is for most of the time exclusively the fierce anticyclonic storm wind directed outward normal to the glacier front. Transportation by running water during the remainder of the year is almost exclusively by meltwater. The volume of this water over the outwash plain fluctuates between trickling braided streams and floods covering the plain on warm days, so that the boulders carried in small icebergs from the glacier front either are buried within the deposits or are left stranded upon the plain. During the longer period of wind dominance there result pebble bands (pavements), ventifacts, etched boulders, dunes, and loess deposits—all of them extramarginal to the glacier.

On the basis of studies made near the border of the Greenland continental glacier I called attention in this *Journal* for May-June, 1931, to the erroneous conception which has long prevailed regarding the conditions which have obtained outside the Pleistocene continental glaciers of North America and Europe.[1] At the Cambridge meeting of the American Association for the Advancement of Science in 1933, I again referred to the Greenland observations and discussed the evidence that the same conditions had been present about the North American continental glaciers of Pleistocene times. The traditional view, based upon studies of mountain glaciers like those of the Alps, had taken no account of the fierce storms which blow outward off the ice of continental glaciers or of the restriction of drainage to a brief warmer season during which meltwater issues from beneath the glacier. Meteoric precipitation is almost exclusively of snow or rime.

As soon as the brief warm season has come to an end, the agent of extraglacial transportation is no longer running water but strong wind currents directed radially outward from the glacier, or nearly

---

[1] "Loess, Pebble Bands, and Boulders from Glacial Outwash of the Greenland Continental Glacier," *Jour. Geol.*, Vol. XXXIX (1931), pp. 381–85; cf. also "The Origin of the Loess Associated with Continental Glaciation Based on Studies in Greenland," *Compt. rend. Congrès internat. géog.*, Vol. II (Paris, 1931), pp. 1–4.

at right angles to the glacier front. During the brief warm season the glacier front becomes more or less undermined by the thaw water which has descended within the marginal crevasses of the ice and issues from beneath the front. Masses of the lower ice layers, loaded with englacial rock debris, including large boulders, become detached as small bergs which are not too heavily burdened to float off in the braided streams on the outwash plain. During exceptionally warm days the streams of outwash may coalesce and become transformed into lakes which may cover most of the plain. When the temperature drops, the flood subsides, and bergs with their enclosed boulders are stranded over the plain away from the braided channels.

When the warmer season ends, no meltwater issues from the glacier, and the entire plain of outwash quickly dries. This marks the beginning of the longer winter season, and wind takes over from the thaw water the work of transportation.

These conditions within the extramarginal zones of the continental glaciers are at this time similar to those within the deserts far removed from glaciers. The rock materials of the outwash, varying in degree of coarseness—rock flour, sand, pebbles, and boulders—are sorted out by the winds. Dust, sand, and the smaller pebbles are lifted and carried away, leaving behind an armor of pebble pavement to protect the materials below, and in every way this pavement is similar to the *sêrir* of low-latitude deserts. The boulders stranded on the plains are now exposed to the driving sands and in consequence become deeply etched, while the larger pebbles and even boulders of uniform hardness become smoothed, planed, and transformed into ventifacts. Those boulders which are crossed by dikes or veins of harder material under this drilling action develop into the "stone lattice" familiar from the low-latitude deserts, and here are found the finest examples anywhere known. Igneous rocks of porphyritic texture show the phenocrysts in strong relief. Conglomerates containing quartzite pebbles show the pebbles projecting from the general surface, whereas pure quartzite rocks may be transformed into smooth ventifacts.

The sand, after accomplishing its work of planing and etching, comes to rest in dunes which under the fierce winds are ridges normal to the glacier front, while the dust is widely spread as loess deposits, sometimes with pebble bands at their base and also in some cases with included etched boulders.

All these types—pebble pavements, sand dunes, loess deposits, deeply etched boulders, pebble bands, and ventifacts—are now known to be widely distributed throughout the glaciated area of North America, but until recently they have found little mention in the literature of the subject. At the time my first paper on the subject was issued (1931), I sent out widely to American geologists and

FIG. 1.—Deeply etched boulders from the glaciated area around Ann Arbor, Michigan

others throughout the glaciated area of North America "A Call for Information concerning Etched Erratic Boulders."[2] This call brought a sufficient number of responses to show that similar etched boulders are common within many parts of the glaciated area of North America, particularly likely to be found selected as curiosities in house yards (see Fig. 1) or even as gravestones in cemeteries. I have myself found such etched boulders by the score within the glaciated area surrounding Ann Arbor, Michigan, and during subsequent travel I have seen them at many other localities.

The restriction of American and European loess deposits to extra-

[2] *Jour. Geol.*, Vol. XLIII (1935), pp. 551–52. The inquiry has not been concluded and further communications are requested for the completion of a map of distribution for the glaciated area of North America which is now in preparation.

marginal zones about Pleistocene continental glaciers had long been recognized, and Dr. George F. Kay has treated of sand dunes, pebble bands, and etched boulders in the loess on Iowan till.[3]

There is need for revision of studies of much of the glaciated areas of North America with this altered picture of extramarginal conditions around the glaciers in mind.

[3] "Origin of the Pebble Band on Iowan Till," *Jour. Geol.*, Vol. XXXIX (1931), pp. 377–80; see also Lincoln R. Thiesmeyer and Ralph E. Digman, "Wind-cut Stones in Kansan Drift of Wisconsin," *Jour. Geol.*, Vol. L (1942), pp. 174–88.

Editor's Comments
on Papers 11, 12, and 13

11   BRYAN
Glacial Versus Desert Origin of Loess

12   SWINEFORD and FRYE
A Mechanical Analysis of Wind-Blown Dust Compared with Analyses of Loess

13   OBRUCHEV
Loess Types and Their Origin

In 1944 the Nebraska Academy of Science organized a symposium on loess; the papers presented were subsequently published, with discussions, in the *American Journal of Science.* Papers were presented by Elias, Schultz and Stout, Bryan, Swineford and Frye, Weaver, Condra, Obruchev, Thorp, Williams, Duley, Bollen, and Watkins. The papers by Bryan, Swineford and Frye, and Obruchev are reproduced here; these relate most closely to the topic of loess lithology and genesis. The Condra paper, listed in the foreword by Elias, was not published.

In his short contribution (Paper II) Bryan makes the suggestion that, although desert conditions may be associated with the formation of loess deposits, glacial or periglacial processes are likely to be involved. Swineford and Frye (Paper 12) used a dust-storm sample to test Russel's contention (see Paper 9) that the wind could not be an effective sorter of loess-sized material. They suggest that the wind could be an effective sorting agent, but their conclusions should be compared with those of Beavers and Albrecht (see Paper 18), who suggested that the wind does not *need* to be an effective sorting agent. Obruchev (Paper 13) argues for a desert source and dismisses Penck's point (Paper 6) about the lack of a loess girdle around a desert such as the Sahara by suggesting that the missing loess material is carried beyond the confines of the continent.

11

Copyright © 1945 by the American Journal of Science (Yale University)

Reprinted from *Amer. Jour. Sci.*, **243**(5), 245–248 (1945)

GLACIAL VERSUS DESERT ORIGIN OF LOESS.

KIRK BRYAN.

ABSTRACT. The main source of loess is the outwash of glacial rivers pulverized by frost action. The main areas of its deposition were marginal to the region most affected by the glacial born anticyclonic winds combined with prevailing westerlies. Steppe loess has its source in the adjacent deserts. The structural similarities in loesses of different origin are due to secondary processes induced by grass vegetation.

Flood plains of Nebraska rivers, which in Pleistocene time were overloaded, glacial-fed streams, were the source of the loess in the State.

THE formation and deposition of wind-borne and wind-deposited silt (loess) is a complex process requiring (1) source of silt; (2) adequate winds blowing predominantly or in net effect from one direction; (3) an adequate place for deposition. This relatively simple chain of thought leads to the conclusion that the flood-ravished outwash plains of glacial rivers made pulverent by frost action form ideal sources of silt. To a lesser extent recently uncovered till plains kept free of vegetation by frost action are also sources. The outblowing or anticyclonic winds of glacial ice masses are combined with the prevailing westerlies to form a system of favorable winds. Places of deposition occur in the tundra-forest border which in middle latitudes once surrounded the Pleistocene ice sheets. Excessive frost-action in these areas was, however, a factor in the removal of wind-borne silt as fast as it was deposited. Large bodies of loess are therefore largely preserved in areas somewhat more marginal to the region most affected by the anticyclonic winds.

The secondary changes in the wind-borne silt give the material its buff color, vertical structure, and in part its calcareous content. These characteristics are induced by the edaphic effect of grass cover which tends to develop on silt in a zone which is generally forested. Loess of glacial origin thus resembles in its secondary features the loess of the steppe zones surrounding desert areas. However, production of dust in deserts is inhibited by the formation of the desert pavement and also by the lack of wind. Only the great trade-wind deserts have strong winds blowing consistently or dominantly in one direction. Thus the Sahara produces dust which is deposited

107

to the South in the Sudan and also on the west in the "Dark Sea" off the shores of Mauretania. Some deserts like that of Chile are, because of calm air, depositories of dust.

In Nebraska the relative influence of glacial and desert sources is acute. It should, however, be borne in mind that the Platte and Missouri Rivers and all their Rocky Mountain tributaries, were in Pleistocene time overloaded, glacially-fed streams. To the extent that their flood plains are the source of Nebraska loess, this loess is essentially a glacial rather than a desert loess.

DISCUSSION.

Between James Thorp and Kirk Bryan.

THORP: 1. Bryan suggests that loess probably has accumulated most abundantly in periglacial areas. I wonder if we could lay a little more stress on the importance of glacial alluvial outwash as a source of the dust, this being, of course, a periglacial phenomenon. In many places the alluvial outwash plains are obviously a source of much of this dust we see accumulated. For example, in Colorado and Wyoming there are islands of loess which occur in association with old outwash plains and alluvial fans. On one of the older of these alluvial-fan remnants in Star Valley, Wyoming, and on adjacent areas of upland which have not been eroded since the time of loess deposition, we find loess deposits and we do not find them on the lower levels. We also find similar islands near Jackson, Wyoming, and assume that outwash from the glacier of that particular area was the source of the loess. The loess is not continuous with the larger deposits farther west.

2. Great loess accumulations southeast of the Ordos Desert in China suggest the desert to be a source of the loess. In that vicinity there is no large alluvial plain that could be interpreted to be the source. In the neighborhood of Loyang in northern Honan Province, I measured a meter thickness of loess over roof tiles, bricks, and other evidences of human culture which could not have been older than about 2000 years. This loess deposit is still accumulating and the dusts of which it is made are blown up from adjacent river valleys (the Loho and Yellow River). Thick clouds of dust can be seen on almost any day from autumn until late spring, borne southeastward by the prevalent northwest winds of the winter

phase of the monsoon. In the Yangtze Valley in the neighborhood of Nanking, there are thick deposits of clayey material of a slightly reddish or light-brown tint which appear to be aeolian in origin. These deposits are closely associated with the flood plains (or former flood plains) of eastern China. Most of the material is now too strongly weathered to bear the name "loess." Deposits of this type near Nanking are known as "Siashu loams."

Another deposit of material, aeolian in origin but not loess in the true sense of the word, was seen in the Szechuan Basin in west China. It occurs on terrace remnants and on adjacent uplands eastward from the Chengtu Plain and elsewhere in the Tibetan borderland.

BRYAN: Regarding the importance of recent alluvium as a source of loess, two questions arise: 1. Whether ordinary alluvium in arid regions, which has no connection with the glacier, forms a source of loess. In any of the generalizations, the exception is always possible and I have no doubt that there are places where dust is being accumulated at the present time. There are also volcanic dusts being accumulated, some of which are mixed with the dusts that are not actually blown out of a volcano but are picked up. This occurs in Mexico where the origin of this theory began. There are no doubt places in China where there is an accumulation of loessial material as of the present day which is not necessarily evidence of the origin of the older and thicker deposits. In many parts of the world, dust which we see is partly man-induced. There probably are exceptions but generally speaking the great bodies of loess are of the past. Are they something which comes out of the desert associated with a dry climate or do they come out of a desert associated with a glacial deposit? In other words, is it not likely that loess deposits from deserts may owe their accumulation to periglacial phenomena?

THORP: Do you think that of the loess that accumulated during the ice age the greater part came from outwash plains or from till?

BRYAN: I would say that in Europe and in the eastern United States the greater part of the loess was obtained from the outwash but one cannot deny that a till plain might be a favorable place to gather dust. That is being studied at the present time.   Relationship of vegetation to frost action is an interesting point to study in this connection.   In Alaska there is a contest between the intensity of frost action and the tundra. If we can determine how close to the continental ice sheet the tundra encroached, we will be able to answer your question. Pollen studies will be one means of determining this.   They will reveal information regarding the status of vegetation upon the continental ice.   A bare area was present and a large tundra area probably went as far south as Missouri and North Carolina.

GEOLOGICAL MUSEUM,
    HARVARD UNIVERSITY,
        CAMBRIDGE, MASS.

12

Reprinted from *Amer. Jour. Sci.*, **243**(5), 249-255 (1945)

A MECHANICAL ANALYSIS OF WIND-BLOWN DUST COMPARED WITH ANALYSES OF LOESS.

ADA SWINEFORD AND JOHN C. FRYE.

ABSTRACT. A mechanical analysis has been made of wind-blown dust collected in September, 1939, from the level of the third floor of the Lakeway Hotel, Meade, Kansas. Comparisons of this analysis with previously published analyses of loess and new analyses of Kansas loess seem to demonstrate that wind can be competent to sort material to the degree represented by some loess deposits.

INTRODUCTION.

RUSSELL (1944, pp. 1-40) recently described loess deposits along the lower Mississippi valley, discussed at some length their possible origin, and concluded that loess is probably of fluvial-colluvial origin. His ideas concerning the origin of loess differ from those held by some geologists. Russell's paper has served to emphasize the fact that the origin of loess is not a closed question, and that some deposits that have been called loess at different localities may have had different origins. The present paper is concerned with only one of the data used by Russell in arriving at his theory of the origin of loess, namely the degree of sorting of the material. He stated (p. 24) that in loess:

"The sorting appears to be too uniform to be the result of direct deposition from a current. It seems improbable that either wind or water could move with the uniform velocity required to permit the accumulation of material so homogeneous that at least a half and ordinarily about three-fourths of its particles (by weight) fall within the limited diameter range 0.01-0.05 mm."

During a dust storm of about two hours duration in September 1939, dust accumulated to an estimated depth of 1/16 inch on surfaces near open third floor windows in the Lakeway Hotel at Meade, Kansas. These windows are about 25 feet above street level. Frye collected a sample of this dust from enameled surfaces which had been cleaned earlier in the day, and stored the sample in a sealed container. The foregoing statement made by Russell prompted the writers to

111

TABLE I. Size distribution of wind-blown dust and Sanborn loess samples.

Sample Number	Location	Size distribution in mm. (per cent by weight).											
		2-1	1-.5	.5-.25	.25-.125	.125-.0625	.0625-.0312	.0312-.0156	.0156-.0078	.0078-.0039	.0039-.00195	.00195-.00098	<.00098
Dust	Meade, Kansas.	0.04	0.19	0.38	1.64	8.45	41.85	24.41	5.63	8.89	5.31	2.67	5.55
1	SE cor. sec. 23, T. 1 S., R. 42 W. Cheyenne County, 30 feet above base of exposure.			0.06	0.80	4.72	60.32	20.55	5.43	2.01	1.30	0.88	4.49
2	SW cor. sec. 4, T. 1 S., R. 41 W. Cheyenne County, road cut.				0.06	3.42	67.25	16.75	3.99	1.82	1.14	0.80	4.78
3	SW sec. 28, T. 3 S., R. 36 W. Rawlins County, 25 feet above base of loess.			0.04	0.07	3.30	55.88	22.40	6.95	2.98	1.70	1.13	5.53
4	NE NE sec. 18, T. 8 S., R. 36 W. Thomas County, in road cut.				0.37	1.45	53.22	22.64	7.27	4.30	2.48	1.49	6.78
5	SW sec. 31, T. 10 S., R. 36 W. Thomas County.		0.09	0.24	0.29	4.41	55.48	22.40	5.69	4.36	1.82	0.97	4.24
6	NE NE sec. 8, T. 8 S., R. 35 W. Thomas County, in railroad cut 7 feet below surface.			0.07	0.30	2.49	58.03	23.81	5.72	2.66	1.46	0.40	5.05
7	SW SE sec. 32, T. 10 S., R. 32 W. Thomas County, 10 feet below surface in road cut.				0.03	2.02	49.41	27.97	8.32	3.70	2.08	1.39	5.09
8	SW SE sec. 32, T. 10 S., R. 32 W. Thomas County, 5 feet below surface in road cut.			0.06	0.06	0.80	53.43	26.65	7.91	3.19	1.53	1.02	5.36
9	NW cor. sec. 30, T. 8 S., R. 31 W. Sheridan County, 4.5 feet below surface in road cut.					0.94	46.01	28.33	9.59	4.80	2.40	1.80	6.14
10	NE sec. 6, T. 3 S., R. 27 W. Decatur County, base of loess.		0.12	0.33	0.33	1.51	37.36	29.81	13.28	6.04	3.62	1.81	5.79
11	NE sec. 6, T. 3 S., R. 27 W. Decatur County, 5 feet above base of loess.					2.62	51.10	25.96	8.13	3.54	1.84	1.05	5.77
12	NE sec. 6, T. 3 S., R. 27 W. Decatur County, 10 feet above base of loess.				0.03	3.75	49.94	24.56	9.55	3.75	1.59	1.25	5.57

analyze this dust in an attempt to test the validity of his conclusion concerning the inability of wind to produce the degree of sorting observed in loess samples.

In order to compare the dust collected at Meade with High Plains loess deposits, size analyses were made of 12 samples of loess collected in 1943 from the Sanborn formation of northwestern Kansas. The location and analyses of these samples are given in Table I.

Udden (1914, pp. 720-726) published analyses of dust, and stated that most eolian deposits are better sorted than most water deposits. Direct comparisons with Udden's analyses have not been made because (1) his analytical methods were quite different from those used in the present study, and (2) the dust samples that he analyzed either were deposited after being transported only a very short distance by wind (such as dust stirred up by a running train and deposited inside the coach) or were collected from the atmosphere where it contained only a small concentration of dust. Thus his samples may not be comparable to the material transported and deposited by western Kansas "dust storms" such as transported and deposited the sample of dust herein described.

ANALYSES.

The dust and loess were prepared for analysis by digesting 30 to 40 grams of sample in hot hydrochloric acid. The acid was removed by six filterings through a Pasteur-Chamberland filter. Each sample was then dried, weighed, dispersed with sodium oxalate solution, and wet-sieved. The fraction caught on a 1/16 mm. sieve was dried, disaggregated with a rubber pestle, and sieved through screens of Wentworth grades. The silt collected in the pan was added to the sodium oxalate solution, and enough solution was added to make a liter. Six portions were withdrawn by a pipette to determine the per cent of material in each Wentworth grade from 1/32 to 1/1024 mm.

COMPARISONS OF SIZE DISTRIBUTIONS.

In order to compare the size distributions of the sample of wind-blown dust with the samples of Sanborn loess and with analyses of Mississippi valley loess published by Russell, plotted cumulative curves of the loess analyses have been superposed (Fig. 1) on the the curve of the dust analysis. The similarity

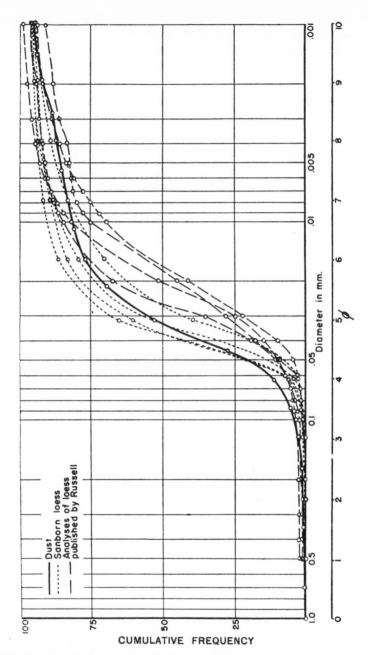

Fig. 1. Cumulative curves of four analyses of Sanborn loess (Sample Nos. 1, 5, 10, 12); four loess analyses published by Russell; and an analysis of modern wind-deposited dust.

of shape of the curves representing the analyses of loess to the curve obtained from the material known to have been transported and deposited by wind is quite apparent. The curves obtained from analyses of Sanborn loess samples indicate slightly coarser material, and Russell's analyses indicate a generally finer material than the 1939 deposit.

The only significant discrepancy between the shape of the curves obtained from analyses of Sanborn loess and the eolian material is in the disproportionately large percentage in the coarse fractions shown by the dust curve. A microscopic examination of the dust, however, reveals that these coarse fractions contain a large percentage of organic material such as fragments of plants and insects and some coal dust. Although such organic materials might have once been in the Pleistocene deposits, they were not preserved. All the distributions, including the dust, fulfill Russell's requirement that at least 50 per cent of the distribution must fall between 0.01 and 0.05 mm. (Russell, 1944, pp. 4, 5).

We have chosen the quartile deviation, defined as half the range of the middle 50 per cent of the frequency distribution, as the measure of sorting for comparison because it is not affected by the ends of the distribution. The data are expressed in terms of phi units, each being equivalent to one Wentworth grade, so that conventional statistical methods can be used.

The quartile deviation for the dust is 0.79 ϕ; that is, the middle half of the sample falls within 1.58 Wentworth grades. The quartile deviations for Russell's four loess samples range from 0.65 ϕ to 0.92 ϕ, and their mean is 0.81 ϕ. The quartile deviations for the 12 samples of Kansas loess range from 0.47 ϕ to 0.90 ϕ, with a mean of 0.67 ϕ. Thus, the Sanborn loess samples seem to be better sorted than both Russell's loess and the dust, but the dust sorting falls within the range of that of the Sanborn loess. Stress has been placed upon the range of quartile deviations rather than upon their standard error because of the small number of analyses involved in each case.

A number of analyses of Peorian loess samples from Iowa were reported in a recent paper by Kay and Graham (1943, pp. 173-183). In order to make further comparisons the quartile deviations were calculated from 18 of these analyses selected at random. The average was 0.75 ϕ with a range from 0.34 ϕ to 1.49 ϕ. This range includes the sample of dust, all of the Sanborn loess samples analyzed, and all of the loess analyses published by Russell.

A comparison of skewness values[1] also shows similarity between the dust and Sanborn loess. The dust skewness is $+0.93$; the average of Sanborn loess analyses is $+0.96$, with a range of $+0.78$ to $+1.14$; and the average skewness of Russell's loess analyses is $+0.41$.[2] The skewness of a "loess-like" material described by Russell as a Pleistocene terrace remnant, is -0.07, which is much smaller and in the opposite direction. Although little is known about the significance of skewness in sediments, analyses of loess samples almost uniformly show a strong positive skewness.

The discrepancies between the sorting and skewness values of the Kansas loess and Russell's loess may be in part the result of differences in preliminary treatment of the samples. The effect of thorough breaking up of the coarse aggregates of the Kansas loess samples by a rubber pestle would be to increase the skewness and improve the sorting. Breaking up of clay aggregates or granules may tend to destroy some characteristics of the loess deposit as it occurs in nature, but it is believed that this did not affect the conclusions. Aggregates of grains were rare in the dust.

CONCLUSIONS.

The comparisons that have been made of the grain size analyses of Kansas Sanborn loess, of loess deposits along the lower Mississippi valley, and of Peorian loess in Iowa with an analysis of material known to have been transported and deposited by wind show that there are no important differences. These data seem to demonstrate that wind can be competent to sort material to the degree represented by some loess deposits.

[1] The skewness formula used is Pearson's $Sk = \dfrac{3(M-Md)}{\sigma}$, where $M =$ arithmetic ϕ mean, $Md = \phi$ median, and $\sigma = \phi$ standard deviation.

[2] In order to compute skewness values from Russell's data it was necessary to estimate hypothetical geometric distributions from his cumulative curves. Inasmuch as the points on the curves were numerous, the error introduced should be small.

REFERENCES.

Kay, F. G., and Graham, J. B.: 1943, The Illinoian and Post-Illinoian Pleistocene Geology of Iowa. Iowa Geological Survey, vol. 38, pp. 1-262.

Russell, Richard Joel: 1944, Lower Mississippi Valley Loess. Geol. Soc. America, Bull., vol. 55, pp. 1-40.

Udden, J. A.: 1914, Mechanical Composition of Clastic Sediments. Geol. Soc. America, Bull., vol. 25, pp. 655-744.

STATE GEOLOGICAL SURVEY, UNIVERSITY OF KANSAS,
LAWRENCE, KANSAS.

13

Reprinted from *Amer. Jour. Sci.*, **243**(5), 256–262 (1945)

LOESS TYPES AND THEIR ORIGIN.

VLADIMIR A. OBRUCHEV.

ABSTRACT. Loess is widespread in the southern part of European USSR and extends through the Kirghiz steppe to Lake Balkhash and farther northwest to Yakutia. Different types of loess are distinguished: primary loess of aeolian origin and secondary loesses, redeposited and originated by other processes. Degraded loess and compact stone-loess are also recognized. Dust from which primary loess originated is believed blown by anticyclonic winds from fluvioglacial alluvium of glaciated areas and foreglacial deserts, and deposited over adjacent prairies. In distinction from this "cold" loess, the other, "warm" primary loess has been blown from the exposed mountain ridges and foothill thallus of Central Asiatic Mountains and adjacent deserts, and deposited in the surrounding prairies, but part of the dust in these areas was also blown in from the distant areas of glaciation.

An alternative, soil-hypothesis of loess ascribes its origin to soil-forming processes from any fine-grained formation deposited by water in fore-glacial and foothill alluvial plains. Comparative petrographic, chemical, and mechanical analyses of different types of loess give little support to this hypothesis, and the true water-laid loesses have greater compaction and offer more resistance to pressure than does the typical, porous, wind-blown loess.

The secondary loesses are more widespread than the primary loess, but are much thinner than the latter.

Richthofen's aeolian theory is emended by elucidation of the part which deserts played in origin of loess dust. Obruchev's critics point out that the Sahara desert is not surrounded by belt of typical loess; however, dust from Sahara is constantly carried away to the Atlantic and the Mediterranean, and is the original material of loess in Algeria, Tunisia and Tripoli. The Nile Valley and Arabia to the east are unsuitable for dust accumulation. The whole of equatorial Africa presents climatic conditions very different from those of east-central Asia, where yellow loess is widespread, and in places is underlain by a more ancient reddish loess, which, beside the color, differs also by having greater clay content and lesser porosity.

LOESS and loess-like rocks are widespread in the south of the European part of the USSR. They cover the Ukraine almost to the shores of Black or Azov seas, and stretch along the northern foothills of the Caucasian Mountains. Toward the north they make tongue-like penetrations toward the towns Vitebsk and Moscow, and also along the right bank of the Volga River to Kazan.

In the Asiatic part of the USSR they cover the southern half of the Kirghiz steppes (prairies) and the foothills of the Altai, and spread out between the Caspian and Aral seas to Kopet-Dagh Range and from there stretch to the plain of Syr-

Daria River and to Lake Balkhash. They cover also the slopes of the Tien-Shan and Pamir-Alai ranges, and spread over Fergana and Ili valleys. They are encountered also in the eastern half of Minusinsk Basin, near Irkutsk, in the southern zone of Trans-Baikalia, and in the north between Nizhni Vilui and Lena Rivers, and westward and eastward from the town Yakutsk on Nizhni Aldan River.

Russian investigators now recognize two kinds of loess, which differ in origin and in characteristic features: (1) typical loess, which could be called *primary loess*, and (2) clay-like loams, sandy loams and sands, which may be grouped as *secondary loess*. The primary loess is an aeolian formation, while the secondary loess is either the result of redeposition of primary loess by water, or is an alluvial and deluvial fine-grained earthy formation, which acquired its loess-like characters by weathering and soil forming processes. The primary loess can, on the other hand, lose its typical properties *in situ*, in which case it is called *altered* or *degraded loess*.

In central Asia an ancient loess has been found; a compact marly rock, frequently schistose, and with insignificant porosity. This rock is called stone-loess.

Most Russian students believe that loess in European Russia and Siberia is a direct consequence of Pleistocene Glaciation. They postulate that widespread ice sheets were centers of anticyclonic winds which blew down the ice-free adjacent plains. Numerous streams and flooding glacial waters ran to these plains, and these carried gravel, sand and mud, and deposited them to form what we call fluvioglacial alluvium. Intense thaw during warm seasons caused their overflow; during cold seasons they were in retreat, and left behind them wide areas covered by loose friable deposits. As the anticyclonic winds blew they were raising from these deposits quantities of fine sand and dust. Similar materials were blown also from exposed grounds and from frontal morains, eskers, kames and outwash plains. From these wide areas of peri-glacial deserts the winds carried the dust far into the adjacent steppe, where it was deposited over the wide watersheds, well covered by herbaceous vegetation; under protection of this vegetation the settled dust became converted into loess formation. Along the glacial streams, which traversed this steppe, banks were exposed at their retreat during cold seasons and these became sources of local dust, to be added to that brought from greater distances. Because of this the loess near such banks is coarser grained than farther away from them. The loess which was thus formed from sources created by glaciation may be called *cold loess*. Another loess type, *warm loess*, was formed in the past, and is still being created by the deserts and semiarid parts of the continent, particularly along vast territories of Central Asia. The ample areas from which wind raises quantities of dust are the bare surfaces of mountain ridges and hills, lose talus along their foothills, salty flats and flat shores of lakes exposed during their drying in summer, fall and winter seasons, and the banks of rivers, which become exposed and dry when waters retract after flowing far into deserts at times of maximum overflow. An important part in raising dust is played by the small blowouts formed in great number in the hot time of day upon the plain. As hot air moves up above them, it sucks the dust from the surface and raises dusty whirlwinds high into the air. The winds which, in the fall, winter and spring blow away from the deserts of Central Asia, are generally directed toward the south, east and west and so they carry the dust to the margins of these deserts. The heavier sand particles are deposited first and form loose sands, while the lighter particles are deposited in the surrounding steppes, where they accumulate under cover of the herbaceous vegetation and form loess. The sandy areas in the marginal parts of the deserts are being continuously stricken by winds and present an additional source of dust, which is raised and carried to the steppes. In this manner were formed the thick beds of primary loess in China and western Manchuria, and on the northern slopes and foothills of Nan-Shan and Kuen-Lung, northern and southern foothills and slopes of Tien-Shan. In all the slopes and foothills of these Ranges and others in Central Asia, materials of "cold" loess play an insignificant rôle at the present, but during glacial epochs much of it was added to the material representing "warm" loess. Thus in the primary loess of these areas certain quantities of local material—from mountain ridges, scattered sands, and fluvial and lacustrine deposits—were added to exogenic material carried in from far away.

In 1915 another hypothesis, the so-called soil-hypothesis, was offered in Russia, which has had and still has many sup-

porters.[1] According to this hypothesis any fine-grained
deposit may be converted into typical loess as a result of
weathering and soil forming processes. The formation thus
produced and simulating loess, consists of an accumulation of
lime acquiring a granular structure in consequence of the
coagulation of minute colloidal particles to larger ones and to
rather stable dusty and fine grained aggregates. Original
stratification disappears and porosity characteristic of loess is
produced. The advocates of this hypothesis hold that the
chief factor involved in transportation and deposition of the
loess-forming material is running water both in peri-glacial
and foothill alluvial plains, where this water carries in and
deposits eluviated material of which typical loess is formed.

Detailed petrographic study in connection with chemical
and mechanical analyses of various loesses showed, however,
that loess-forming processes can neither produce the calcium
saturation characteristic of loess, nor explain the inertness
of colloidal (alumosilicate) and carbonate parts to each other
in the same loess. Besides, as erection of large buildings upon
loess in the Ukraine shows, the loess, when saturated with water,
becomes condensed under load and subsides, causing deformation in the heavy buildings. The fact that this tendency
toward subsidence is observed only in primary loess proves that
they were not deposited by water and never were thoroughly
and intensely soaked in it—that is they were accumulated
gradually in dry steppes from dust brought in by winds and
not by water.

In regard to origin it would be necessary therefore to distinguish (1) primary (or typical) loess, cold and warm, an
aeolian product, the material for which was brought in from
periglacial and inner deserts, and deposited upon dry steppes;
and
(2) Secondary loess, consisting of different loess-like clays,
loams, sandy loams, which represent alluvium, talus, proluvium,
as well as primary loess redeposited by water; all of these were
to greater or less extent a subject of loess forming processes
and acquired certain, but not all, characteristics of loess.

Secondary loesses are much more widespread than the primary ones. They are frequently encountered in the zonal devel-

[1] This explanation has recently been advocated by R. T. Russell for the
loess of the lower Mississippi Valley. See Geol. Soc. Amer., Bull. vol. 55,
144, pp. 1-40. Ed.

opment of primary loess alongside with the latter, but their thicknesses are limited to several meters, while that of primary loess is usually up to ten, twenty, and not infrequently to forty or fifty, and in China even to one or two hundred meters. According to Richthofen, the originator of the aeolian hypothesis, Central Asia was almost completely covered with loess, which filled the valleys and hollows between the mountains, and was produced as a result of the weathering in the mountains, the loess-forming material from these being moved downward by water and wind. Obruchev has observed that in central Asia there is no loess: it appears only in the marginal zones. His observation of the distribution of loose sands and loess along these marginal zones made it possible to emend the theory proposed by Richthofen and so to elucidate the part played by the desert as a source of dust and sand. Antagonists of this theory point out that the vast Sahara desert is not surrounded by a belt of typical loess. They forget, however, that the red dust from the Sahara is carried away by winds blowing westward to the Atlantic (as it has been observed to settle on sailing ships since long ago), and northward to western Europe, as well as forming loesses in Algeria, Tunisia and Tripoli. In the East the desert is bordered by the Nile Valley where dust blown in becomes mixed with alluvium and cultural soils. Still farther eastward lie the Red Sea and the Arabian desert, which are unfavorable for dust accumulation. Southward from the Sahara stretches the equatorial zone which is rich in atmospheric precipitation. In general, the climatic conditions in northern Africa are quite different from those observed in Asia. Regions convenient for accumulation of a thick dust formation, so common there, are almost absent here. Nevertheless, here too loose sands occupy great areas in the western, northern, and eastern margins of the desert.

In China the beds of primary yellow loess are underlain in places by beds of reddish loess, which is distinguished from the yellow variety by greater clay content and lesser porosity. These beds are frequently covered by a layer of gravel. This older loess was probably deposited in early Quaternary time when in Central Asia took place an extensive outwash and blow-out of the higher territory beds, where the red beds of the upper Cretaceous and Tertiary continental deposits occupied considerable areas in hollows between the higher ground. From these areas the reddish dust was brought to China and accumulated to produce the older loess, which since has become a subject of intense degradation *in situ*.   Locally it is underlain by darker red clays with Pleistocene faunas.

V. A. OBRUCHEV,
  ACADEMICIAN,
    ACADEMY OF SCIENCES OF U.S.S.R.

DISCUSSION.

THORP: Regarding the loess in China, some of the reddish color in the older loess is probably due to the soil-forming processes that were active after older beds of loess were formed and before younger deposits were made.

BRYAN: Obruchev made argument for formation of loess in northwest China, southeast of the Ordos Desert, but I believe it is essentially glacial in character.   The loess blown from the Ordos Desert was probably due to periglacial climatic effects (i.e., strong anticyclonic winds).   There are known to have been more extensive glaciers in the mountains of Tibet than exist there at the present time.

TO THE ANSWER BY THORP TO BRYAN'S REMARK: The loess-like rocks of southern China seem more likely to represent not the recent, but rather the older "cold" loess, which is connected with the ancient glaciation of Tibet, and subsequently degraded under the very moist recent climate of southern China.   At such climatic conditions the eolian loess cannot be formed, but instead various kinds of red soils (krasnoziem) are usually developed.

ELIAS: Some processes, generally called "soil processes," are supposed to be responsible for both building up and deterioration of loess.   To say "soil processes" is not enough; we must be more specific.

OBRUCHEV'S ANSWERS TO REMARKS.[2]

TO JAMES THORP: If we assume that the reddish color of the more ancient Chinese loess resulted from some soil processes then we should postulate for the time of its origin much warmer climate accompanied by greater humidity.   However, a simpler explanation of the reddish color of the ancient loess is fur-

[2] Submitted in written form after the meeting.

nished by the wide development of the red Tertiary and upper Cretaceous deposits in central Asia, which are now largely covered by the yellow Quaternary deposits, and which were the source of the ancient loess.   Besides, moist climate is generally unfavorable for the origin of loess.

To Kirk Bryan: The greater thickness (to 400 meters?) of the loess, which is to the south of Ordos, is directly at its border, that is to the south of the sands, and diminishes farther to the South.   If this loess were a "cold" one and had originated in connection with the glaciation of Tibet, which, by the way, is fairly distant from this country, then its thickness would have been on a decrease in the opposite direction: from the south northward.   Besides, this hypothesis leaves without explanation the occurrence of the large area of Ordos sands to the north of the thick loess.   In European Russia the sands are located between the southern border of glaciation and the area of the development of the loess, which is a natural consequence of the proximal deposition of the coarser material by the anticyclonic winds, while the dust has been carried farther away.   Thus in China the sands should have been deposited to the south of the loess, closer to the border of Tibet, if it were a "cold" loess.

To M. K. Elias: In order to explain what is meant by "soil processes" a special report would be necessary, which the author could not furnish because of not being a specialist on soils.   Much attention has been devoted to this question in the recent Russian literature.

Editor's Comments
on Paper 14

14   ZEUNER
        Excerpts from *Frost Soils on Mount Kenya, and the Relation
        of Frost Soils to Aeolian Deposits*

Zeuner was more interested in loess stratigraphy than loess sedimentology, but he produced a theory to account for the formation of fine loess material and thus tackled a problem other investigators had almost totally failed to appreciate. He has been described as a "one-time colleague and follower" of W. Soergel, who more or less invented loess stratigraphy and certainly had a considerable influence on Zeuner. One of the few books on loess is Soergel's *Lösse Eiszeiten und paläolithische Kulturen*, published in Jena in 1919; his later studies on the loess at Mauer, where the skull of Heidelberg man was found, established loess stratigraphy as a valuable source of data for Pleistocene dating attempts. Soergel (1925) proposed a detailed general subdivision and correlation of European continental glacial and interglacial deposits. This showed two phases each for the first three main glaciations and three for the last, these *Stadia* being interrupted by minor retreat phases called interstadials (Cornwall, 1970, p. 15).

Twenty years later, the whole subject was summarized and interpreted by Zeuner in his book *The Pleistocene Period* (1945; 2nd ed., 1959) with special reference to chronology. He described loess as a "most important climatic deposit" and defined it as "wind-blown dust which is finer than sand, but coarser than clay." The weakness of his theory lies in the fact that he requires thermal stresses to operate effectively in small monomineralic rock fragments. It seems unlikely that this occurs, and Bagnold (1945, p. 8) has suggested that such thermal fracture will not occur in a fragment with a diameter less than about 1 cm.

REFERENCES

Bagnold, R. A. 1945. The physics of blown sand and desert dunes. Methuen, London. Reprinted in 1954 with new preface.

Cornwall, Ian. 1970. Ice ages—their nature and effects. John Barker, London, 180p.

Soergel, W. 1925. Die Gliederung und absolute Zeitrechnung des Eiszeitalters. Fortschr. Geol. Palaeontol. Berlin 13, 125–251.

14

Reprinted by permission of the Clarendon Press, Oxford, from *Jour. Soil Sci.,* **1**(1),
25–30 (1949)

FROST SOILS ON MOUNT KENYA, AND THE RELATION OF FROST SOILS TO AEOLIAN DEPOSITS

F. E. Zeuner

[*Editor's Note:* In the original, material precedes this excerpt.]

Following Richthofen's observations on loess in China [23, p. 74], the theory that loess is wind-transported dust has become widely accepted,

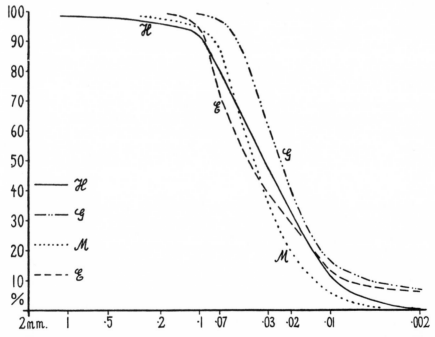

TEXT-FIG. 2. Mechanical analyses of true, unweathered loesses. *H*: Hárshegy, Budapest, Hungary. *G*: Gaza Pit, Gnadenfeld, Upper Silesia. *M*: Carrière Chemin-de-fer, Montières near Amiens, France. *E*: 'Upper middle loam', Ebbsfleet near Gravesend, Kent.
Analyses by the author.

and it has been applied to the vast deposits of Pleistocene loess in Europe, which extend from southern England to the Caspian Sea.

Now, loess exhibits the same predominance of the silt grade which we have observed in mature frost soils [5, 21]. Four examples are shown in Text-fig. 2, of loess from Hungary, Silesia, France, and England. They resemble strikingly the mature frost soil from Spitsbergen, Text-fig. 1, *S*. It has been suggested that such resemblance is not accidental but that frost-weathering provided the vast amount of dust deposited in the loess belt of Europe. If this is right, the puzzle of the provenance of the European loess would be solved, since it has always been one of the difficulties of the aeolian theory that bare gravel surfaces, moraines, hill-sides, &c., would not have been able to supply all the material deposited as loess.

One might even be inclined to discard aeolian transport and to assume that loess is a soil formation resulting from physical weathering *in situ*. This view would not be identical with that of the Russian geologist Leo Berg [24], who holds that loess is formed *in situ* by chemical weathering and coagulation of particles of the clay grade by calcium carbonate. Both views, however, would require formation *in situ* without transport of the particles, and geological opinion is almost unanimously agreed that loess is a genuine sediment. It appears to me that the most conclusive evidence against formation of loess *in situ* is the large number of sections containing several horizons of loess separated by fossil soils of the podzol, brown-earth, or chernozem type. It is impossible for a new layer of fresh loess to form by any suggested weathering process on top of an ordinary weathering soil developed on some older loess, without obliterating the soil profile.

If, then, it is necessary to assume transport, the agent suggested by mechanical analysis is wind. In Text-fig. 3 a few undoubtedly wind-borne dusts are shown. The most instructive is the recent dust from Breslau, Silesia (Text-fig. 3, *D*), which was collected after a strong wind had carried dust from the town and deposited it on snow. It contained a large amount of particles of typically urban origin (slag, brickdust, plant-fibre, soot, &c.). This dust shows unmistakably the grading of a 'loess'. Similarly, the cryoconite from Spitsbergen is a wind-borne dust deposited on the surface of a glacier. Again its grading is of the loess type, but, of course, also of the type of mature frost soils from which the material may well be derived.

It appears, therefore, that physical weathering, as demonstrated in the present paper for frost-weathering, produces a dust of a grain size which happens to be eminently suitable for transport by air currents. If one raises the question why this range from 0·1 to 0·01 mm. should be so favoured, it is not difficult to explain the upper size limit. According to Kölbl [25], the rate of sedimentation from air rises with increasing size of the particles so rapidly that grains of more than 0·06 mm. diameter are unlikely to be carried far in suspension by air currents. Their rate of sedimentation is 50 cm./sec. Above this size the sedimentation rate increases very rapidly, being 167 cm. for particles of 0·1 mm. diameter, and 250 cm. for those of 0·2 mm. Udden [26] calculated the distance over which particles of different sizes are liable to be carried by wind. He found that grains of 0·06–0·03 mm. would still be carried some 200 miles, but grains of 0·13–0·06 mm. only a 'few miles'. Above this size no more than local shifting is normally achieved by the wind. A rapid decrease in the effectiveness of wind transport occurs between 0·06 and roughly 0·1 mm. This is indeed reproduced by the summation

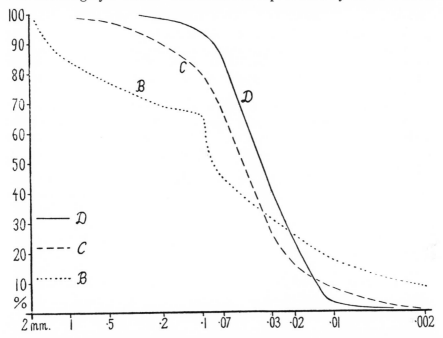

TEXT-FIG. 3. Mechanical analyses of wind-blown dusts. *D*: Dust collected on snow, east of the city of Breslau, Silesia, after a spell of west winds carrying dust from the city. *C*: Cryoconite, from Green Bay Glacier, Spitsbergen. *B*: Volcanic ash from northern Tanganyika, probably from Lemagrut volcano in the Ngorongoro chain, deposited on the edge of the Balbal depression. Wind is constantly playing with this material, and numerous dust-devils are seen. The curve suggests that wind-sorting is only beginning and, since the material has not been transported far, a fair amount of particles larger than 0·1 mm. is present.
Analyses by the author.

curves for loesses, Text-fig. 2. Bagnold [27, p. 6] approached the problem from the other side, that of the sand. According to this author the tendency of the particles to be carried up into the air and to be scattered as dust increases rapidly below 0·2 mm. In the finest wind-blown sands the predominant diameter is never less than 0·08 mm. These figures again suggest that the limit between wind-blown sand and dust lies, very approximately, near 0·1 mm., which value also is the mean of those accepted by Kölbl (0·06 mm.) and Bagnold (0·2 mm.), respectively.

The lower limit observed in wind-blown dust near 0·01 mm.* appears

---

* Since the analyses are intended to reveal the grading of the deposits as they occur in nature, all were carried out on samples prepared by shaking only. No dispersing agents were applied in order to avoid changes in the natural grading of the material.

to be determined by the co-operation of two factors. The first is the sedimentation rate which, according to Kölbl [25], is very slow for particles of less than 0·02 mm. diameter. It is conceivable, therefore, that dust particles much below this size remain suspended in air almost indefinitely, and that they are scarce in loesses for this reason. On the other hand, a varying proportion of 'clay-grade' particles *is* found in

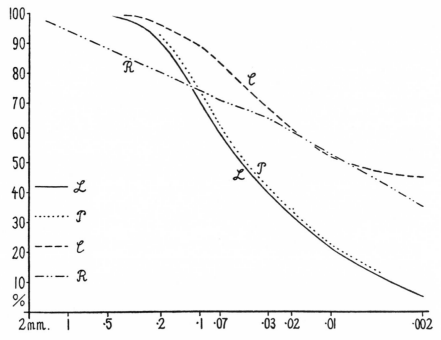

TEXT-FIG. 4. Mechanical analyses of floodloams and other deposits superficially resembling loess. *L*: Floodloam of the Thames, Lavender's Pit, Shepperton, Middlesex. *P*: Floodloam of a small river at Psychod, Upper Silesia. (Both *L* and *P* may contain particles derived from loessic deposits.) *C*: Floodloam of the Neckar at Cannstatt near Stuttgart, west Germany. *R*: 'Terra Bruna' of Grotta Romanelli, Apulia, south Italy, probably blown into the cave by wind, but derived from soils of the Terra Rossa type found in the neighbourhood, and mixed with blown sand.
   Analyses: *C*, Laufer and Hötzel in Soergel [29]; *L*, *P*, *R* by the author.

loesses, especially in weathered loesses from which the calcium carbonate has disappeared. Grahmann [28] emphasizes that certain Chinese loesses contain up to 40 per cent. of particles below 0·01 mm., but it may be assumed that the samples were dispersed with hydrochloric acid prior to analysis. Unweathered loesses which have not been dispersed artificially rarely contain more than 20 per cent. of the grades below 0·01 mm., and usually much less. In practice this affords a means of distinguishing certain types of flood loams from loesses (Text-fig. 4, *C*).
   The second factor which tends to reduce the amount of particles below 0·01 mm. appears to be coagulation by calcium carbonate. Very few unweathered loesses are known which do not contain at least a few per cent. of calcium carbonate, and Berg [24] has pointed out (though in

order thus to explain the origin of the loess altogether) that the presence of calcium carbonate increases the amount of particles above o·o1 mm. at the expense of the clay. The two factors in combination, therefore—namely, permanent suspension of the finest particles in the air, and coagulation of such particles by calcium carbonate—appear to explain the lower limit of the 'loess grade' in a satisfactory manner.

Thus one arrives at the conclusion that the identity in grading of fine-earths from mature frost soils and of wind-borne dust deposits like loess is a coincidence. This coincidence explains why there are enormous dust deposits, i.e. loess, in a belt bordering the formerly glaciated areas of the northern hemisphere. Here frost-weathering prepared enormous quantities of material which could readily be carried away by the wind.

In the border zones of hot deserts, dust deposits are rare, and one is inclined to interpret this as due to the absence of frost-weathering. But dust is by no means absent, and the question whether insolation weathering (heat-fracturing) produces similar grades as does frost-weathering remains to be investigated.

Summary

1. Structure soils due to frost action occur on Mount Kenya at about 13,000–14,000 ft. altitude.

2. The stone nets of Mount Kenya belong to the group of frost soils formed by the action of needle ice.

3. The grading of the fine-earth shows that the Mount Kenya structures are immature, and that the disintegration of the crystals of anorthoclase produces certain irregularities in the grading.

4. Fine-earth from mature frost soils is predominantly composed of the silt-grade, from about o·o1 to o·1 mm.

5. Wind-borne dust has the same grading, but from different causes.

6. The coincidence of 4 and 5 explains the wide distribution of loess in the periglacial areas of the northern hemisphere.

REFERENCES

1. GREGORY, J. W. 1930. (a) On frost stone-packing on the shores of Loch Lomond. Q. J. Geol. Soc. London, **86**, cxliv–v. (b) Stone polygons beside Loch Lomond. Geogr. J. London, **76**, 415–8.
2. HAY, T. 1936. Stone stripes. Geogr. J. London, **87**.
3. HOLLINGWORTH, S. E. 1934. Some solifluxion phenomena in the northern part of the Lake District. Proc. Geol. Assoc. London, **45**, 167–88.
4. BESKOW, G. 1935. Tjälbildningen och tjällyftningen med särskild hänsyn till vägar och järnvägar. (English summary: Soil freezing and frost heaving.) Sver. Geol. Unders. Stockholm, Ser. C, No. 375, 242 pp.
5. ZEUNER, F. E. 1945. The Pleistocene Period. 322 pp. Quaritch, London.
6. BRYAN, K. 1946. Cryopedology—the study of frozen ground and intensive frost-action with suggestions on nomenclature. Amer. J. Sci. **244**, 622–42.
7. SHARP, R. P. 1942. Soil structures in the St. Elias Range, Yukon Territory. J. Geomorph. **5**, 274–301.
8. BESKOW, G. 1930. Erdfliessen und Strukturböden der Hochgebirge im Licht der Frosthebung. Geol. Fören. Stockholm Förh. **52**, 622–38.
9. ELTON, C. S. 1927. The nature and origin of soil-polygons in Spitsbergen. Q. J. Geol. Soc. London, **83**, 163–94.
10. HÖGBOM, B. 1914. Über die geologische Bedeutung des Frostes. Bull. Geol. Inst. Univ. Uppsala **12**, 257–390.

11. HUXLEY, J. S., and ODELL, N. E. 1924. Notes on surface markings in Spitsbergen. Geogr. J. London, **63**, 207–29.
12. PATERSON, T. T. 1940. The effects of frost action and solifluxion around Baffin Bay and in the Cambridge District. Q. J. Geoi. Soc. London, **96**, 99–130.
13. POSER, H. 1933. Das Problem des Strukturbodens. Geol. Rundschau Berlin, **24**, 105–21.
14. STOLTENBERG, H. 1935. Der Dauerfrostboden. Ibid. **26**, 412–23.
15. SMITH, W. CAMPBELL. 1931. A classification of some rhyolites, trachytes, and phonolites from part of Kenya Colony, with a note on some associated basaltic rocks. Q. J. Geol. Soc. London, **87**, 212–58.
16. HEILIG, H. 1931. Untersuchungen über Klima, Boden und Pflanzenleben des Zentralkaiserstuhls. Z. Bot. **24**, 225–79.
17. TABER, S. 1929. Frost heaving. J. Geol. Chicago, **37** (5), 428–61.
18. —— 1930. The mechanics of frost heaving. Ibid. **38** (4), 303–17.
19. HAMBERG, A. 1915. Zur Kenntnis der Vorgänge im Erdboden beim Gefrieren und Auftauen sowie Bemerkungen über die erste Kristallisation des Eises in Wasser. Geol. Fören. Stockholm Förh. **37**.
20. NANSEN, F. 1921. Spitzbergen. Leipzig (pp. 111–12).
21. DÜCKER, A. 1937. Über Strukturböden im Riesengebirge. Z. Deutsch. Geol. Ges. Berlin, **89**, 113–29.
22. ZEUNER, F. E. 1933. Die Schotteranalyse. Ein Verfahren zur Untersuchung der Genese von Flusschottern. Geol. Rundschau, **24**, 65–104.
23. RICHTHOFEN, F. VON. 1877. China. Vol. i, 758 pp. Berlin.
24. BERG, L. S. 1932. Löss als Produkt der Verwitterung und Bodenbildung. Trans. 2nd Int. Conf. Assoc. Study Quartern. Per. Moscow–Leningrad, **1**, 57–62.
25. KÖLBL, L. 1931. Über die Aufbereitung fluviatiler und ˙scher Sedimente. Min. Petr. Mitt. Leipzig, **41** (2), 129–44.
26. UDDEN, J. A. 1898. The mechanical composition of wind deposits. Augustana Library Publ. Ref.: N. Jahrb. Min. Stuttgart, 1900.
27. BAGNOLD, R. A. 1941. The Physics of Blown Sands and Desert Dunes. 265 pp. London.
28. GRAHMANN, R. 1932. Der Löss in Europa. Mitt. Ges. Erdk. Leipzig, 1930–1, 5–24.
29. SOERGEL, W. 1929. Das Alter der Sauerwasserkalke von Cannstatt. Jahresber. Mitt. Oberrhein. Geol. Ver. 1929, 93–153.

Editor's Comments
on Papers 15, 16, and 17

The Russell paper (Paper 9) provoked many replies; Chauncey Holmes (1944) was very quickly off the mark and Russell (1944) replied to his criticisms. In Europe, and some time later, Doeglas published his criticisms of the Russell *in situ* approach in a short but classic paper in which he compared Mississippi Valley loess with Dutch loess (Paper 15).

The Doeglas argument is mainly mineralogical, and there is a short skirmish with van Rummelen (Paper 16) about epidote. Holmes also made some mineralogical comments and those concerning calcite were very pertinent:

> The occurrence of calcite grains and their relations to carbonate enrichment raise questions not answered in the paper. Calcite seems to be listed among the detrital minerals supposedly segregated from the parent material which, however, is said to be noncalcareous and to have undergone weathering. As their origin under these conditions seems impossible, perhaps they were developed as a consequence of carbonate enrichment. If so, what is the process by which discrete calcite grains develop in a porous matrix? If they are concretionary forms, at which stage of enrichment do they appear? What is the grade-size distribution of calcite in the deposit?

Paper 18 continues the mineralogical discussion.

REFERENCES

Holmes, C. D. 1944. Origin of loess—a criticism. Amer. J. Sci. 242, 442–446.
Russell, R. J. 1944. Origin of loess—reply. Amer. J. Sci. 242, 447–450.

15

Reprinted from *Jour. Sed. Petrol.*, **19**(3), 112–117 (1949)

LOESS, AN EOLIAN PRODUCT

D. J. DOEGLAS

Agricultural University, Wageningen, Holland

ABSTRACT

Petrographic characteristics and field relationships of Dutch and of Mississippi Valley, United States, loess deposits are described. The evidence indicates that the loess of both regions is eolian in origin.

Whether loess is an eolian product or a water-laid deposit still seems to be an unsettled problem. Although the eolian genesis probably is more widely accepted, other explanations for its origin continuously appear.

R. J. Russell (1944) refers to van Rummelen who defended the non-eolian theory of the Dutch loess in many papers. In the meantime, however, the eolian origin of the Dutch loess has been proved by van Doormaal (1945).

In 1927 Druif pointed out that the mineralogical composition of the Dutch loess was different from that of the underlying formations. Edelman confirmed this on a quantitative base. Van Doormaal continued these studies on a larger scale. He gives extensive petrological descriptions of many sections of the loess of the southern part of the province of Limburg in the Netherlands. These profiles are very similar to those of the Mississippi River Valley. The author will give two of them for comparison with those of R. J. Russell. (See "Profiles of Dutch loess.")

All features given by Russell are found also in the Dutch loess. In some layers gravel occurs and creep (solifluction) can be observed especially well in the older loess (zone IV). The deposit is evenly distributed over nearly the entire area. In the valleys the thickness is increased, in some places, due to solifluction.

Two high points, however, are not covered with loess and these are always used as proof against the eolian origin.

Both localities, however, have definitely eroded surfaces. Van Doormaal showed that there exists a transition between the loess-covered and loess-free areas. The wind-blown sands which cover the loess-free spots get more and more mixed with loess material toward the typical loess-covered parts.

The loess overlies Cretaceous limestones, Oligocene and Miocene sandstones, and Pleistocene river terrace deposits.

The mechanical composition is given in figure 1. It is very homogeneous. The grade $50-10\mu$ makes up 60–80 per cent of the deposit. The grades $>50\mu$ are less than 5 per cent and those $<10\mu$ less than 30 per cent. The size frequency distribution is similar to that of the Mississippi River Valley loess.

According to unpublished data of Dewez, a gradual change in the mechanical composition exists between the loess and fine, slightly loamy sands toward the north. These sands form a nearly continuous, thin cover over a large part of the Netherlands and are supposed to be wind-blown.

The heavy mineral composition of the loess is entirely different from that of the underlying formations and the recent deposits of the Meuse River. Table 1 gives a selection out of more than 300 samples investigated for heavy minerals by Druif, van Baren, Edelman, Van Doormaal. Muller, and Zonneveld.

The theory of van Rummelen that the Dutch loess is a weathering product of

Profiles of Dutch loess

Quarry, Brick factory at Beek

Zone I 0–20 cm.
Gray loam, free of lime.
Zone II
(a) 20–150 cm.
Light brown to reddish or chocolate brown loess, free of lime, many large root tubes, reddish brown spots of $Fe(OH)_3$ and black spots of Fe- and Mn-concretions. A few small yellow silt lenses. Bottom layer lighter brown.
(b) 150–240 cm.
Light brown loess with many horizontal, yellow lenses up to 2 cm thick and maximum 200 cm long. The lenses follow shape of pockets in zone III. Many small root tubes. Structure gets looser downward. Non-calcareous.
(c) 240–270 cm.
Light brown loess without lenses, small root tubes. Color gets more yellow. Loose structure. Free of lime.
Zone III 280–665 cm.
So called "Eerdmergel" (earthy marl), calcareous. Top well marked, undulating. From 280–470 cm, an even straw yellow color, with root tubes. Between 330 and 470 cm many small veins of $CaCO_3$. From 470–665 cm horizontal lamination of more or less compact layers. In bottom layers brown spots of $Fe(OH)_3$ due to action of ground water.
Zone IV 665–695 cm.
Calcareous, compact, greenish yellow loam. Solifluction phenomena.
Zone V below 695 cm.
Terrace gravel.

Graetheide, Welzenheuvel, old lignite quarry

Zone I 0–25 cm.
Grayish yellow loam, top layer a black forest soil mixed with lignite dust and Miocene sand, white sand grains (podsol). Lower layer sandy loam, with pieces of brick. Loose structure.
Zone II 25–325 cm.
Non-calcareous loess, leached.
(a) 25–145 cm.
Yellowish white to yellowish brown loam getting more compact to 75 cm and looser further downward. Varying colors with white spots down to 90 cm. Below 90 cm even yellow color with light yellow, sandy spots.
(b) 145–275 cm.
Alternating yellowish white and brownish yellow, horizontal layers of varying thickness. Their lengths may be many meters. The thickness decreases downward.
(c) 275–325 cm.
Non-laminated loam, yellow with orange yellow spots. Transition to zone III is sharp, undulating.
Zone III 325–450 cm.
Calcareous loam, slightly lighter than overlying zone and locally with orange yellow spots. Lime concretions. Between 425 and 450 cm folded due to solifluction. Concretions on surface of laminae.
Zone IV 450–575 cm.
Heavy, slightly plastic loam, orange to brownish red, non-laminated, slightly calcareous. 450–525 cm even color with blackish brown spots. 525–575 cm slightly laminated, brownish red to whitish yellow. Some pieces of flint. Probably a "gley" horizon.
Zone V Below 575 cm.
Terrace sand and gravel.

the Cretaceous should involve an increase of the number of mineral species with epidote and hornblende during the weathering. The Tertiary formations do not contain these minerals either.

Zonneveld informed the author that 8 river terraces can be distinguished by means of heavy minerals. He examined more than 100 samples. They contain less than 1 per cent epidote and garnet. A brown hornblende variety which sometimes is present, does not occur in the loess. The terrace deposits contain a high percentage of zircon, tourmaline, staurolite, kyanite and sometimes chloritoid.

The mineralogical similarity between the loess and the glacial deposits in the north is striking. The loess components, however, could have been derived from these glacial formations only if they had been transported by the wind.

The investigation of van Doormaal proves the eolian origin of the loess in the Netherlands. His work furthermore shows that loess studies should be made with mechanical analyses and mineralogical investigations of many samples of

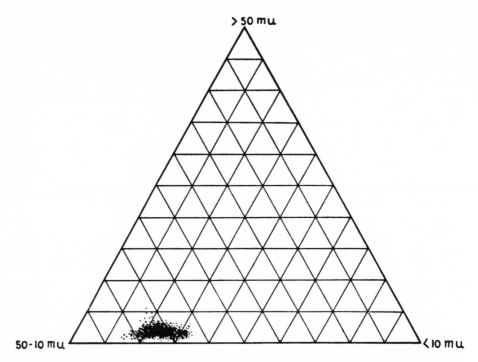

FIG. 1.—Mechanical composition of Dutch loess samples.

the deposit itself and the underlying formations, combined with a study of the regional distribution of the deposit.

In 1946 the author was able to collect a few samples of the loess and underlying formations during an excursion with Prof. R. J. Russell along the Mississippi River Valley between Doloroso and Vicksburg, Mississippi.

Table 2 gives the heavy mineral data of these samples. The microscopic determinations were made by Miss S. M. van der Baan. Heavy mineral data of recent sediments of the Mississippi River shown in table 2 are taken from the study of Dana Russell. The percentage of the heavy non-opaque minerals have been calculated from his paper.[1]

For the Mississippi River Valley the difference in mineralogical composition

[1] Dutch sedimentary petrologists always give the mutual percentage of the non-opaque minerals and, in addition, the opaque minerals as percentage of the total heavy fraction.

between the loess and the underlying river terrace sands and upper Tertiary formation is also evident. The upper Tertiary and the terrace sands contain staurolite-kyanite-zircon assemblages, the loess has a garnet-epidote-hornblende association. Epidote, garnet and hornblende are absent in the older formations. Only the uppermost terrace sand on highway 61, one mile south of Doloroso, contains 4 per cent of epidote and a few grains of hornblende. This may be due to admixture of loess material during its deposition or to contamination with loess at the surface of the exposure.

The loess samples 13 and 15 contain a lower hornblende, epidote and garnet content; 13, furthermore, contains a higher percentage of staurolite, kyanite and zircon. These samples also contain gravel and show creep phenomena. During the solifluction terrace material moved also and mixed with the loess. This caused a lowering of the percentage

TABLE 1.—*Minerals of loess and other sediments from the Netherlands*

Sediment type and locality	Opaque	Tourmaline	Zircon	Garnet	Rutile	Brookite	Anatase	Titanite	Staurolite	Kyanite	Andalusite	Sillimanite	Chloritoid	Epidote	Hornblende	Glaucophane	Augite	Hypersthene	Enstatite	Saussurite	Picotite	Diaspore
Loess, Quarry Beek																						
Zone I, 0–20 cm	45	2	31	13	13	2	2			3	3	1		25	5							
Zone II, 20–45 cm	34	1	29	21	11		1	1		2	2			28	4							
Zone II, 155 cm	40	5	21	16	11		2	2		3	2	1		32	5							
Zone II, 210 cm	41	1	29	7	16	3	3			3	5			25	7	1						
Zone III, 320 cm	43	1	26	29	15	1	2	1		1	2	1		18	2							
Zone III, 430 cm	40	1	24	11	19		3				8			29	5							
Zone III, 650 cm	36	1	20	19	14	1	4		3	1	4	1		24	6					2		
Zone IV, 685 cm	49	3	35	16	6		2	2	3	4	1	1	1	19	7							
Loess, Greatheide																						
Zone I, 4–25 cm	36	2	39	10	15	1	5			2	1	2		21	2							
Zone II, 145 cm	43	1	30	17	10		6			1	55			24	5							
Zone II, 300 cm	46	2	31	25	10	1	4			1	3			18	7							
Zone III, 440 cm	50	1	34	11	19	2	8			3	2	1		19								
Zone IV, 535 cm	45	1	32	3	14	4	9		3	4	4	1		25								
Mixed Loess, Heerlen																						
Zone III	67	8	48	3	17	3	4	1		5	3	1		4	1					2		
Cretaceous	59	7	66	1	20	2	2	1			1											
Oligocene	57	14	38	3	25		6	1		6	5	2										
Miocene	47	11	40	3	18	1	4			9	10	3					1					
Pliocene	52	20	33		26		1			5	8	5		2								
Pleistocene Terrace	43	18	48	2	20	1	1				3	3	1	3								
Recent Maas River sand (16 samples)	45	16	10	22	3					7	2	2		27	2	2		4		3		
Average Glacial deposits N. Netherlands	30	2	12	18	8			1		2	2	1		40	13					1		

‹f the characteristic loess minerals and an increase of those of the terrace material. This also has been observed by van Doormaal in the Dutch loess. The mechanical analyses of such samples generally give a higher content of grades $>50\mu$ and a too high or too low percentage of the grade $<10\mu$.

The few samples of the Mississippi Valley loess which have been investigated indicate that they can not have been derived from the terrace deposits or Tertiary formations. The loess does not contain pyroxenes which are present in about the same percentage as hornblende in the Recent Mississippi River sands (Dana Russell).

A more regional study of the Mississippi River Valley loess, combined with mechanical and mineralogical analyses of many samples has to be made in order to ascertain the dissimilarity of the loess and the underlying formations and to make sure that its composition is uniform over a large area.

The theory of the non-eolian origin of the loess and its derivation from backswamp deposits, pushed forward by R. J. Russell, has not been confirmed by mineralogical evidence. It seems rather certain that the Mississippi loess is of eolian origin just as well as the Dutch loess.

REFERENCES

DRUIF, J. H., *1927*. Over het Ontstaan der Limburgsche Löss in Verband met haar Mineralogische Samenstelling (The origin of the Limburg Loess based on its mineralogical composition): Thesis, Univ. Utrecht, Holland, 330 pp.

EDELMAN, C. H., *1931*. Over de Mineralogische Samenstelling van de Linburgsche Loess en haar Ontstaan: *Nederlandsch Aardr. Gen. Amsterdam Tijds.*, vol. 48, pp. 442–446.

———, *1933*. Petrologische provincies in het Nederlandse Kwartair (Sedimentary petrological provinces of the Dutch Quaternary): Thesis, Univ. Amsterdam, Holland, 104 pp.

———, and van Baren, F. H., *1935*. La petrographie des sables de la Meuse Neerlandaise: *Med. Landbouwhogeschool*, Wageningen, vol. 39, no. 2, pp. 1–15.

D. J. DOEGLAS

TABLE 2.—Minerals of loess and other sediments from the Mississippi River Valley, U.S.A.

Sediment type and locality	Opaque	Tourmaline	Zircon	Garnet	Rutile	Brookite	Anatase	Titanite	Staurolite	Kyanite	Andalusite	Sillimanite	Chloritoid	Epidote	Hornblende	Glaucophane	Augite	Hypersthene	Enstatite	Saussurite	Picotite	Diaspore
Loess (weathered), Highway 61, 12 miles S. of Vicksburg, Miss.	35		20	23			12	1						18	25						1	
Loess, North of Highway 61, 12 miles S. of Vicksburg, Miss.	61	2	25	12			6		2	1				25	29							
Loess, South of Highway 61, 12 miles S. of Vicksburg, Miss.	33		18	16	4		8	1		1				19	30						1	
Loess, Highway 61, 4 miles S. of Natchez, Miss., 3' above low. layer	51		19	18	1		5			3				24	30							
Loess, Highway 61, 4 miles S. of Natchez, Miss., 3' above low. layer	18		16	14			1	2		1				29	39							
Loess, Highway 61, 4 miles S. of Natchez, Miss., lower layer	34	1	10	20	2		5	1	2	2				10	47							
Loess, Highway 61, 4 miles S. of Natchez, Miss., lower layer, mixed	59	2	30	5	9		8		10	15				16	5							
Loess, Highway 61, 4 miles S. of Natchez, Miss., low. layer (with gravel)	49	3	31	12	6		9	1	1	3				17	17							
Loess, Highway 61, one mile South of Dolorosa, Miss.	32		18	14	3		3							19	42						1	
Terrace sand, Highway 61, one mile South of Dolorosa, Miss.	67	4	50	1	3		10		6	26				4								
Terrace sand, Highway 61, one mile South of Dolorosa, Miss.	58	6	48		5		3		11	18					1							
Williana Terrace sand, half mile S. of Dolorosa, Miss.	60	5	34	1	7		1		16	33		3										
Williana Terrace sand, 2 miles North of Washington, Highway 85	78	6	38		5		8	2	16	18	2	4										
Upper Tertiary, Highway 61, one mile South of Dolorosa, Miss.	39	6	62		8		11		6	6		4										
Upper Tertiary, Highway 61, one mile North of Dolorosa, Miss.	42	10	57		13		11		6	3												
Mississippi River sand near Vicksburg, Miss., Recent		*	11	22	*			6	*	*		*		22	25		14	*				
Mississippi River sand near Natchez, Miss., Recent		*	2	21	*			2	*	*		*		22	33		18	2				
Mississippi River sand near Natchez, Miss., Recent		*	4	14	*			4	*	*				19	37		21	1			*	

* Mineral present below 1 percent.

138

HOLMES, C. D., *1944*. Origin of loess—a criticism: *Am. Jour. Sci.*, vol. 242, pp. 442–445.

JONGMANS, W. J., and VAN RUMMELEN, B. H., *1937*. De bodem van Zuid Limburg (The geology of South Limburg): Zeist, Holland, 79 pp.

MULLER, J. E., *1943*. Sediment-petrologie van het Dekgebergte in Limburg (Sedimentary petrology of the Post-Carboniferous formations in Limburg): *Netherlands Geol. Stichting*, Med. s. C-II, no. 2, pp. 1–62.

RUSSELL, R. D., *1937*. Mineral composition of Mississippi River sands: *Geol. Soc. America Bull.*, vol. 48, pp. 1307–1348.

RUSSELL, R. J., *1944*. Lower Mississippi Valley loess: *Geol. Soc. America Bull.*, vol. 55, pp. 1–40.

———, *1944*. Origin of loess—reply: *Am. Jour. Sci.*, vol. 242, pp. 447–450.

VAN BAREN, F. A., *1934*. Het voorkomen en de beteekenis van kalihoudende mineralen en Nederlandsche gronden (The occurence and significance of potash-bearing minerals in Dutch soils): Thesis, Agr. Univ., Wageningen, Holland, 109 pp.

VAN DOORMAAL, J. C. A., *1945*. Onderzoekingen betreffende de Lössgronden van Zuid Limburg (Loess soils of South Limburg, the Netherlands): Thesis, Agr. Univ., Wageningen, Holland, 94 pp.

VAN RUMMELEN, B. H., *1942*. Bijdragen tot de Kennis van het Ontstaan der Lössoiden (The origin of the Lössoid soils): Med. Jaarverslag over 1940 en 1941, Geol. Stichting, *Geol. Bur. voor het Nederlandsche Mijngebied*, Heerlen, pp. 73–84.

ZONNEVELD, J. C. S., *1947*. Het Kwartair van het Peelgebied en de Naaste Omgeving (The Quaternary of East Brabant, Holland): *Netherlands Geol. Stichting*, Med. s. C-VI, no. 3, pp. 1–223.

16

Reprinted from *Jour. Sed. Petrol.*, **21**(3), 183–184 (1951)

SOME REMARKS ON THE MINERAL EPIDOTE IN CONNECTION WITH THE LOESS PROBLEM

A reply to D. J. Doeglas, "Loess, an eolian product"

F. F. F. E. VAN RUMMELEN[1]

University of Indonesia, Bogor, Indonesia

In the paper, "Loess, an eolian product," Doeglas (1949) compares the "loess" of southern Limburg (the Netherlands) with loess-products of the Mississippi Valley. Van Rummelen, Sr. (1950) has disputed the views of Doeglas, but has not presented quantitative results of a conclusive nature concerning the epidote content of "loess."

The present author has no intention of going into the problem of genesis of the loessoïdes of Limburg. That matter has been the subject of many publications by van Rummelen, Sr. However, one of the basic arguments in the problem has been that of high epidote percentage. Van Rummelen, Sr., not being a mineralogist, has not been able to answer such arguments. The high epidote percentage in loess is a fact not in dispute.

Doeglas uses the high epidote percentage to prove that loess could not be derived locally, from older sediments. His table 1 is presented in an attempt to prove that the Cretaceous, Oligocene, Miocene, and Pliocene contain practically no epidote, whereas loess contains a high percentage. If true, the theory of van Rummelen, Sr., is contradicted, and the theory of eolian-glacial origin, as advocated by Doeglas, appears more reasonable. The views of Doeglas, as presented in his table 1, gives basis for a belief that van Rummelen, Sr., lacks factual support for his theory of loess origin.

No evidence was presented by Doeglas as to the number of samples upon which the idea was advanced that the Cretaceous and Tertiary formations lack epidote. That the notion is incorrect is hereby presented on the basis of 751 samples, all of which were analyzed mineralogically by the author and others. The present author has analyzed more samples than any previous worker in the Netherlands. Epidote percentages were determined, in the heavy mineral content, and were averaged by formations, with the following result:

TABLE 1

Formation age	Epidote percentage	Number of samples
Amstelian	0.9	119
Poederlian	2	127
Upper Diestian	2	114
Middle Diestian	11	20
Lower Diestian	1	41
Upper Miocene	22	11
Middle Miocene	6	41
Braunkohlform.	10	7
Upper Oligocene	14	60
Middle Oligocene	9	162
Lower Oligocene	24	48
Maestrichtian	14	1

The mean value for the terrestrial Pliocene is 5 per cent, based on 421 samples.

[1] This paper has been edited by R. J. Russell. Although the exact wording is not that of the author, the factual data and general character of discussion are accurately presented. J.L.H.

140

A comparison of values with those of Doeglas is given below:

TABLE 2

Age	Epidote percentage	
	Doeglas (1949)	Various authors
Pliocene	2	5
Miocene	—	13
Oligocene	—	16
Maestrichtian	—	14

It is impossible to accept the interpretations of Doeglas. They must be based either upon too few samples or else upon selected samples. The mean values of epidote percentages based on the work of the author and others is entirely objective. Doeglas' statement that the Tertiary formations of Limburg are deficient in epidote is incorrect. Both hornblende and garnet are also present in those formations.

There is, therefore, no reason at all for believing that the loessoïdes of Limburg could not have been derived from Tertiary formations present in the region.

REFERENCES

DOEGLAS, D. J., *1949*. Loess, an eolian product: *Jour. Sedimentary Petrology*, vol. 19, pp. 122–117.

MULLER, J. E., *1943*. Sedimentpetrologie van het dekgebergte in Limburg: *Med. Geol. Stichting*, serie C-II-2-2.

VAN RUMMELEN, F. F. F. E., *1950*. Plioceen in Zuid- en Midden Limburg (manuscript).

VAN RUMMELEN, F. H., *1950*. Erratica in de loessoïden van Ransdaal (met medewerking van F. F. F. E. van Rummelen): *Nat. Hist. Maandbl.* 39e, Jrg. no. 5-6-7-8.

LOESS, AN EOLIAN PRODUCT

Correction and a reply to the remarks of F. F. F. E. van Rummelen

D. J. DOEGLAS

Agricultural University, Wageningen, Netherlands

In the author's paper on the origin of loess in the *Journal of Sedimentary Petrology* of December 1949 a regrettable error has been made. According to table I the epidote percentage of the Oligocene was zero and that of garnet only 3 per cent. The author used mineral data from the continental Oligocene in N.E. Belgium. In the Dutch South Limburg, however, the Oligocene is marine and contains an average of 12 per cent epidote and 14 per cent garnet. The other Tertiary formations in this area are continental and are practically free of epidote.

In December 1950 F. F. F. E. van Rummelen discussed this error in a Dutch popular scientific paper and in the March, 1951 number of the *Journal of Sedimentary Petrology* a translation of this paper appeared. The tables I and II of these papers by van Rummelen not only show epidote in the Oligocene but also in the other Tertiary series of South Limburg. As van Rummelen referred to a large number of his own analyses, present already in manuscript form, the author waited to answer him until this paper appeared.

The second paper of van Rummelen (1951), however, forces the author not only to correct his error but also to prove that the average epidote percentages given by van Rummelen give a wrong impression of the mineral associations of the Cretaceous, Miocene and Pliocene.

The percentages in the author's paper of 1949 had been estimated. In the present paper (table 1) the percentages of garnet, hornblende and epidote have been calculated. The number of samples used for these calculations have been indicated and the authors from which the data have been taken are mentioned. The mineral tables in the papers used are indicated in the bibliography. The percentages of epidote given by van Rummelen and by the author in 1949 are given too.

The average percentages of epidote calculated by the author and those given by him in 1949 only deviate for the Oligocene and a thin basal layer of the Middle Miocene which according to Mueller (1943) consists of reworked marine Oligocene. The discrepancies between the average percentages of epidote of van Rummelen and those calculated by the author, however, are large except for the Oligocene. The number of data used has no influence: van Rummelen 751—Doeglas 566.

When the percentage of epidote of the loess is compared with those calculated by the author for the underlying formations, it is clear that only the base of the Middle Miocene and the Oligocene could have contributed epidote to the loess. The average percentage of epidote in these formations, however, is about half that of the loess.

The percentage of hornblende gives a more striking difference. The loess contains an average of 7.2 per cent and the average percentage of the underlying formations is less than 0.1. Only the Pleistocene terraces contain a high hornblende percentage. The hornblende variety in the terraces, however, is brown green and that of the loess blue green.

The garnet of the loess (average 16 per cent) also could have been derived from the basal Miocene and the Oligo-

TABLE 1

	Doeglas, 1951				v. Rummelen 1951		Doeglas 1949
	Number of samples	Garnet	Hornblende	Epidote	Epidote	Number of samples	Epid · te
Loess (Edelman)	16	16	5.0	24.0			
Loess (Doormaal)	36	16	8.2	28.7			
Loess—average	(52)	16	7.2	27.3			
van Baren, Zonneveld							
Middle terrace	39	1.4	27.0[1]	3.6[2]			3
Upper terrace	213	1.4	11.4[1]	2.3[2]			
Muller							
Pliocene							
Kiezeloölite terrace	49	2.4	0.04[3]	1.1	2[4]	421	2
Upper Miocene	0				22	11	
Middle Miocene							
garnet-free zone	34	0.7	0.0	0.1			
garnet-rich zone	48	19.5	0.04	0.12	6.6	48	0
basal marine zone	15	27.4	0.6	9.3[5]			
Upper Oligocene	24	22.2	0.04	15.4	14	60	
Middle Oligocene	136	12.4	0.02	9.0	9	162	
Lower Oligocene	49	17.2	0.18	17.2	24	48	
Total Oligocene	(209)	14.0	0.06	12.0	12.8	(270)	0
Senonian, S. Limburg	39	3.9	0.0	0.25	14	1	0
Senonian, E. Brabant	72	3.5	0.0	0.04			
Edelman-Doeglas							
Pliocene	9	1.3	0.0	0.1			
Miocene	5	0.6	0.0	0.0			
Upper Oligocene	12	5.0	0.0	4.6			
Middle Oligocene	66	15.4	0.15	2.3			
Lower Oligocene	8	10.7	0.4	8.5			

[1] Hornblende in Pleistocene terraces is brown-green, in loess blue-green.

[2] Includes saussurite.

[3] Muller did not analyse samples from Upper Miocene.

[4] van Rummelen calculated 5 per cent epidote. According to his own table I it is only 2 per cent.

[5] The garnet- and garnet-free zones are 60–100 meters thick, the marine base of the Middle Miocene is 10–20 meters and represents reworked Oligocene according to Muller. Lower Miocene is absent.

cene. In the southern part of South Limburg and in Belgium the Miocene is absent and the Oligocene only occurs in small patches. The loess overlies the Senonian or Pleistocene terraces.

The Senonian of South Limburg (39 well samples) and of areas farther North investigated by Muller (1943), a total of 111 well samples, contains less than 0.25 per cent epidote. Van Rummelen gives only one sample with 14 per cent epidote, analyzed by himself.

The basis of the author's paper on the eolian origin of the loess minerals, there-

52 DISCUSSION

fore, has not been weakened. The mineral association of the loess in South Limburg could not have been derived from the underlying formations and as it is very similar to the mineral association of the glacial deposits in the central part of the Netherlands an eolian transport is necessary.

REFERENCES

BAREN, F. A. VAN, *1934*. Het voorkomen en de beteekenis van kalihoudende mineralen in Nederlandsche gronden: Thesis, Wageningen, Netherlands, tables V and VI.

DOEGLAS, D. J., *1949*. Loess, an eolian product: *Jour. Sedimentary Petrology*, vol. 19, pp. 112–117.

DOORMAAL, J. C. A. VAN, *1945*. Onderzoekingen betreffende de loessgronden van Zuid-Limburg: Thesis, Wageningen, 88 pp.

EDELMAN, C. H., *1933*. Petrologische Provincies in het Nederlandsche Kwartair, Thesis, Amsterdam, table XXIII.

EDELMAN, C. H., and DOEGLAS, D. J., *1933*. Bijdrage tot de petrologie van het Nederlandsche Tertiair: *Verhandelingen Geologisch-Mijnbouwk*. Genootschap, 10, 1, pp. 1–38, tables IVb, Va–b and VI.

MULLER, J. E., *1943*. Sediment-petrologie van het dekgebergte in Limburg: *Mededelingen Geologische Stichting*, serie C-II-2-2, tables II, VII–XV, XVII, XVIII.

RUMMELEN, F. F. F. E. VAN, *1950*. Enige opmerkingen omtrent het mineraal epidoot naar aanleiding van het loessprobleem: Natuurhistorisch Maandblad, 39, 12, pp. 139–140.

RUMMELEN, F. F. F. E. VAN, *1951*. Some remarks on the mineral epidote in connection with the loess problem: *Jour. Sedimentary Petrology*, vol. 21, pp. 183–184.

RUMMELEN, F. H. VAN, *1950*. Erratica in de loessoïden van Ransdaal (met medewerking van F. F. F. E. van Rummelen): *Natuur-Hittorisch Maandblad*, 39, 5–8.

ZONNEVELD, J. I. S., *1949*. Zand-petrologische onderzoekingen in de terrassen van Zuid-Limburg: *Mededelingen van de Geologische Stichting*, no. 3, pp. 103–123.

Editor's Comments
on Papers 18 and 19

18   **BEAVERS and ALBRECHT**
Composition of Alluvial Deposits Viewed as Probable Source of Loess

19   **BEAVERS**
Source and Deposition of Clay Minerals in Peorian Loess

Beavers and Albrecht (Paper 18) present critical data concerning the source of loess material related to the Mississippi system. They suggest that the wind acts essentially as a transporting agent rather than as a sorting *and* transporting agent. The implication is that the critical material-forming factors were in operation earlier in the sequence of loess-forming events. They propose that the Missouri-Mississippi river system is bringing about uniformity of its sediment throughout its course. They make passing reference to the variation in thickness of loess with distance from river source (a topic discussed later in this collection with relation to Papers 28 and 29).

The authors cite Chamberlin's 1897 paper (Paper 4), and the data that they present tend to confirm his suggestion that the Mississippian loess is first aqueous and second eolian, although they do not mention Chamberlin's further observation about glacial origin. Much of the Mississippi loess material must previously have been incorporated in Great Plains loess deposits, the material itself existing as a result of glacial action (this is discussed further in Paper 46).

The glacial action theory provides a mechanism to supply abundant silt-sized quartz material, but the origin of carbonates and clay minerals is a bit more obscure. Beaver's ingenious proposal about the supply of clay minerals (Paper 19) supplies a possible solution to one of these problems. The observations of Beavers and Albrecht suggest that carbonate material can be blown into the deposit region with the quartz particles.

18

This is Soil Science Society of America reprint.

Copyright © 1948 by the Soil Science Society of America

Reprinted from *Soil Sci. Soc. America Proc.*, **13**, 468–470 (1948)

Composition of Alluvial Deposits Viewed as Probable Source of Loess[1]

ALVIN H. BEAVERS and WM. A. ALBRECHT[2]

LOESS, as a geological deposit, is characterized by its silty texture and diversity of its component minerals. The distribution of the large loess deposits over extensive areas has been attributed to the wind as the agency for transport. However, the large body of loess deposit in central United States is very closely associated with the Mississippi river and its tributaries. Such rivers have water flow with periods of high and periods of low volumes. Consequently, during the former or for a part of the season, they may carry, mix, and deposit large quantities of sediment. This sediment, exposed on the flood plain during periods of low water, is subject to drying, wind removal, and deposition on the adjoining areas. It suggests itself as the logical source material of the loessial soils.

Although the unweathered loess has been shown to be relatively uniform in chemical and mineral composition over broad areas (7)[3], the reason for this uniformity has never been fully established. If flood plain material is the source of loess, then these river sediments must likewise show uniformity in mineral and chemical composition over extended distances of the river flood plains.

The study reported herewith was a test of the hypothesis that some loess might have its origin in the river flood plains. Mineralogical and chemical studies were made of the flood plain material of the Missouri-Mississippi river system to learn how uniform these sediments are.

REVIEW OF LITERATURE

Theories aiming to explain the formation of loess deposits have been numerous. One particular feature seems to be common to all of these theories; namely, the pulverized mineral and rock material, called loess, is believed to originate in an arid climate or in an area of glaciation. According to Free (2) the primary American deposit of loess may be considered as made up of glacial silt and wind-borne rock debris. These two materials, separated or mixed, have been collected either by aeolian action or by streams on the flood plains of large but sluggish rivers. These large rivers, which periodically pour muddy floods over their plains, are in this way depositing flood plain materials as potential loess subject to the activities of the aeolian agencies operating during low water stages. Free has pointed out that the very finely comminuted particles of loess have been the product of aqueous, sub-aerial, or glacial action. Whatever may have been the source of this material, it was not produced where it is now found.

Chamberlain (1) locates and describes the loess in central United States. The loess of the Mississippi River basin is distributed along its leading tributaries. "Along these valleys the loess is thickest, coarsest, and most typical on the bluffs bordering the rivers and grades away into thinness, fineness, and nontypical nature as the distance from the rivers increases."

Guy D. Smith (4) of Illinois studied the variation in particle size and the thickness of the loess deposit with distance from the river channel. He found that the thinning and the texture of the loess bear a close linear relation to the logarithm of the distance from the river flood plain.

Smith and Norton (5) prepared a map showing the distribution and thickness of the loess mantle in the State of Illinois. The depth of the loess was found to be directly related to the present major stream channels. The thickest deposit is adjacent to the wide bottoms of the Mississippi River. From this study, Smith and Norton attributed the loess deposits as being the result of wind action on the barren river bottom during low water level.

Vanderford and Albrecht (7) made a study of the loess on the river bluff from Iowa to Vicksburg, Miss. They found that the unweathered loess (C-horizon) was similar in chemical composition over the entire distance between these two points.

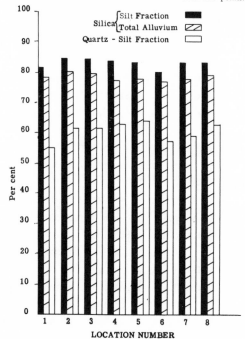

FIG. 1.—Quartz in the silt fraction and silica in both the silt fraction and total alluvium in relation to location.

---

[1] Contribution from the Department of Soils, Missouri Agricultural Experiment Station, Columbia, Mo. Journal Series No. 1126.
[2] Graduate Assistant, and Professor of Soils, respectively.
[3] Figures in parenthesis refer to "Literature Cited", p. 470.

However, the soil developed from this loess reflected the climate of the particular location.

While the loess and the soils developed from it have had extensive research consideration, the material of the flood plains as potential loess has not. For that reason mineralogical and chemical studies of this flood plain material in the Missouri–Mississippi river system were made.

PLAN AND METHODS

Samples were collected from the river flood plains to represent the recently deposited alluvia in the Missouri–Mississippi river system from Rockport, Mo., to Vicksburg, Miss. The samples were collected over an extended period of time and do not represent the sediments from one flood. They were taken from the river bars so as to represent material of recent deposition. They were collected near the following locations in Missouri: No. 1, Rockport, No. 2, Oregon, No. 3, Lexington, No. 4, Glasgow, No. 5, McBaine, No. 6, St. Marys, No. 7, Cape Girardeau; and in Mississippi: No. 8 near Vicksburg. Only these corresponding numbers instead of the names of the locations will be used in the discussion, table, and figure which follow.

ANALYTICAL METHODS

Since loess consists dominantly of the silt separate, the silt fraction with equivalent diameters of 0.05 to 0.005 mm was separated from the total sample of the river alluvium. The total sample and its silt fraction were fused in duplicate with sodium carbonate and the content of silica determined by the method of Triebold (6). Each silica-free sample was then made up to a definite volume and further analyses made on aliquots of this solution. Calcium, potassium, and phosphorus were determined by standard methods. The quartz in the silt fraction was determined by the method described by Knopf (3). The pH of the total alluvium was measured with the Beckman potentiometer using a ratio of 1:1 of water to alluvium.

RESULTS AND DISCUSSION

Silica and Quartz Contents.—The silica contents of the alluvia and of the silt fractions are shown in Fig. 1. The mineral quartz contents of the silt fractions are also given. The variation in the silica between all samples is so slight that this uniformity indicates that these river sediments are of nearly constant composition regardless of location. The amounts of the mineral quartz in the alluvium exceed those of all other minerals combined. Therefore a quantitative measure of the variation of this mineral should serve as an important index of the variation of the reciprocal or non-quartz minerals. Any variation therefore in the general mineral content should be reflected by the variation in the quartz content.

Phosphorus Content.—The phosphorus content of any one of the different samples of the total alluvium does not vary from that of any other by a figure of more than 0.013%, according to the data in Table 1. Therefore, the phosphorus-bearing minerals must be uniformly distributed in these river deposits over the area investigated.

TABLE 1.—*Chemical and mineral composition of alluvial deposits at various locations in the Missouri–Mississippi River system. (Data are percentages of the dry sample.)*

Location	SiO$_2$	Mineral quartz	Phosphorus	Potassium	Calcium	pH
No. 1 Rockport, Missouri	*A 78.1		0.045	1.38	2.00	7.04
	†B 81.4	54.8			0.71	
No. 2 Oregon, Missouri	A 80.1		0.044	1.42	1.21	6.80
	B 84.7	61.7			0.47	
No. 3 Lexington, Missouri	A 79.5		0.044	1.38	1.55	8.01
	B 84.4	61.4			0.51	
No. 4 Glasgow, Missouri	A 77.3		0.045	1.38	1.59	8.02
	B 83.7	62.6			0.43	
No. 5 McBaine, Missouri	A 77.7		0.045	1.49	1.53	6.62
	B 83.3	63.8			0.73	
No. 6 St. Marys, Missouri	A 76.7		0.050	1.25	1.86	7.79
	B 80.0	57.1			1.62	
No. 7 Cape Girardeau, Missouri	A 77.8		0.055	1.62	1.84	7.70
	B 82.8	58.7			1.31	
No. 8 Vicksburg, Mississippi	A 79.1		0.042	1.22	1.37	6.68
	B 82.6	62.3			0.90	

* Total alluvium.
† Silt fraction.

Potassium Content.—The uniformity of the content of potassium in the total alluvium, as shown in Table 1, points out that also this element is evenly distributed along the channels of the Missouri and Mississippi rivers.

Calcium Content.—The calcium contents of both the total sample and its silt fraction, as shown in Table 1, indicate a relative uniformity of this mineral over the distance of the river course investigated.

pH—Degree of Acidity.—The pH of the total alluvium varied between 6.08 and 8.02 with a mean value of 7.33.

The data characterizing the recently deposited alluvium indicate that this material to be relatively uniform in mineral and chemical composition throughout the river length investigated. Therefore, if the wind-borne material constituting the loess is picked up from the flood plain of the river and deposited on the adjacent area, it must also be uniform throughout the course of the river.

SUMMARY AND CONCLUSION

That the alluvial deposits on the flood plains of the Missouri–Mississippi rivers between Rockport, Mo., and Vicksburg, Miss., are highly similar in mineral and chemical composition is pointed out by the studies of the silica, quartz, phosphorus, potassium, and pH of the alluvium.

As a collector, mixer, and depositor of the material which it carries, according to the evidence here, the Missouri–Mississippi river system is bringing about uniformity of its sediment throughout its course. There-fore, the barren river flood plain, swept by winds that pick up certain separates of the sediments and re-deposit them on the upland would make such deposits, of necessity, consist of material that is relatively constant in mineral and chemical makeup.

According to these data, then, the relative mineral and chemical uniformity of loess over wide areas is not due to any specific part played by the wind. Rather, it seems plausibly explainable as the result of the action of a river, in bringing about uniformity of its sediments left in its flood plains to be picked up by the winds and deposited as loess. Soils developed from loess, then, would vary according to the climatic factors operating in any specific location.

LITERATURE CITED

1. CHAMBERLAIN, T. C. A supplementary hypothesis respecting the origin of the loess of the Mississippi valley. Jour. Geol., 5: 795–802. 1897.
2. FREE, E. E. The movement of soil material by the wind. U. S. D. A. Bureau Soils Bul. 68. 1910.
3. KNOPF, A. The quantitative determination of quartz (free silica) in dusts. Public Health Reports, 48(8): 183–190. 1933.
4. SMITH, GUY D. Illinois loess-variation and distribution: A pedologic interpretation. Ill. Agr. Exp. Sta. Bul. 490. 1942.
5. SMITH, R. S., and NORTON, E. A. Parent material of Illinois soils in "Parent materials, subsoil permeability and surface character of Illinois soils". Ill. Agr. Exp. Sta. and Ext. Ser. 1935.
6. TRIEBOLD, H. O. Quantitative analysis of agricultural and food products. New York: D. Van Nostrand Co., Inc. 1946.
7. VANDERFORD, HARVEY B., and ALBRECHT, W. A. The development of loessial soils in central United States as it reflects differences in climate. Mo. Agr. Res. Bul. 345. 1942.

19

Reprinted from *Science*, **126**(3286), 1285 (1957)

Source and Deposition of Clay Minerals in Peorian Loess

Loess is one of the most remarkable of the Pleistocene deposits. It is associated with and covers to varying depths and extent most of the major sheets of glacial drift. The origin of loess, however, has been debated. Scheidig (*1*) lists some 20 hypotheses that have been advanced at one time or another to explain its presence and distribution. Chamberlain (*2*) has advanced the most widely accepted theory to explain the origin of loess in the upper Mississippi Valley. He considered the loess as a wind deposit emanating from the flood plains of the major Pleistocene rivers. His concept was that proloess materials were deposited from glacial melt waters on the flood plains of the rivers. After drying, these materials were picked up by strong winds and redeposited as loess on the adjacent uplands.

Chamberlain's theory implies that the mineralogy of the unaltered loess and that of the associated unaltered tills should be similar, including the clay minerals. Studies of clay minerals (*3–5*), however, have shown that the principal type of clay in calcareous Peorian loess in Illinois, Kansas, Nebraska, Iowa, and Missouri is montmorillonite, whereas, illite and some chlorite are the principal clay minerals in tills of Wisconsin age over a broad area (*3, 6, 7*). Some explanations given to account for this difference in mineralogy follow: (i) the montmorillonite clay now found in Peorian loess resulted from weathering of the illite and chlorite in the calcareous material after deposition; (ii) the clay and silt minerals of Wisconsin age weathered to form montmorillonite before and/or during transport; and (iii) the montmorillonite was differentially picked up by the silt particles from the river flood plains, thus concentrating this type of clay in Peorian loess.

On the basis of studies of clay mineral in soils developed entirely from tills of Wisconsin age, explanations i and ii above were considered unlikely. For instance, Beavers et al. (*3*) found that only small amounts of montmorillonite (maximum 10 percent) had formed in soils developed from Tazewell and Cary age tills and that no montmorillonite had formed in calcareous tills of the same age. Similar results were found by Bidwell and Page (*6*). Explanation iii cannot be ruled out, although I believe that the bulk of the sediments carried by the Illinois and Wabash rivers during the time of deposition of Peorian loess were of Wisconsin age and that illite was the principal clay mineral in the sediments. The influence of local flood-plain clay sediments is indicated by the tendency of illite clay to concentrate in calcareous Peorian loess in Illinois (5 to 20 percent) near the major rivers (*3, 5*).

I postulate that the bulk of the clay minerals in Peorian loess did not come from local flood plains but that these minerals were carried in by strong winds from widely scattered sources throughout the central United States. The problem is essentially one of deposition of the fine clay. I suggest the following as a possible mechanism that may account for the deposition of fine clays carried from afar, along with local flood plain silts. The air-borne clay minerals were electrostatically attracted and adsorbed onto the larger silt-sized particles that were blown from local flood plains, and then the clays and silts were deposited together.

Charge spectrometer studies of quartz and standard clay minerals, as well as of clays and silts from Peorian loess, show that these materials have a tendency to take on strong electrostatic charges (*5*). It is well established that dust storms are highly electrified, the friction of the particles providing a source of electricity. Boning (*8*) advanced the theory that a part of the charge developed in dust clouds was the result of friction between particles of different sizes. That particles of silt and clay minerals have different electrostatic charges is suggested not only by the fact that the two kinds of particles are different in size but also by the fact that their crystalline structure and dielectric properties are different.

Dallavalle (*9*) states: "Fine dust particles may be swept upward by turbulent wind and kept in motion by it so that the effect of gravity is nullified." Even today, Illinois receives clay from western storms that occasionally cause the sun to appear hazy. When these fine air-borne clays are brought down by snow or rain, they fall in sufficient concentrations to cover clean surfaces with buff-colored clay particles. We also know that fine clay-size material from bomb blasts and volcanoes is carried long distances by wind, even across continents and oceans.

A unique property of loess is its unstratified nature. Thin sections of Peorian loess adjacent to the Wabash, Mississippi, and Illinois rivers show that the materials possess a fine porous fabric with the larger silt-sized grains connected with intergranular braces of a light ocher color consisting of very fine silt with clay minerals evenly disseminated throughout. A homogeneous and unstratified deposit would not be expected to result from the normal settling of silts and clays. Here again it appears that some mechanism other than the normal settling forces were operative and that the silt and clay did not settle independently.

The electrostatic adsorption and deposition of fine clay by local flood plain silts could explain the distribution of the montmorillonitic type of clay throughout the Peorian loess area as well as the unstratified nature of the loess deposit.

A. H. BEAVERS
*Department of Agronomy,
University of Illinois, Urbana*

References and Notes

1. A. Scheidig, *Der Loss* (Steinkopff, Dresden, Germany. 1934).
2. T. C. Chamberlin, *J. Geol.* 5, 795 (1897).
3. A. H. Beavers *et al.*, *Natl. Acad. Sci.–Natl. Research Council Publ. No. 395* (1955), p. 356.
4. A. Swineford and J. C. Frye, *J. Sediment. Petrol.* 25, 3 (1955); E. P. Whiteside and C. E. Marshall, *Univ. Missouri Agr. Expt. Sta. Research Bull.* 386 (1944).
5. A. H. Beavers, unpublished.
6. O. W. Bidwell and J. B. Page, *Soil Sci. Soc. Am. Proc.* 15, 314 (1950).
7. J. B. Droste, *Bull. Geol. Soc. Am.* 67, 911 (1956).
8. P. Boning, *Z. tech. Physik* 8, 385 (1927).
9. J. M. Dallavalle, *Micromeritics* (Pitman, New York, ed. 2, 1948).

12 August 1957

Editor's Comments
on Paper 20

20 DAVIDSON and SHEELER
   Cation Exchange Capacity of the Loess in Southwestern Iowa

Davidson and Sheeler have each made considerable contributions to the study of loess, particularly within the state of Iowa and with respect to engineering problems. The study of cation exchange capacity has, however, much wider implications, and this is the reason for the inclusion of Paper 20. In 1951, Davidson et al. published a loess bibliography that had fairly general coverage of the literature but was essentially restricted to English-language papers.

Sheeler and Davidson (1957) have correlated the consistency limits of Iowa loess with the clay content, and in 1968 Sheeler published a summary of the engineering properties of loess in the United States. This review covers identification, mechanical analyses, specific gravity, Atterberg limits, permeability, density, shear strength, consolidation, bearing capacity, natural moisture content, and erosion. Paper 20 gives some results for Iowa loess. The techniques used have been described in detail by the authors in their contribution to ASTM STP-142 (see Paper 20, reference 2, for full citation).

Engineering investigations on the Iowa loess are continuing and Handy (an associate of Davidson and Sheeler) has recently published an interesting paper on collapsing loess in Iowa (Handy, 1973).

REFERENCES

Davidson, D. J., Chu, T. Y., and Sheeler, J. B. 1951. A bibliography of the loess. Eng. Rept. Iowa Eng. Expt. Station No. 8, Ames, Iowa, 15p.

Handy, R. L. 1973. Collapsible loess in Iowa. Soil Sci. Soc. America Proc. 37, 281–284.

Sheeler, J. B. 1968. Summarization and comparsion of engineering properties of loess in the United States. Highway Res. Record No. 212, p. 1–9.

——, and Davidson, D. T. 1957. Further correlation of consistency limits of Iowa loess with clay content. Proc. Iowa Acad. Sci. 64, 407–412.

20

Reprinted from *Iowa Acad. Sci. Proc.*, **60**, 354–361 (1953)

Cation Exchange Capacity of the Clay Fraction of Loess in Southwestern Iowa

By D. T. Davidson and J. B. Sheeler

The cation exchange capacity of clay-size material extracted from soil with a low organic matter content is largely dependent on the kinds of clay minerals present. If the extracted clay is composed mostly of one kind of clay mineral, the exchange capacity will indicate what that mineral is. This paper reports on cation exchange capacity determinations made on the minus 2 micron clay-size range of selected samples of loess from the southwestern Iowa area shown in Figure 1.

WHOLE LOESS SAMPLES

The origin, distribution, and property variations of the Wiscon-

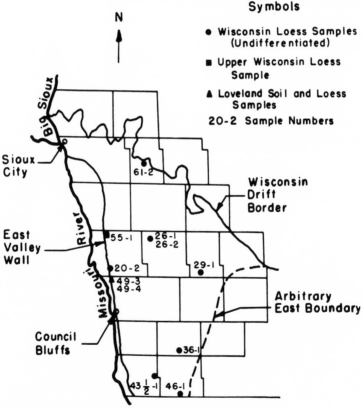

Figure 1. Distribution of sampling locations in southwestern Iowa loess area. Symbols indicate age classification of loess sampled.

sin* loess which mantles much of the southwestern Iowa area shown in Figure 1 have been discussed in previous papers (1, 2, 3). The minus 2 'micron clay fractions used in the study reported in this paper were extracted from eleven samples of whole loess which were selected as representing the range in properties of the more than 150 loess samples that have been tested in the property variation studies of the Iowa Engineering Experiment Station.

Descriptive information of the locations from which the whole loess samples were taken are given in Table I, and the distribution of sampling locations is shown in Figure 1. Sample 55-1 is Upper Wisconsin or Cary-Mankato loess from the Pisgah road section identified by Ruhe (4). The other Wisconsin age loess samples are undifferentiated because the buried (Brady) soil (6) which separates the upper and lower Wisconsin components was not present. The samples of Loveland soil† and of Loveland loess from the type section at Loveland, Pottawattamie County, were included in the study for comparative purposes.

Table II gives some properties of the whole loess which are indicative of the cation exchange material in the samples. The Wisconsin loess samples are arranged in this table in the order of increasing clay content. The range in clay content of the Wisconsin loess in southwestern Iowa is shown by these samples. Whether the type section Loveland samples used herein are representative of Loveland soil and loess exposed elsewhere in southwestern Iowa is not as yet known. The Loveland soil had a higher clay content than the underlying loess, and both Loveland samples had lower clay contents than some of the more plastic Wisconsin loes samples.

Sample 26-2 is from the same location as sample 26-1 but was taken at a greater depth in the Wisconsin loess section. Both samples contain practically the same amount of clay. However sample 26-2 is unoxidized and unleached, and sample 26-1 is oxidized and leached. Sample 26-2 was included in the study to determine what effect oxidation and leaching might have on the clay mineral present in the loess. This was the only unoxidized sample used in the study.

The organic matter content of all samples was low, and for this reason the inorganic clay minerals are considered to be largely responsible for cation exchange. Carbonate contents were variable, being relatively high for unleached samples taken near the major source areas of the loess (3). The distribution of carbonates in

---

*Also referred to as Peorian loess in the geological literature.
†Sangamon soil profile on Loveland loess.

Table 1
Loess Sampling Locations in Southwestern Iowa.

Sample No.	Material	Age classification	Sampling depth[c] (ft.)	County	Section	Township North	Range West	Soil series
55-1	Loess	Upper Wisconsin[b]	2½–3½	Harrison	SW/c,S-8	81	44	Hamburg
20-2	"	Wisconsin (Undifferentiated)	39–40	Harrison	S-15[f]	78	43	Hamburg
61-2	"	"	17–18	Ida	NW¼,S-9	87	40	Monona
26-1	"	"	4–5	Shelby	SE¼,S-21	81	40	Monona
26-2	"	"	10–11	"	"	"	"	"
29-1	"	"	5–6	Audubon	NW/c,SW¼,S-13	78	36	Marshall
36-1	"	"	5½–6½	Montgomery	SE¼,NE¼,S-14	72	38	Marshall
43½-1	"	"	5–6	Fremont	NW/c,S-36	68	40	Marshall
46-1	"	"	5–6	Page	NW¼,S-30	67	37	Marshall
49-3	Soil[a]	Loveland	55[d]	Pottawattamie	SE/c,NW¼,S-3[g]	77	44	Hamburg
49-4	Loess	"	65[e]	"	"	"	"	"

[a] Sangamon soil profile on Loveland Loess.
[b] Also called Cary-Mankato loess and Bignell loess. Sampled from Pisgah road section (4).
[c] Measurements are from earth's surface.
[d] Sampled from about middle of 10 ± ft. Sangamon soil profile.
[e] Sampled in Loveland loess about 3 ft. above slump.
[f] Sampled from bluff behind third ward school in city of Missouri Valley.
[g] Sampled from type section of Loveland loess at northeast edge of town of Loveland (5).

Table 2

Some Properties of Whole Loess Samples.

No.	Age Classification	Textural Composition[a]				Organic matter (%)	Carbonates (%CaCO₃)	Oxidation	pH	Cat. Ex. Cap (m.e./100g)	Plasticity Index (%)	B.P.R.[b] Classification
		Sand (%)	Silt (%)	Clay (%) -5μ	Clay (%) -2μ							
55-1	Upper Wisconsin	4.0	82.6	13.4	12.0	0.24	11.0	Oxidized	8.4	11.2	2.3	A-4(8)
20-2	Wisconsin (Undifferentiated)	1.4	78.8	19.8	16.0	0.17	10.2	Oxidized	8.7	13.4	6.2	A-4(8)
61-2	"	5.2	70.8	24.0	19.8	0.15	11.5	Oxidized	8.3	14.2	10.8	A-6(8)
26-1	"	2.0	70.6	27.4	22.4	0.18	1.4	Oxidized	7.0	18.2	12.5	A-6(9)
26-2	"	0.9	69.8	29.3	23.1	0.17	8.7	Unoxidized	8.3	17.9	17.8	A-6(9)
29-1	"	1.0	67.9	31.1	25.0	0.25	2.9	Oxidized	8.3	19.5	18.0	A-6(11)
36-1	"	0.8	63.2	36.0	28.9	0.21	1.8	Oxidized	6.7	21.0	20.7	A-7-6(13)
43½-1	"	0.4	60.2	39.4	33.0	0.37	0.5	Oxidized	6.7	24.4	33.4	A-7-6(18)
46-1	"	0.8	55.7	43.5	36.2	0.30	1.5	Oxidized	6.3	22.6	32.7	A-7-6(19)
49-3	Loveland (soil)	2.7	61.1	36.2	31.4	0.19	1.3	Oxidized	8.2	22.7	24.6	A-7-6(15)
49-4	Loveland	5.3	65.9	28.8	24.6	0.11	7.0	Oxidized	8.2	16.6	15.3	A-6(10)

[a]Sand—2.0 to 0.05 mm., silt—0.05 to 0.005 mm. One micron equals 0.001 mm.
[b]Bureau of Public Roads Soil Classification System. Also referred to as the Highway Research Board System or the American Association of State Highway Officials (AASHO) System (7).

the different particle-size fractions of the loess has not as yet been determined. The pH values varied from 6.7 to 8.7 or from near neutral to alkaline. The variations in whole loess cation exchange capacity, in plasticity index, and in B.P.R. classification are mainly due to the variation in clay content (2).

CLAY FRACTION

The minus 2 micron portions of the whole loess samples were used for the exchange capacity determinations because practically all of the cation exchange material, the clay minerals, occur in this particle-size range. Complete separation of the clay minerals from other substances such as quartz and carbonates is difficult, but only very small amounts of such substances are commonly found in the minus 2 micron soil fraction.

The separation of the minus 2 micron clay material from the whole loess was by means of a sedimentation procedure which is the subject of another paper now in preparation. In this procedure neither the whole loess nor the separated clay was given hydrochloric acid or hydrogen peroxide treatments. When a deflocculating agent was needed to prevent flocculation, 0.1 N sodium hydroxide was used.

DETERMINATION OF CATION EXCHANGE CAPACITY

Cation exchange capacity can be determined by a number of methods, most of which involve leaching the soil sample with a salt solution containing known cations followed by analysis either of the resulting soil or of the solution for the amount of cations exchanged. Because of solubility and decomposition effects Kelly and Brown (8) recommended the determination of the cations absorbed by the soil rather than of the amount of cations that are brought into solution from the soil.

The cation exchange capacity of a soil denotes the total amount of cations that can be exchanged under a given set of conditions and not necessarily the amount that could be exchanged under other conditions. The determination is particularly sensitive to the pH of the salt solution; the common practice is to use a neutral (pH = 7) solution. Neutral normal ammonium acetate has been found to be a salt solution especially well adapted to the exchange capacity determination (9). With this solution the exchange capacity of inorganic soils can be determined with reasonable accuracy even when the soil contains soluble salts and calcium carbonate.

Test methods used in determining cation exchange capacities of whole Wisconsin loess samples have been previously presented by the authors (2).   A step-by-step summary of the procedure used in the present study is as follows:

1. Weigh out about one gram (accurate to 1 mg) of representative air-dry clay and place in a 300 ml centrifuge bottle. (A similar sample should be weighed out for the hygroscopic moisture determination needed to convert air-dry weight to oven-dry weight.)
2. Add 10 g of fine Ottawa sand to the clay in the centrifuge bottle. (Ottawa sand is inert and increases the permeability for the purpose of filtration in Step 7.)
3. Add 250 ml of neutral normal ammonium acetate to the contents of the centrifuge bottle and shake for 3 min. (Higher normalities and increased shaking times were experimented with but did not significantly affect results.)
4. Centrifuge at 2000 RPM for 10 min.
5. Decant the clear supernatant liquid.
6. Repeat Step 3.
7. Filter the contents of the centrifuge bottle with a Buchner funnel containing two fine filter papers.
8. Wash the material retained on the filter paper with 150 ml of neutral 70 percent (by volume) methyl alcohol to remove the excess ammonium acetate trapped in void spaces.
9. Determine the amount of ammonia held in the exchange positions of the clay by a modified Kjeldahl nitrogen determination (2).
10. Calculate the cation exchange capacity in milliequivalents per 100 g of oven-dry clay (2).

CATION EXCHANGE CAPACITIES

Cation exchange capacities of minus 2 micron clay fractions are given in Table III.   Since the clay minerals are the primary seat of cation exchange, the uniformity of the exchange capacity data

Table 3

Cation Exchange Capacities of Minus 2 Micron Clay Fractions

Sample no.	Material	Age Classification	Cation Exchange Capacity[a] (m.e./100g)
55-1	Loess	Upper Wisconsin	59.3
20-2	"	Wisconsin (Undifferentiated)	58.6
61-2	"	"	52.5
26-1	"	"	62.8
26-2	"	"	59.0
29-1	"	"	62.2
36-1	"	"	63.1
43½-1	"	"	59.9
46-1	"	"	57.4
49-3	Soil	Loveland	63.4
49-4	Loess	"	54.4

[a]Values reported are the average of two determinations.

indicates that there is little variation in the kinds of clay minerals in the Wisconsin Loess. The data further indicates that there is little difference between the clay fractions of the Wisconsin loess and the Loveland Loess. The results of differential thermal analyses on whole loess samples (3) substantiate this interpretation.

Slight variations in the cation exchange capacity values in Table III may or may not be significant. Further studies are in progress to determine whether they are due to experimental factors or to slight variations in mineral composition.

The clay fractions extracted from soils are rarely composed of a single kind of clay mineral but usually contain two or more mixed with other substances from which complete separation is difficult. The exchange capacity of the minus 2 micron particle-size range for this reason can at best be used only to estimate roughly the predominant kind of clay mineral present. The estimation can be made by comparing the determined value with the exchange capacities of comparatively pure clay minerals. Grim (10) gives cation exchange capacities of the common clay minerals as:

> Montmorillonite.............................60-100 m.e./100g
> Attapulgite....................................25-30
> Illite..20-40
> Kaolinite......................................3 -15
> Halloysite.....................................6 -10

A comparison of the data in Table III with Grim's values indicates that the loess clay fractions contain a predominance of montmorillonite group minerals. Differential thermal curves for samples of whole loess from southwestern Iowa have indicated the presence of illite (3). This mineral is often associated with montmorillonite in soils, and its presence would tend to lower the exchange capacity (9).

According to Jackson et al (11), montomorillonite can develop from illite by weathering, which may be the reason why the Loveland soil has a higher exchange capacity than the loess on which it developed. The same explanation may apply to the difference in the exchange capacities of samples 26-1 and 26-2. The oxidized and leached sample (26-1) had a slightly higher exchange capacity than the unoxidized and unleached sample (26-2).

ACKNOWLEDGMENT

The subject matter of this paper was obtained as part of the research being done under Project 283-S of the Iowa Engineering Experiment Station of Iowa State College. This project, entitled "The Loess and Glacial Till Materials of Iowa; An Investigation of

Their Physicial and Chemical Properties and Techniques for pro-
cessing Them to Increase Their All-Weather Stability for Road
Construction" is being carried on under contract with the Iowa
State Highway Commission and under the sponsorship of the Iowa
Highway Research Board. It is supported by funds supplied by the
Commission and the U. S. Bureau of Public Roads.

References Cited

1. Davidson, D. T. and Sheeler, J. B. 1951. Studies of the clay fraction in engineering soils: III Influence of amount of clay on engineering properties. Proc. Hwy. Res. Bd. 31:558-563.

2. Davidson, D. T. and Sheeler, J. B. 1952. Cation exchange capacity of loess and its relation to engineering properties. Symposium on Exchange Phenomena in Soils, Am. Soc. Testing Mat. Special Tech. Pub. 142. pp. 1-19

3. Davidson, D. T. and Handy, R. L. 1952. Property variations in the Peorian loess of southwestern Iowa. Proc. Iowa Acad. Sci. 59:248-265.

4. Ruhe, R. V. 1949. A Bignell (?) loess section in western Iowa. Proc. Iowa Acad. Sci. 56:229-231.

5. Kay, George F. et al. 1944. The Pleistocene geology of Iowa. Spec. Rpt., Iowa Geologic Survey.

6. Ruhe, R. V. 1952. Classification of the Wisconsin glacial stage. Jour. Geol. 60:398-401.

7. American Association of State Highway Officials. 1950. Standard specifications for highway materials and methods of sampling and testing, Part I. Specifications. The Association, Washington 25, D. C.

8. Kelly, W. P. and Brown, S. M. 1924. Replaceable bases in soils. California Agr. Exp. Sta. Tech Paper. 15:1-39.

9. Kelly, W. P. 1948. Cation Exchange in Soils. Reinhold Publishing Corp., New York, N. Y.

10. Grim, R. E. 1942. Modern concepts of clay materials. Jour. Geol. 50:225-275.

11. Jackson et al. 1948. Weathering sequence of clay-size minerals in soils and sediments. Jour. Phys. and Colloid Chem. 52:1237-1260.

IOWA ENGINEERING EXPERIMENT STATION

IOWA STATE COLLEGE

AMES, IOWA

Editor's Comments
on Papers 21 and 22

21 **DYLIK**
The Problem of the Origin of Loess in Poland

22 **TOKARSKI** et al.
Remarks on the Loess

Dylik quotes the remark by Tokarski to the effect that every geological structure bears the imprint of its genetic process and that this message can be read if adequate methods are applied. The nature of loess material and the structure and position of loess deposits should tell us about the mode of formation both of the material and deposits, and modes of formation should characterize the material. Dylik implies that loess definitions should include mechanism.

Paper 21 is essentially a review of the loess formation problem from a Polish viewpoint. It includes references to Polish papers, which have perhaps not had the general impact that they deserve and which should be better known. The paper appeared in the first issue of the journal *Biuletyn Peryglacjalny,* the first journal devoted to the periglacial environment. In another paper in the same issue, Jahn (1954) discussed the work of Walery Lozinski and his role in initiating periglacial studies. Lozinski (1912) made some very pertinent observations on loess, in particular regarding the formation of particles by mechanical action and the role of the wind as a transporting agent in periglacial regions. It is interesting that the Eurpoean and American loess deposits have long been associated with glacial action and discussed in terms of periglacial phenomena, and yet the proposed desert origin of the large Chinese deposits was general accepted.

Paper 22 summarizes many years of effort by a leading Polish loess investigator. The term "anemoclastic," which appears to be unique to Tokarski, comes from two Greek roots meaning "wind" and "broken." Tokarski is committed to the eolian theory of deposit formation and indicates very clearly the importance of examing loess material. Like Lugn

(1962) and Swineford and Frye (see Paper 12), he presents dust-storm data; he also considers variations in loess properties with distance from source, as Waggoner and Bingham (Paper 28) and Frazee et al. (Paper 29) have done. He was also one of the few loess investigators to use thermogravimetric techniques, and his application can be usefully compared with that of Seppälä (Paper 36). The text figures have been reproduced in this shortened translation, but the photographic plates were not considered essential and have been omitted.

It would have been useful to include the well-known paper by Jahn (1950), but lack of space meant that the Polish contribution had to be curtailed in some way; it seemed more desirable to include the more recent papers.

REFERENCES

Jahn, A. 1950. Loess, its origin and connection with the climate of the Glacial Epoch. Acta Geol. Polonica 1, 257–310.

———. 1954. Walery Lozinski's merits for the advancement of periglacial studies. Biul. Perygalc. 1, 5–16 (Polish text), 117–124 (English text), 16–18 (references).

Lozinski, W. 1912. Die periglaziale Frazies der mechanischen Verwitterung. C. R. 11th Intl. Geol. Congr. Stockholm, 1910, 2, 1039–1053.

Lugn, A. L. 1962. The origin and sources of loess. University of Nebraska Studies No. 26, Lincoln, Neb., 105p.

21

Reprinted from *Biul. Peryglacjalny*, No. 1, 28–30, 125–131 (1954)

THE PROBLEM OF THE ORIGIN OF LOESS IN POLAND

Jan Dylik

Abstract

Recent years have witnessed an animation in the discussion of the problem of loess. The development of periglacial investigations 'has greatly contributed to this increase of interest in loess-problematics, foremost as concerns the question of the origin of loess.

The most important achievements of the newest Polish research work in this field are: the passage from the adaptation of general theories to more direct, and thorough investigations; the distinction of a whole set of active processes which contributed to the formation of loess; the recognition of a periglacial environment in which these processes occurred; the investigation of the parent-loess material and the development of scientific methods.

Among the problems which still remain open thus giving rise to new scientific needs the most important ones are: the problem of „typical sub-aerial loess", that of the eolian process and of its exclusivity as well as the necessity of a fuller knowledge of the morphogenetic characteristics of periglacial environment.

Although its history dates from some 130 years back the problem of loess is still the object of animated discussion all over the world. The eolian theory though preponderant in Western science is still far from being unanimously accepted. The majority of Soviet experts such as B e r g (*3,4*), G e r a s i m o v and M a r k o v (*15, 16*) take the view of weathering theory of loessification. I. P. G e r a s i m o v (*14*) believes the eolian theory to be purely academic and one-sided. In the West, objections against the eolian theory were raised by G a n s s e n (*13*), M ü n i c h s d o r f e r (*26*) and recently by R. J. R u s s e l l (*35*) and F i s k (*12*).

The last post-war years have witnessed a revival of interest in the problem of loess; numerous contributions testify to the fact (*1, 5, 6, 12, 18, 19, 20, 24, 27, 29, 32, 41, 42, 43, 45*).

One of the reasons which brought about this animation of the discussion was the original theory of R. J. R u s s e l l. His colluvial hypothesis of loessification aroused many criticisms and largely contributed to the growth of loess-literature in the United States (*12, 20, 24, 42, 43*).

The development of periglacial studies creates a new, more general and more urgent need of a revision of the problem of loess. The investigations conducted during the last few years have endavoured to determine with ever-increasing precision the periglacial, climatic environment as well as the assemblage of processes which were active within its limits. Being indeniably a periglacial phenomenon, loess should be studied in intimate connection with the entire set of other periglacial facts (*9*, p. 10).

Since the war, the problem of loess was treated in Poland in a considerable number of publications, though in diverse connections (*11, 21, 22, 25, 30, 31, 33, 34, 39, 40*).

In the intricate problem of loess, the most important question is that of its origin. The discussion to follow — which is of a surveing character — will be focussed upon this question. It refers to the general genetic conception and to several subordinate problems involved: such as climatic environment, origin of the loess-material, the active processes which contributed to its formation, as well as the processes of transportation and accumulation of loess.

The authors of the papers published in Poland all, save R o k i c k i (*34*) take the view-point of the eolian theory. M a l i c k i (*25*) has set forth an original conception that of a modified eolian hypothesis. He holds that eolian transport was but a short, local process, confined to the transportation of particles from the closest neighbourhood.

R o k i c k i devoted his attention to loess-like formations such as dusty clays and pulverulent sandy formations which he thinks, are partly local weathering products, partly — deposits due to water action. The writer mentions also eolian loess in the area of the Trzebnica Hills, which constituted the field of his investigations. He failed however to give them closer attention. Anyhow, the criticism made by R o k i c k i seems to raise doubts as to the eolian origin of this loess (*34*, p. 7).

R o k i c k i's critical attitude with regard to these formations, recalls the vital problem of what should be called loess-formation. Apart from the general term of loess, other more precise definitions are being used such as loess, loess-like formations and single loess-facies.

All the authors distinguish typical subaerial loess. K l i m a s z e w s k i (*22*) mentions moreover striated loess and loess-like clay due to weathering. M a l i c k i (*25*) distinguishes stratified loess-like formations swept down from higher spots and forming a bottom of loess on the edge of the Carpathians, fluvial deposits created by the overflow a re-deposition of weathering products from small-grained rocks, slope-loess consisting of the weathering products of calcareous Flysch rocks and of the weathering products of uncalcareous, small-grained carpathian rocks. P o ż a r y s k i (*31*) distinguishes typical subaerial loess, slope - (soliflual) loess and lake-loess. In connection with typical loess, R o k i c k i investigated loess-like pulverulent clays and sandy dusts. J a h n (*21*) recognizes three different kinds of eolian loess, one formed in the upland-facie, the second — in the valley-facie and a third, consisting of sandy loess characterized by the presence of sandy intercalations in the loess.

The differentiation of the general notion of loess into typical subaerial loess and loess-like formations was operated on the support of a number of criteria, among which foremost that of the microtexture visible in the character of the mechanical composition.

This differentiation is mainly based upon observations relating to structure. The majority of the authors holds that „typical loess" is neither stratified nor bears any traces of stratification (*25*). The capacity of spliting along vertical joints is supposed to be one of the most striking features of typical loess. The presence of calcareous concretions is also considered as one of the main characteristics of this formation. M a l i c k i finds also that fauna, molluscian as well as mammalian appears rather in loess-like formations than in typical loess.

The list of characteristic traits includes also the topographical situation and the thickness. Typical loess occurs in the highest points of uplands whereas loess--like formations are to be 'found on slopes and in valleys (*25*). This distinction is not however generally respected since e. g. S a w i c k i's (*39*, *40*) typical loess appears on slopes and K l i m a s z e w s k i mentions typical loess in valleys.

R o k i c k i holds that typical loess exhibits a greater thickness than loess-like formations.

On the support of extensive studies by Soviet experts, I. P. G e r a s i m o v believes the wide spread opinion of the occurrence under the shape of covers to be entirely devoid of foundations (14, p. 103).

Save R o k i c k i, all the authors seem to be aware of the fact that loess originated in periglacial environment. However, the role of this environment and the relationship of its characteristic phenomena with the process of loess-formation have been given a largely varying attention.

The majority of authors — also out-side of Poland –– confines the correlation between loess and periglacial conditions merely to the statement of the periglacial situation of this formation. The term *periglacial* acquires thus only a spatial connotation.

K l i m a s z e w s k i and M a l i c k i also fail to give any closer attention to the essence of periglacial environment in their discussion of the problem of loess. Among the diverse phenomena characteristic of this environment they solely stress weathering and wind-action.

The works by J a h n and S a w i c k i distinguish themselvès in that respect. These writers regard the loess problem as intimately connected with a number of processes proper to periglacial environment. It should be also stressed that the relation of loess-origin to other periglacial phenomena is based by these authors on direct observations of the formation and not on general principles. They both rely upon data derived from the study of periglacial structures in loess and draw their conclusions on this support. These conclusions refer to the process itself as well as to the time and to the conditions both climatic and floral.

There are however marked differences between the individual lines of interest in periglacial environment studied in its connection to loess. S a w i c k i who has the merit of having been the first, apart from K r u k o w s k i (23), to appreciate in a large measure and still before the war, the importance of periglacial environment for the problem of loess, accentuates in his latest work the morphogenetic processes. J a h n is mainly concerned with climatic conditions and chronology.

The origin of the material plays a predominant part in the problem of the genesis of loess. R o k i c k i regards it as a residual or else as a weathering displaced product. M a l i c k i sees the source of loess material in river-alluvial which partly represent an accumulation of periglacial weathering-products. According to his view, areas of eolian deflation and accumulation are rather topographically than geographically differentiated.

J a h n, K l i m a s z e w s k i and S a w i c k i point to the geographical specifity of the terrains of deflation and accumulation. S a w i c k i does not pay any closer attention to the question. Neither does K l i m a s z e w s k i who discusses the area of Podlasie and the adjacent northern tracts, where loess-material was mainly prepared by periglacial weathering.

J a h n demonstrates the peculiarity of the areas of deflation and accumulation on the evidence of the belt-like occurrence of loess and on the support of the analysis of the loess-material. His interpretation of the role of heavy minerals differs from that by M a l i c k i (11); instead he demonstrates the presence of northern material. Deflation occurred in lowland terrains which provided material directly represented by glacial and fluvial deposits as well as by products of periglacial weathering. He assumes besides the participation of local weathering formations.

The research by the present writer and K l a t k a (*7, 8, 10*) are devoted to areas of deflation. They brought evidence of the presence on these terrains of loess-like formations and discussed in detail the mechanism of the processes involved in the formation of the loess-fraction.

In Poland as well as in other countries, the attention of workers dealing with the problem of loess is mainly focussed upon such processes as transportation and accumulation. Transportation may be eolian, fluvial, or due to mass-movement.

Eolian transport is not as a rule a matter of direct observation. Conjectures are based upon assumptions relating to the distribution of the atmospheric pressure prevailing at the time of the formation of loess and on the resulting wind-directions as well as on the geographical and topographical situation of the loess-deposits. Conclusions directly derived from loess-deposits are also known. They are deduced from such features as the structure, texture and origin of the material. Such evidence however is frequently of too general a character, it lacks the necessary precision and even clarity and thus leads to divergent interpretation. The „massive-structure" of eolian loess and the mineralogical composition of loess-material — may serve as examples (*11; 21*, p. 280). The most direct argument may be found in the works by S a w i c k i which concludes to the eolian transport on the evidence of the characteristics of the grains of sand found in loess and of the morphology of the layers in the formation (*40*).

Among pre-war contributions, J. T o k a r s k i's study (*44*) deserves special mention. It is mainly devoted to the genesis of loess. The author's conviction of the eolian origin of loess is based upon direct and detailed inguiries into its texture. His investigations followed the line indicated by his own basic methodological principle. For, T o k a r s k i is quite right in saying that every geological structures bears the imprint of its genetic processes, the legability of which depends upon the application of adequate methods (p. 21).

The part played by water in the accumulation of loess in a general sense is almost universally accepted. Evidence of the action of running water is more direct than the assumption of eolian transportation. Attention has been attracted to river--waters (*25*) and unorganized waters. K l i m a s z e w s k i explicitly refers to a process of down wash. In other papers the character of this process lacks precise definition.

The same may be said of mass-movement the character of which requires penetrating investigations in order to become precisely determined. The research by H a l i c k i and S a w i c k i (*17*), J a h n (*21*), P o ż a r y s k i (*31*) and S a w i c k i (*40*) led to the recognition of congelifluction in loess deposits. The correlation of this process with the problem of the origin of loess is discussed by these authors mainly from the view-point of climatic conditions and of chronology.

Among the achievements of the most recent loess-literature the first place should be given to the noticeable change in scientific approach. J a h n remarks that earlier investigations were chiefly concerned with the problem of stratigraphy. Genesis was considered in Poland as a matter of general knowledge the explanation of which could be found elsewhere. In face of such attitude the problem of the origin of loess was threatened by the danger of becoming a matter of dead routine.

In a number of the most recent papers (*21, 25, 34, 39, 40*) the problem of the genesis of loess assumes an outstanding position; the writers are trying to solve it by means of diverse methods. They show besides a marked tendency to base this solution upon direct observation and not upon the application of general ideas.

This dominant scientific attitude raises the problem of the method of work. The largest application should be given to such methods which arise from the necessity of a possibly intimate and direct study of the loess-deposits. S a w i c k i is right in saying that (40) natural exposure or even artificial ones, if not produced for scientific purposes, are irrelevant. Observations derived from the surface of an exposure which has been especially prepared by careful cleansing of the wall and those made without such previous operation, are largely divergend. The legitimity of this assumption is confirmed by research of J a h n and R o k i c k i, but foremost by that of S a w i c k i himself, who owing to the continual application of his basic and simple method succeeded in proving the complexity of the set of processes which contributed to the accumulation of loess.

The importance of this preparatory method consists in the fact that it represents a necessary condition of the recognition of structure and texture of the loess. Thorough inquiry into the structure of the loess-deposit leads to the recognition of the processes which contributed to its accumulation and to its subsequent transformations. It throws light upon the climatic conditions under which these processes operated and permits to draw conclusions relating to chronology.

The application of this method in the areas which may be assumed to be the parent-areas of the loess-material permitted to recognize the processes involved in the formation of the corresponding fraction of the grains which may have undergone eolian transportation (7, 8, 10).

Thorough investigation of the loess-deposits led directly to the recognition of their disturbance by frost-action and yielded direct evidence for the inference of their relation to periglacial environment, in which the process of loess-formation took place. Owing to the research by H a l i c k i and S a w i c k i (17), J a h n (21) and S a w i c k i (40) who discovered in loess the presence of congelifluction, frost--heaving, frost-fissures and ice-wedges the idea of periglacial environment in the problem of the genesis of loess has assumed a new and fuller sense. For it thus ceases to be a mere spatial notion determined by general climatic coditions. These writers have stressed the morphogenetic characteristics of this environment, especially in connection with the problem of the origin of loess.

The statement of the correlation of loess with other periglacial phenomena is to be considered as the greatest achievement in the study of this problem. It opens a new line of investigation which cannot be neglected in further study. The meaning of this remark refers both to the study of loess-deposits and to that of the terrains of origin of its material.

Closer, more penetrating investigation of loess especially with regard to structure has created favourable conditions for a more precise differentiation of the loess--deposits. Apart from the so-called „typical loess" several categories of loess-like formations (25, 31, 34) were distinguished as well as diverse loess-facies (21). The merit of these distinctions is to have widened the knowledge of these processes and of the conditions under which they were active.

The survey above has outlined the present-day problematics of the genesis of loess in Poland as well as the progress achieved in that line. But at the same time, on that background all the problems which are still open, which require further study, greater precision and insight, are being brought to light.

9 — Biuletyn Nr 1

The notion of „typical subaerial loess" raises many objections. The very basis of the distinction of this formation seems doubtful. The criteria applied are not characterized by indisputable exclusivity; besides, not all the workers apply them with the same meaning. Finally one may have some suspicions — which seem to be not altogether devoid of foundations — as to whether the recognition of these characteristic traits is invariably based upon thorough investigations.

The very structure of „typical subaerial loess", seems rather enigmatic. The majority of the authors are in favour of the massive structure which shows neither stratification, lamination nor their traces. S a w i c k i states however the presence of layers and laminations in the loess of Zwierzyniec which he regards as indisputably „typical subaerial loess". So does P r ó s z y ń s k i (*33*, p. 329) while describing the bedding of typical loess. However lack of stratification appears in some sedimentary rocks cf remarkable thickness when the material is homogeneous and identical conditions of sedimentation were longlasting.

All these authors confine the occurrence of „typical" loess of massive structure to top-portions reaching about 2 m. from the surface. The present writer examined about 60 exposures in the region of Lublin, Sandomierz, Miechow as well as in the Carpathians and in Silesia. He nowhere noticed loess that would be unstratified, unlaminated or devoid of pseudo-layers, save in top-portions of a thickness which practically never exceeded 1,5 m.

The capacity to split along vertical   joints is generally   believed to be the characteristic and distinctive feature of „typical" loess. One should however remember that this feature characterizes also such deposits which, on account of their stratification, lamination and other properties, are being regarded as loess-like formations. Moreover, vertical jointing may be found also in boulder clays, fine sands and other fine-grained formations. This characteristic grows together with the increase of $CaCO_3$. R u s s e l l (*35*, p. 17) holds that vertical jointing is induced by tensional stresses associated with down-slope creep and settling.

The criterion of mechanical composition does not seem more decisive for it is mainly based upon the analysis of samples of such material which had been beforehand classified according to the insufficiently distinctive characteristics mentioned above. Besides the material hitherto published is still too scanty.

Hence, one may venture to express some doubts as to the existence of „typical subaerial loess". Its characteristics seem to be delusive and the best one may say is that „typical" loess i. e. such which displays all the features that are believed to characterize it, is extremely rare. Would this not prompt the paradoxal conclusion that really typical is the „non-typical" loess? Would it not thus suggest the inference that the accumulation of loess is a complicated process and that, accordingly, the different forms under which loess occurs (together with loess-like formations) illustrate the predominance of the individual agents which took part in this intricate process?

The answers to these questions are intimately connected with the basic problem of the eolian process and of its exclusivity in the origin of loess. The structure of the loess deposits — as far as it is known to-day — does not provide any foundations which would permit to conclude to eolian accumulation. The belief in such accumulation is rather derived from general principles which, are moreover, not always formulated with the same meaning. It is as yet impossible to know whether accumulation resulted from suspension in tranquil air or if loess material was deposited by wind. If the latter of these two hypotheses were the right one, loess should display a structure reminiscent of that at dunes with only those modifications which would result from the lower value of the angle of repose of the material.

The question has not as yet been discussed from this view-point. The only exception is in that respect the study by S a w i c k i who concludes to wind-action 'on the support of the direct evidence of flutes and grooves he had opportunity to observe (40). The line of investigation thus indicated is highly interesting from the methodological view-point and deserves to be followed. Nontheless, the inferences drawn raise some doubts. There is no certainty as to whether these forms are really eolian. Similar forms arise in result of the action of running water. This can be decided only after closer study of these microforms, of their size and foremost of their direction, which the author failed to note.

However readily legible traces of other processes are generally inscribed into the other loess-exposures. These are slope-processes, foremost congelifluction. In the profile of Zwierzyniec, which was already published one may clearly see a series of congeliflual horizons, which S a w i c k i points out in the text. The action of down-slope processes is still more visible in the profile which runs vertically to the published one and follows the sloping. The influence of congelifluction and other down-slope processes, mainly down-wash is also distinctly marked in many other exposures, provided the walls have been carefully prepared and cleaned.

Without entering for the moment upon a discussion on the question of eolian origin, one is bound to state, that the present-day loess deposits bears foremost traces of others processes. Examination of loess-exposures has not led hitherto to conclusions confirming with absolute certainty the theory of eolian accumulation of loess in the places it occupies at present [1]).

Further studies should follow the line laid down by previous research and try to free themselves more and more from all routine and rigid dogmatism. They must try to devise and develop methods of thorough investigation of tangible evidence. Apart from the necessity of perfectioning and of a broader application of petrographic methods, e.g. T o k a r s k i 's paper (44), the vital thing is to gain fuller knowledge of the morphogenetic characteristics of periglacial environment.

The importance of the above mentioned lines of study with regard to the problem of loess is not only due to the fact that loess represents a specific instance of petrographic research, but foremost to its being a spectacular document of the complexity of morphogenetic processes. Hence the study of loess requires geomorphological methods, and, in turn, the study of this phenomenon assumes the role of an active process which contributes to the formation of the geomorphologist.

Translated by T. Dmochowska

---

[1]) Together with professor A. J a h n the present writer conducted investigations in a series of exposures near Ostrowiec and Sandomierz. On having discussed these exposures they both agreed upon the above conclusion.

BIBLIOGRAPHY

1. A m b r o ž V. — Spraše pahorkatin (summary: The loess of the hill countries). · *Sborník Státního Geol. Ústavu Rep. Československé*, sv. 14, 1947.
2. B e r g L. S. — O proischożdienii lessa (The origin of loess). *Izw. Gieogr. Obszcz.*, 1916.
3. B e r g L. S. — O poczwiennoj tieorii obrazcwanija lessa (About the soil theory of the origin of loess). *Izw. Gieogr. Inst.*, 1926.
4. B e r g L. S. — Less kak produkt wywietriwanija i poczwoobrazowanija. *Trudy II Mieżdunarodn. Konf. A.I.C.P.J.*, 1932.
5. B r y a n K. — Glacial versus desert origin of loess. *Am. Jour. of Sci.*, vol. 243, 1945.
6. B ü d e l J. — Die Klimazonen des Eiszeitalter. *Eiszeitalter u. Gegenwart*, Bd. 1, 1951.
7. D y l i k J. — The loess-like formations and the wind-worn stones in Middle Poland. *Bull. Soc. Sci. Lettr. Łódź*, vol. III, 3, 1951.
8. D y l i k J. — Głazy rzeźbione przez wiatr i utwory podobne do lessu w środkowej Polsce (summary: Wind-worn stones and loess-like formations in Middle Poland). *Biul. Państw. Inst. Geol.*, 67, 1952.
9. D y l i k J. — Periglacial investigations in Poland. *Bull. Soc. Sci. Lettr. Łódź*, vol. IV, 2. 1953.
10. D y l i k J., K l a t k a T. — Recherches microscopiques sur la désintégration périglaciaire. *Bull. Soc. Sci. Lettr. Łódź*, vol. III, 4, 1952.
11. D o b r z a ń s k i B., M a l i c k i A. — Rzekome loessy i rzekome gleby loessowe w okolicy Leżajska (Summary: Pseudo-loesses and pseudo-loess soils in the environment of Leżajsk). *Ann. Univ. M. Curie-Skłodowska*, sectio B, vol 3, 1948.
12. F i s k H. N. — Loess and Quaternary geology of the lower Mississippi valley. *Jour. Geol.*, vol. 59, 1951.
13. G a n s s e n R. — Die Entstehung und Herkunft des Löss. *Mitt. Preuss. Geol. L. A.*, 1922.
14. G i e r a s i m o w I. P. — K woprosu o gienezise lessow i lessowidnych otłożenii (summary: On the question of the genesis of the loess and the deposits of the loess type). *Izw. Ak. Nauk SSSR, s. Gieogr. i Gieofiz.*, 1939.
15. G i e r a s i m o w I. P., M a r k o w K .K. — Czetwierticznaja gieołogia (Quaternary geology). Moskwa 1939.

16. G i e r a s i m o w J. P., M a r k o w K. K. — Liednikowyj pieriod na tieritorii SSSR (Ice age in the area of USSR), 1939.

17. H a l i c k i B., S a w i c k i Ludw. — Less nowogródzki (Loess of Nowogródek). *Zbiór prac poświęcony E. Romerowi*, Lwów 1934.

18. H o b b s W. H. — Wind, the dominant transportation agent within extra marginal zones to continental glacier. *Jour. Geol.*, vol. 50, 1942.

19. H o b b s W. H. — Glacial studies of the Pleistocene of North America. Ann Arbor, 1947.

20. H o l m e s Ch. D. — Origin of loess — a criticizm. *Am. Jour. of Sci.*, vol 242, 1944.

21. J a h n A. — Less, jego pochodzenie i związek z klimatem epoki lodowej (summary: Loess, its origin and connection with climate of the glacial epoch). *Acta Geol. Polonica*, vol. 1, 1950.

22. K l i m a s z e w s k i M. — Polskie Karpaty Zachodnie w okresie dyluwialnym (Polish West Carpathian during Ice Age). *Prace Wrocł. Tow. Nauk.*, Ser. B, nr 7, 1948.

23. K r u k o w s k i S. — Paleolit Polski (Paleolith of Poland). *PAU, Encyklopedia Polska*, t. 4, cz. 1, dz. 5, 1939-1948.

24. L e i g h t o n M. M., W i l l m a n H. B. — Loess formation of the Mississippi valley. *Jour. Geol.*, vol. 58, 1950.

25. M a l i c k i A. — Geneza i rozmieszczenie loessów w środkowej i wschodniej Polsce (summary: The origin and distribution of loess in central and eastern Poland). *Ann. Univ. M. Curie-Skłodowska*, sectio B, vol. 4, 1950.

26. M ü n i c h s d o r f e r F. — Der Löss als Bodenbildung. *Geol. Rundschau*, Bd. 17, 1926.

27. O b r u c z e w W. A. — Loess types and their origin. *Am. Jour. of Sci.*, vol. 243, 1945.

28. O b r u c z e w W. A. — Less kak osobyj wid poczwy, jego gienezis i zadaczi jego izuczenia. *Woprosy Gienezisa i Gieografii Poczw.*, Akad. Nauk SSSR, 1948.

29. P o s e r H. — Die nördliche Lössgrenze in Mitteleuropa und das spätglaziale Klima. *Eiszeitalter u. Gegenwart*, Bd. 1, 1951.

30. P o ż a r y s k a K. — Stratygrafia plejstocenu w dolinie dolnej Kamiennej (summary: Pleistocene of Lower Kamienna). *Biul. Państw. Inst. Geol.*, nr 52, 1948.

31. P o ż a r y s k i W. — Plejstocen w przełomie Wisły przez Wyżyny Południowe (summary: The Pleistocene in the Vistula gap across the Southern Uplands). *Prace Inst. Geol.*, t. 9, 1953.

32. P r o š e k F., L o ż e k V. — Výzkum sprašového pokryvu v Sedlci u Prahy. *Věstník Ústředn. Ústavu Geol.*, r. 27, 1952.

33. P r ó s z y ń s k i M. — Spostrzeżenia geologiczne z dorzecza Bugu (summary: Notes sur la géologie du bassin de la rivière Bug). *Biul. Państw. Inst. Geol.*, nr 65, 1952.

34. R o k i c k i J. — Less i utwory pyłowe Wzgórz Trzebnickich (summary: Loess and pelitic deposits of Trzebinica Hills). *Biul. Państw. Inst. Geol.*, nr 65, 1952.

35 R u s s e l l R.J.—Lower Mississippi valley loess. *Bull. Geol. Soc. Am.*, vol. 55, 1944.

36. S a m s o n o w i c z J. — O loessie wschodniej części Gór Świętokrzyskich (summary: Sur le loess dans la partie orientale des montagnes de S-te Croix). *Wiad. Archeologiczne*, vol. 9, 1924—1925.

37. S a m s o n o w i c z J. — Objaśnienia arkusza Opatów. *Państw. Inst. Geol.*, 1934.

38. S a w i c k i Ludwik — Sur la stratigraphie du loess en Pologne. *Roczn. P. T. Geol.*, t. 8, 1932.

39. S a w i c k i Ludwik — Les conditions climatiques de la période d'accumulation du loess supérieur aux environs de Cracovie. *Sédimentation et Quaternaire, France*, 1949.

40. S a w i c k i Ludwik — Warunki klimatyczne akumulacji lessu młodszego w świetle wyników badań stratygraficznych stanowiska paleolitycznego lessowego na Zwierzyńcu w Krakowie (summary: Les conditions climatiques de la période de l'accumulation du loess supérieur aux environs de Cracovie). *Biul. Państw. Inst. Geol.*, nr 66, 1952.

41. S c h ö n h a l s E. — Über fossile Boden im nichtvereisten Gebiet. *Eiszeitalter u. Gegenwart*, Bd. 1, 1951.

42. S w i n e f o r d A., F r y e J. C. — A mechanical analysis of wind-blown dust compared with analyses of loess. *Am. Jour. of Sci.*, vol. 243, 1945.

43. S w i n e f o r d A., F r y e J. C. — Petrography of the Peoria loess in Kansas. *Jour. Geol.*, vol. 59, 1951.

44. T o k a r s k i J. — Fizjografia lessu podolskiego, oraz zagadnienie jego stratygrafii (in German: Physiographie des podolischen Lösses und das Problem seiner Stratigraphie). *Mém. Acad. Polonaise d. Sci. et Lettr.*, Cl. Sci. Math. Nat., S.A., 1936, no 4, Cracovie 1936.

45. U r b á n e k L. — Aeolické sedimenty katastru kolínského (summary: Les dépôts éoliens du cadastre de Kolín). *Sborník Státního Geol. Ústavu Československé Rep.*, sv. 16, 1949.

22

Reprinted from *Rocznik Pol. Tow. Geol.*, 31(2–4), 250–258 (1961)

REMARKS ON THE LOESS

Julian Tokarski

in collaboration with W. Parachoniak, W. Kowalski, A. Manecki and B. Oszacka

Abstract. Anemoclastic loess is an important historical document not only of Polish diluvium. The dust of this rock, blown by the wind from above the glaciers, was deposited in periglacial zones in secluded places where the strength of the wind dropped to its minimum. Its origin is connected with the barometric gradient caused by the formation of the glacier plateau. It follows logically that the presence of each glacier is connected with the formation on its foreland of loess deposit. In other words the amount of loess depositions corresponds to the amount of glaciation on a given area.

The author began a broad methodical investigation on Polish loess. In the first phase of research the 1928 phenomenon of dust falling which had been carried by east winds from the Azov steppes towards the eastern territories of Poland was used. At that time both the kind of falling matter and its granulometric differentiation were determined according to the distance from its starting place. Characteristic isarythms of the size of the dust grain, called „isograns", were drawn on a special map.

In the regional research of the Podole loess, using the methods already established, a result determining the presence of three kinds of loess in the area was obtained. The measurement of their grain-size variation made it possible to draw on the map isograns determining the diminishing of the average grain in the lower part of the outcrop in the direction from NW towards SE. Isograns of the upper level of loess showed a similar decrease from W to E. These differentiations of isograns could be caused only by the change of the wind direction. Numerous detailed analyses — chemical, microscopic and of heavy minerals — determined exactly the quality of loess material.

In the western part of Poland the Cracow loesses were analysed in the same way. This material appeared to be identical with the loess of Podole. From all the investigations resulted the main conclusion that p r i m a r y loess is characterized by extremely typical grain-size variation which can be determined by microscopic measurements. The quantity of the grain-size variation is illustrated by the „summation" curve shown mathematically by means of appropriate equations. Secondary loesses show in this respect different relations.

INTRODUCTION

Anemoclastic loess is — as is known — an important historical document not only of Polish diluvium. Long lasting disputes concerning its origin, at first based only on its megascopic description or on academic discussions, once caused confusion in the opinions as to its origin (Scheidig (1934), Quiring (1934), Vendel and others). At last, with the increase in the amount of field observation the thesis on the eolian origin of this rock was slowly established. The first to point to it on the basis of long studies in Central Asia was the German scientist Richthofen. The hypothesis was reinforced by the Russian scientist Obrutchev and some others so that at present any controversy about the eolian origin of loess is invalid. Nowadays it is generally accepted that loess material is rock dust blown out from desert centres on their foreland which may sometimes lie at a considerable distance.

The problem of the mechanics of loess sedimentation as an eolian deposit could be solved ultimately only on the basis of analysis of the most precise and various kind performed on the loess matter itself. There is no doubt that each fragment of any rock contains in its mineral and chemical composition and structure the history of its origin. The reading of this history is only a matter of using the correct methods of investigation. The methods in the field of petrography of all rock types have been sufficiently set for a number of years so that they might be applied to all types of materials classified on various tables.

In Poland loess appears — in accordance with its origin — chiefly in the southern areas. No doubt its original cover was fairly uniform. During the interglacial periods erosion and denudation factors managed to destroy it in many places owing to the known „flimsy" structure of loess rock which could not resist denudation.

In Poland this cover really forms only large and small „remnants" which do not always consist from primary loess or have not remained unchanged in its full profiles. If we add that loess sedimentation was several times re-laid in the course of diluvium history, and erosion of their profiles independently of this reached various depths, in the historical analysis of this rock there occurred a new difficulty in distinction, as for example older loess from younger. Therefore in order to solve the problem very precise quantitative and qualitative methods had to be used to determine that in any case we are without doubt dealing with material of primary loess or with a loess of first, second and third sedimentation. Without this approach to the problem it would be difficult to solve the historical geological role of the rock during diluvium.

A good field for research of loess cover was the area of Podole bounded in the north by the so called „Podolian border" and in the south by the river Dniestr. The glaciers never moved into this high area, dominating with rising land over the northern depression of the river Bug. This country was then a foreland of glaciation untouched by glacial erosion, and its loess profiles did not suffer such radical destruction as in the north and west, where the continental glaciers moving southwards repeatedly partially or even completely destroyed the deposits over which they moved.

In the whole of Podole loess covered the surface of older prediluvial beds and its thickness often reaches here 30 m (e. g. in Halicz). An important impulse in explaining the origin of the rock in Podole was following accident.

On 29th April 1928 the inhabitants of Lwów witnessed a mysterious and powerful phenomenon of nature. Throughout the day a brownish dust fell on their heads and even caused panic here and there because of the extraordinary solar eclipse. At the beginning two of the professors of the University (R o m e r and A r c t o w s k i) tried to explain the phenomenon in the newspapers by the eruption of a volcano not very far from the town. This of course increased the anxiety. Their opinion was not supported by the investigation of the material in question. An immediate analysis made by K. S m u l i k o w s k i, assistant at the Chair of Mineralogy at the Lwów Technical College showed in a few hours that the dust matter falling from the air on the area of Podole was brown soil which had been blown by eastern winds and which reached our territories as the result of a powerful hurricane on the Azov steppes. Three important facts were then determined: first, that the matter was of anemoclastic origin, second, that the place it came from was recognized, and third, that the dust settled on Podole at a distance which could be precisely estimated in kilometres from the place where it was picked up by the winds. These very facts appearing as *experimentum cruçis* caused the author to explain the mechanics of dust falling and at the same time gave the natural bases for study on loess. A special expedition was organized which collected samples of the dust from 72 places, exactly determining these places topographically; the samples underwent two analyses — mechanical and physicochemical. The former by microscopic measurements determined the average size of dust grains (chiefly of quartz), the latter the quantities of colloidal components (clay) by adsorption of methylene blue. Finally two types of isarythms were drawn. One, called isograns showed distinctly the decrease of dust grains westwards (1 micron per 50 km). It was the expression of logical consequence which proved the decrease in the average grain with the increase in the distance from its starting place. The isarythms of adsorption of methylene blue called isoadsorbents behaved contrary to isograns i. e. the quantity of colloidal components in the blown soil, also in accordance with the logical postulate, increased westwards. The final result of the investigation on dust in Podole in 1928 was that a fairly precise method of investigation of anemoclastic matter was determined which in consequence indicated even the direction of winds (J. T o k a r s k i 1936). No wonder then that the carrying out of the analytical experiment with dust was immediately followed by a regional and quantitative elaboration of the physiography of Podole loess.

PODOLE LOESS

In order to solve rationally the problem of loess it was necessary to collect samples from loess profiles consistently in a uniform way from various places in Podole. It is obvious that the samples had to correspond to primary loess. Because of the fact mentioned above that such loess undoubtedly underwent to a certain degree mechanical and chemical degradation, it was necessary when collecting samples to find places where the degradation was very slight. It was logically deduced that places where loess was to be found in its primary form would appear in morphological culminations of the area. There, though the chances of degradation of loess matter were the greatest, the rock could form „remnants" of the primary deposit. Preliminary field investigation confirmed this opinion. The profiles performed in the culminations, even quite deep ones, sometimes disclosed the primary form of the rock with all its typical characteristics, i. e.: uniform grain, lack of stratification or any addition of sand etc. Only eight such profiles were investigated in detail because of the lack of means.

In order to establish the methods of investigation an 11 metre primary profile of loess in Grzybowice near Lwów was chosen in the preliminary research. The results of the methodical work were as follows:

1. Loess samples collected at 1 metre intervals along profiles were analysed microscopically, mechanically, and chemically.

2. Qualitative microscopic analysis determined their mineral composition in general. Quantitative microscopic analysis showed the mechanical composition of the samples (characterized by the measurements of the greatest diameters of grains which appeared in the field of vision of the microscope) by means of an appropriate summation curve for which the following general equation was determined in the form (Fig. 1):

$$y = A - a \cdot x + B (x - 3)^n$$

where $A = 100,3$, $a = 14.8$, $B = 3.864$, $n = 1.38$.

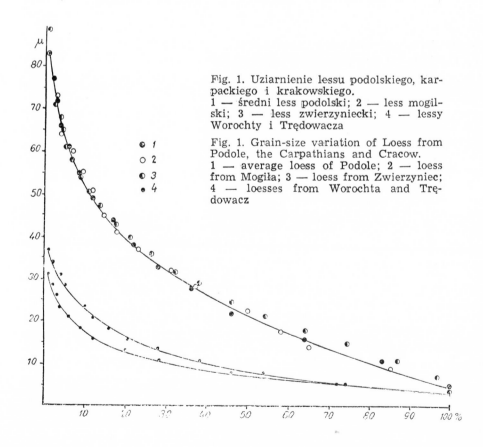

Fig. 1. Uziarnienie lessu podolskiego, karpackiego i krakowskiego.
1 — średni less podolski; 2 — less mogilski; 3 — less zwierzyniecki; 4 — lessy Worochty i Trędowacza

Fig. 1. Grain-size variation of Loess from Podole, the Carpathians and Cracow.
1 — average loess of Podole; 2 — loess from Mogiła; 3 — loess from Zwierzyniec; 4 — loesses from Worochta and Trędowacz

3. Chemical brutto and fractional analyses of ten samples characterized exactly the investigated zones of the rock.

4. The loess of Grzybowice cannot be genetically connected with tertiary sand from the vicinity of Lwów as was shown by comparative analyses.

5. Heavy minerals contained in the loess of Grzybowice analysed in ten zones showed the presence of a great quantity of three most important components: iron and titanium oxides, garnet, and zircon. The quantity relation of these components in the profile was variable. Analytic details of these investigations may be found in the author's paper (J. Tokarski 1935).

After having established the efficiency of the investigation methods on the basis of loess from Grzybowice, the regional investigation of the rock of Podole near Lwów began. The mechanical composition was investigated in eight places at one metre intervals by means of the microscopic method. It soon became evident that on certain zones of investigated profiles which were not of equal thickness characteristic culminations appeared of the size of an average grain; lover culminations showed larger grains in general (with the exception of the profile in Halicz). As the lower zones of the investigated profiles were the oldest and

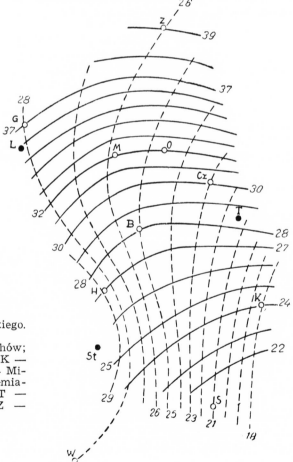

Fig. 2. Izograny lessu podolskiego. Isograns of Loess of Podole.
B — Brzeżany; Cz — Czerniechów; G — Grzybowice; H — Halicz; K — Kopyczyńce; L — Lwów; M — Mitulin; O — Opaki; S — Siemiakowce; St. — Stanisławów; T — Tarnopol; W — Worochta; Z — Zboryszów

least degraded — which was logical — their culminations were put to-gether into the first picture of isograns. The second culmination was treated in the same way. The third culmination could not be presented consistently by means of a graph of isograns as the corresponding upper zones of loess underwent erosion in most cases. On the included table (Fig. 2) are shown the isograns for the first culmination (from the bot-tom) and of the second one. The former shows distinctly the decrease of the grain from the NW towards the SE. This extremely characteristic differentiation in the size of loess grain in this direction could be caused, as in the dust grain of 1928, by the winds blowing from the NW and weakening towards the SE. The second culminations assembled in the isograns give a different picture. Here the decrease of the grain appears in the direction from W to E.

The following conclusions result from the entire regional investiga-tion of the Podole loess:

1. The characteristic physiography of the primary loess of Podole was established on the basis of 23 precise chemical analyses of various sam-ples. Besides this the mineral nature of the clayey component isolated in its pure form was precisely analysed, and on the basis of microscopic measurements of the diameters of quartz grain carried out on 114 samples the characteristic dispersion of the size of its quartz components was shown. In the course of research about 9 million planimetric lines were counted in the microscope. The control of the mineral composition was made by means of adsorption of methylene blue. All the various kinds of investigation determined without doubt the mineralogical nature of the loess rock of Podole.

It should be pointed out that these tiring and precise petrographic investigations had no precedent. As anemoclastic material loess of Podole could be formed by the blowing of dust components out of the sands gathered on its hinterland over a large area.

2. The methods used for investigation made it possible to distin-guish exactly primary from the secondary loess. Primary loess may be precisely differentiated from the secondary matter (loess-like) already on the basis of microscopic determination of the mechanical composition.

3. The regional investigation of the loess rock in Podole facilitated the exact determination of the direction of diluvial winds. It was model-led on the investigation of the dust fall in 1928.

4. The presence in the loess profiles of Podole of three culminations of the average size of grain confirmed the conclusion concerning the presence of three loesses of Podole parallel to three glaciations.

5. In the investigated eight loess profiles of Podole „fossil soil" was not found. This is quite accidental. These soils, which in the interglacial periods should be formed on the loess cover, were in a certain already cold period destroyed by erosive factors.

The details concerning the research on loess in Podole were monogra-phically elaborated in the PAU publications (J. Tokarski 1936).

All the above mentioned results of the research on loess encouraged without doubt the extending of the investigations to the territories outside Podole. The collection of materials on the territories west of Podole was entrusted to Professor A. Malicki, a co-worker of the Depart-ment of Mineralogy and Petrography at the University of Lwów. Pro-fessor Malicki collected extensive material from these areas which.

though assembled in the laboratories could not be analysed because of the outbreak of the second world war.

The Lwów investigations determined the following important physiographic characteristics of the primary loess:

1. The loess rock contains an extremely characteristic and constant individual mechanical composition which could be shown exactly in the microscopic measurements by means of a summation curve. This was possible because the investigation of the characteristic in this way determined a uniform dispersion of grain size in numerous samples collected in many places in Podole. Thus the statement in a given case whether we are dealing with a sample of primary loess should be made on the basis of determination — an easy one — in the microscope of a sufficient number of grain sizes. All the loesses on which erosive factors have worked lose to a certain extent various fractions (especially the dusty ones), thus giving a different curve of grain dispersion.

2. The chemical composition of the loess of Podole, investigated precisely in 23 samples, showed the following variability in percentages of five main components, calculated into 100:

$$SiO_2 = 81.05 — 89.00, \quad Al_2O_3 = 4.37 — 8.65, \quad Fe_2O_3 = 1.21 — 4.95,$$
$$CaO = 1.17 — 6.24, \quad MgO = 0.64 — 1.75.$$

In the analyses not calculated in 100 K_2O varied from 1.35 to 2.14, $Na_2O = 0.77 — 0.96$. The full average analysis of this loess gave the following figures:

$$SiO_2 = 71.67, \quad R_2O_3 = 10.35, \quad CaO = 5.73, \quad MgO = 1.12, \quad K_2O = 1.83,$$
$$Na_2O = 0.84, \quad \pm H_2O = 2.74.$$

3. The rational chemical analysis performed on the sample of loess from Grzybowice (zone 1) showed that in hydrochloric acid 5.59% components dissolve in weight percentages, of which:

$$SiO_2 = 1.33, \quad Al_2O_3 = 1.54, \quad Fe_2O_3 = 2.18, \quad CaO = 0.24, \quad MgO = 0.30,$$

and in sulphuric acid 2.48%, of which

$$SiO_2 = 0.06, \quad Al_2O_3 = 1.56, \quad Fe_2O_3 = 1.83, \quad CaO = 1.10, \quad MgO = 0.13.$$

The alkalis in the amount: $K_2O = 1.31$, $Na_2O = 0.78$ also passed into the solution. As the quantity of components soluble in muriatic acid from any clastic rock throws some light on its genetic relations (hydration and hydrolysis of the primary material) the given results of the rational analysis of loess here characterize this rock.

4. The average size of loess grain in the investigated profiles in Podole consequently decreases from bottom to top though not constantly. The profiles investigated for mechanical composition are divided into sections in which the size of grains increase and after reaching culminations at a certain point consistently decrease. These culminations are extremely characteristic for the primary loesses. When all the investigated profiles of Podole showed the same characteristic in this respect the phenomenon could be caught synthetically by drawing the isarythms of selected culminations. These isarythms — as was mentioned above — were called isograns. From the dispersion of isograns it appeared that the loess culminations in various profiles increase in the lower parts of the profile from NW towards SE and in the upper (culmination 2) from W towards E. The explanation of this phenomenon was based on the pre-

vious investigation of the dispersion of dust grain from the year 1928. The conclusion was that only the wind direction could decide on the regional variability of grain culmination.

The following important conclusion concerning the mechanical composition of loess appears after these considerations. The investigation in Podole established that the average size of grains of the loess rock varies in the c u l m i n a t i o n s from 39 to 18 microns. It is clear that in the course of the given profile most of the averages contain smaller grains. One should bear in mind that one micron is one thousandth of a millimetre. Thus the mass of one grain is very small. We know well that dust can float even in calm air for quite a long time (the eruption of the volcano Krakatau in 1883). The smallest dust will fall on the earth's surface only when, for instance, its electrically active surfaces forced by the humidity of the medium join into bigger or smaller aggregates. If the speed of air current is for instance only 6 cm/sec. loess dust cannot yet fall to the ground. Such „silence" in the air approaches the „absolute". In other words it means that loess dust in the periods of glaciation could fall on the foreland of the glaciers in the regions of almost „absolute silence". The zones of atmospheric „silence" are well known in hot climates, but we have not much knowledge about such conditions of atmosphere in the polar climate. The falls of loess dust are documentary proof that such „silence" in climates which were decided by glaciation are possible.

The problem mentioned above is undoubtedly connected with the interpretation of falls of plant „dust" which when blown out from the centres of phytosociological media had probably to follow the same laws of eolian dynamics as loess dust. The falls of such „dust" were widely considered recently as paleobotanical documents. Here we shall only point out that this organic „dust" has a considerably smaller specific gravity than the loess grains and that they can make use of various means of flying which must have caused the differentiation of their concentration in the places of falling.

[*Editor's Note:* Material has been omitted at this point.]

REFERENCES

K l i m a s z e w s k i M. (1948), Polskie Karpaty Zachodnie w okresie dyluwialnym. *Pr. Wrocł. TN. Ser. B* nr 7, Wrocław.

K u ź n i a r Cz. (1912), Less w Beskidzie Galicji Zachodniej. Sur les loess dans les Beskides de la Galicie occidentale. *Kosmos* 37 (za rok 1912), str. 671, Lwów.

O b r u c z e w (1933), Problema lessa. *Tr. Mieżdunarod. Konfer. Assocj. Izucz. Czetw. Per. Europy* 2, wyp. 2.

Q u i r i n g H. (1934), Die Unterscheidung von Löss und Hochflutlehm. *Z. prakt. Geol.* Jg. 42, H. 10 Halle.

S a w i c k i L. (1952), Warunki klimatyczne akumulacji lessu młodszego w świetle wyników badań stratygraficznych stanowiska paleolitycznego lessowego na Zwierzyńcu w Krakowie. *Biul. Państw. Inst. Geol.* 66. Z badań czwartorzędu w Polsce. T. 2, Warszawa.

S c h e i d i g A. (1934). Der Löss und seine geochemischen Eigenschaften. Dresden, Leipzig.

T o k a r s k i J. (1935), Studia nad lessem podolskim. I. Analiza petrograficzna profilu lessowego z okolic Grzybowic koło Lwowa. Studien über den podolischen Löss. I. Petrographische Analyse eines Lössprofiles aus Grzybowice bei Lwów. *Bull. intern. Acad. Pol. Sc.* nr 5 — 6 a, Kraków.

T o k a r s k i J. (1935), Beitrag zur Kenntnis der Hydroklastischen Elemente des Czeremoszgebietes. *Bull. intern. Acad. Pol. Sc.,* nr 1 — 2 a Kraków.

T o k a r s k i J. (1936), Fizjografia lessu podolskiego oraz zagadnienie jego stratygrafii. Physiographie des podolischen Lösses und das Problem seiner Stratigraphie. *Mém. Acad. pol. Sc.* nr 4, Kraków.

T o k a r s k i J. (1947), Ciężkie minerały jako wskaźniki stratygraficzne serii fliszowych. *Nafta* nr 9, Kraków.

T o k a r s k i J. & O l e k s y n o w a K. (1951), Specjalna analiza lessu okolic Mogiły pod Krakowem. Special Analysis of Löss from the environs of Mogiła near Cracow. *Bull. intern. Acad. Pol. Sc.,* Kraków.

T o k a r s k i J. (1951), Zasady termicznej analizy gleb. Principles of the thermal analysis of soils. *Bull. intern. Acad. Pol. Sc.,* Kraków.

T o k a r s k i J. (1954), Zagadnienie koloidów glebowych. *Post. Nauk. roln.* nr 5, Kraków.

T o k a r s k i J. (1955), Ilościowa charakterystyka składu mineralnego niektórych gleb lekkich okręgu krakowskiego. *Post. Nauk. roln.,* Kraków.

T o k a r s k i J. (1957), Minerały ilaste jako możliwe wskaźniki paleogeograficzne. *Biul. III Wydz. PAN.*

T o k a r s k i J. (1957), Studia nad koloidami gleb lekkich w Polsce, cz. I. *Rocz. Nauk roln.* T. 76-A-3, Warszawa.

T o k a r s k i J. (1960), Ilościowa charakterystyka mineralogiczna polskich gleb piaszczysto-ilastych cz. I. *Rocz. Nauk roln.* T. 80 A 4, Warszawa.

T o k a r s k i J. (1960), Sprawność nowej termoanalizy wagowej w ilościowych badaniach materii glebowej. *Rocz. Nauk roln.* T. 82-A-1, Warszawa.

W o l d s t e d t P. (1929), Das Eiszeitalter Grundlinien einer Geologie des Diluviums, s. 15, Stuttgart.

Editor's Comments
on Papers 23 and 24

Paper 23, dealing with the petrographic examination of loess material, is very important in that it is one of the few published papers to give very detailed comparisons of widely separated loess deposits. There is still a pressing need for this type of comparative work to be extended to other deposits. A very useful introduction to loess petrography is that by Guenther (1961); his book is essentially a practical manual and deals with techniques in sedimentary petrography applied to loess. The author is basically more interested in loess stratigraphy than loess lithology and genesis, and the petrographic information is sought with stratigraphic ends in mind. The book is one of a series of monographs called *Fundamenta* that deal with prehistory rather than geology.

Another German book (Richter et al., 1970) should be noted at this point. Loess investigations have been pursued in the German Democratic Republic but because of the lack of communication between East and West have been appreciated in the English-speaking world even less than the West German contributions. Supplement 274 to *Petermanns Geographische Mitteilungen* deals largely with the loess in the German Democratic Republic and gives some very useful petrographic information, and also detailed stratigraphic data.

Paper 24 is coupled with Paper 23 because it was published at around the same time and illustrates in a very graphic manner the difference between the Russian and American approaches to loess. The roots of the Russian *in situ* approach are shown by Lysenko to go back to Dokuchaev, whose prestige and influence in the Soviet Union are enormous. He set the scene for the *in situ* approach and its development was inevitable. There is only room for a short extract from Lysenko's important paper;

a full-length translation has been published by the British Library Lending Division (Boston Spa, England; translation RTS 8864).

REFERENCES

Guenther, E. W. 1961. Sedimentpetrographische Untersuchung von Lössen, Teil 1: Methodische Grundlagen mit Erläuterung an Profilen. Böhlau Verlag, Cologne, 91p.

Richter, H., Haase, G., Lieberoth, I., Ruske, E., and Ruske, R. (eds.). 1970. Periglazial-Löss-Paläolithikum im Jungpleistozän der Deutschen Demokratischen Republik. VEB Hermann Haack, Gotha, 422p.

23

Copyright © 1955 by the Society of Economic Paleontologists and Mineralogists

Reprinted from *Jour. Sed. Petrol.*, **25**(1), 3–7, 12, 21–23 (1955)

PETROGRAPHIC COMPARISON OF SOME LOESS SAMPLES FROM WESTERN EUROPE WITH KANSAS LOESS

ADA SWINEFORD AND JOHN C. FRYE[1]
State Geological Survey, University of Kansas

ABSTRACT

Fifteen samples of loess collected in 1952 in Italy, France, Germany, and Belgium were studied by mechanical and chemical analysis, X-ray diffraction, and light and electron microscopy. These data are compared and contrasted with similar data from 47 samples from Kansas and Nebraska. It is shown that in spite of the similarities of the loess of the two regions in field appearance, relationship, buried soil profiles, texture, and stratigraphy, the mineralogy of the European samples contrasts strongly with that of the Kansas samples. Abundant montmorillonite and volcanic ash shards in the Kansas samples are the most striking points of difference, but strong variations are also noted in other minerals of the coarse fraction as well as in other of the clay minerals.

INTRODUCTION

Loess, or massive well-sorted silt with some clay and very fine sand, is perhaps the most widespread of surficial deposits in the northern hemisphere. Although there is lack of general agreement concerning its origin—it is probable that deposits included by various workers under the name "loess" represent deposits of several origins—western Europe, and particularly the Rhine Valley, has long been a classic region for study of these massive silts. The central Great Plains of the United States (Kansas and Nebraska) contain more widespread and thicker deposits of surficial silts that bear a striking superficial similarity to the European deposits. We have had opportunity during the past decade to study in some detail the loess deposits of the central Great Plains (Swineford and Frye, 1951; Frye and Leonard, 1951); it is our purpose here to present petrographic data on samples from western Europe and to compare and contrast their properties with typical loess from Kansas and Nebraska.

Opportunity to examine loess deposits

and to collect samples in western Europe was made possible to Frye during the summer of 1952 by a grant from the University of Kansas Endowment Associa-

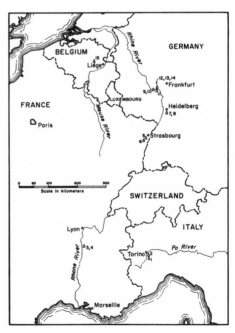

FIG. 1.—Map of a part of western Europe showing location of loess samples described.

[1] Present address: Illinois State Geological Survey, Urbana, Illinois.

tion and the generosity of Prof. Raymond C. Moore of the University of Kansas with whom he traveled. Data on 15 samples from 10 localities in Italy, France, Germany, and Belgium are reported here (fig. 1); data on 47 samples from 42 localities in Kansas and Nebraska have previously been published (Swineford and Frye, 1951). Field guidance in the case of all but one of the localities described here was in company of local authorities; the single exception was found by following maps and notes furnished by them. Ages were assigned, and collateral data in many cases were furnished by them, and their assistance is hereby acknowledged. In the Rhine Valley in the vicinity of Wiesbaden, Germany, we were guided by Prof. Dr. Franz Michels and Dr. Ernst Schonhals, Hessisches Landesant für Bodenforschung, Wiesbaden; at Mauer, Germany, by Dr. Arthur Rosler, Geologische-Paläontologische Institut, Heidelberg; in the vicinity of Strasbourg, France, by Prof. Dr. Georges Dubois and Dr. Paul Wernert, Laboratoire de Géologie, Université de Strasbourg; in the vicinity of Torino, Italy, by Prof. C. Socin, Istituto di Geologia, Torino; and in the vincity of Landen, west of Liége, Belgium, by Prof. F. Gullentops, Laboratoire de Géologie, Université de Louvain. Prof. F. Schaub, Naturhistorisches Museum, Basel, Switzerland, furnished maps and notes for the location of exposures in the vicinity of St. Vallier, France, along the Rhone Valley.

General Stratigraphy

Loess deposits of the northern hemisphere are predominantly of late Pleistocene age. In the Great Plains region loess deposits are Illinoian and Wisconsinan in age and are subdivided by the use of buried soils and molluscan faunas into three stratigraphic units: the Loveland, Peoria, and Bignell silt members of the Sanborn formation (Frye and Leonard, 1951; 1952). Of these three the Peoria is by far the thickest and most extensive; all samples used for comparison are from this early Wisconsinan unit.

Stratigraphic placement of the samples from western Europe is with respect to the standard section of the Alpine region. Most of the European loess samples described here are assignable to the Würm; however, sample 10 from the Rhine Valley is of a Chernozem soil developed in loess tentatively assigned to the Riss; sample 6 from Achenheim is assigned to the late Mindel or Riss; and samples 3 and 4 along the Rhone Valley are dated only by their position above a deep, mature soil developed in the top of deposits that have yielded diagnostic Villafranchian vertebrate fossils.

In general loess samples studied from both North America and Europe are of corresponding age, and are assigned to the last major episode of continental glaciation.

Laboratory Procedure

Mechanical analyses of the 15 samples were made by the same method used for the Kansas samples in 1950 (Swineford and Frye, 1951) in order to facilitate comparison. The results are shown in table 1, and a few cumulative size-frequency curves are plotted in figure 2.

Chemical analyses of the 15 European samples were made under the direction of Russell T. Runnels in the laboratories of the State Geological Survey of Kansas, and are shown in table 2, along with calcium carbonate content computed by Runnels.

The size fraction between 44 and two microns (in all samples except those from Torino) consists predominately of clay-coated quartz grains, plus calcite and feldspars in those samples which also contain calcite and feldspars in the coarser fractions. Micas and rock fragments are less abundant than in the fractions coarser than 44 microns. Most of the heavy minerals are within the 20–44 micron fraction.

The silt and sand fractions coarser than about 20 microns were examined with binocular and petrographic microscopes. Detailed observations were restricted to

TABLE 1.—*Mechanical analyses of samples of loess from western Europe and selected samples from Kansas and Nebraska*

No.	Location	Size distribution in phi units (percent by weight)														M_ϕ	σ_ϕ	Md_ϕ	QD_ϕ	Sk_ϕ	PD_ϕ
		−2−−1	−1−0	0−1	1−2	2−3	3−4	4−4.5	4.5−5	5−6	6−7	7−8	8−9	9−10	10+						
1	Italy				0.1	1.0	24.8	27.1	19.5	15.8	4.9	2.3	1.6	1.1	1.9	4.79	1.411	4.45	0.56	0.723	1.35
2	Italy			0.0	0.1	1.2	14.1	18.5	19.0	18.3	8.7	7.2	4.9	3.3	4.7	5.55	1.892	4.90	1.06	1.031	2.35
3	France		0.1	0.1	0.0	0.4	10.7	21.6	25.7	21.1	7.5	4.3	3.0	2.2	3.5	5.32	1.633	4.85	0.68	0.863	1.82
4	France			0.0	0.4	1.1	6.7	12.6	17.7	18.9	9.1	7.6	7.0	5.6	13.0	6.33	2.298	5.50	1.72	1.084	3.20 ++
5	France				0.1	0.2	0.6	3.5	8.3	26.7	16.5	4.8	3.7	3.1	28.4	6.55	2.274	6.55	2.62	1.082	3.58 +
6	Germany			0.2	1.0	1.7	1.1	7.2	21.7	34.1	11.8	4.8	3.7	3.1	9.7	7.37	1.982	5.45	0.85	0.878	2.73
7	Germany		0.7	0.7	0.4	0.5	2.2	10.8	20.1	32.2	11.9	6.8	5.0	3.7	5.9	6.03	1.959	5.45	0.92	0.965	2.75
8	Germany			0.6	0.3	1.1	3.4	12.1	19.3	31.7	14.1	6.2	3.2	4.3	5.0	5.45	1.785	5.50	0.91	0.840	2.27
9	Germany		0.2		2.4	2.3	4.3	9.5	21.2	27.5	10.5	4.9	4.9	2.4	18.2	6.00	2.175	5.00	0.94	1.048	2.83
10	Germany	0.1			1.4	3.8	1.6	14.2	16.1	21.6	10.7	6.2	3.1	2.4	12.7	5.76	2.482	5.70	1.88	0.919	3.43 ++
11	Germany		2.3	1.4	1.7	5.7	8.2	14.8	20.7	29.0	9.8	4.0	3.1	1.9	8.9	6.46	2.536	5.30	0.94	0.591	3.33 +
12	Germany	1.1	0.2	0.8	2.1	2.5	9.2	11.6	18.2	25.7	8.0	3.0	2.0	2.2	9.7	5.80	2.131	5.03	0.82	0.718	2.92
13	Germany	0.0	0.1	0.3	0.9	0.1	6.1	12.4	17.2	23.5	8.3	3.7	2.8	1.9	12.7	5.54	2.226	4.95	0.89	0.674	3.15
14	Germany				0.0	1.6	0.9	32.2	33.7	28.5	10.6	4.6	2.8	0.4	5.1	5.96	2.206	5.35	1.02	0.830	3.88 +
15	Belgium						18.9	28.9	22.6	34.5	6.5	2.5	0.6		1.7	5.54	1.568	5.05	0.52	0.938	1.58
	Nebraska[1]						3.1	11.9	28.0	16.4	4.2	1.5	1.7	1.7	1.7	4.69	1.164	4.45	0.46	0.619	1.16
	Kansas[1]						0.4	31.8	21.4	23.2	7.2	2.8	1.1	1.4	1.4	5.15	1.323	4.79	0.56	0.816	1.28
	Kansas[1]						7.5	17.6	30.7	38.3	14.8	5.8	1.7	1.2	1.7	5.72	1.376	5.43	0.68	0.632	1.52
	Kansas[1]						0.6		36.2	19.7	4.3	2.0	0.6		2.1	4.97	1.226	4.67	0.43	0.734	1.00
	Kansas[1]					0.1				34.1	5.6	2.1	0.5		1.8	5.22	1.176	4.94	0.44	0.663	0.92
	Kansas[1]				2.2	21.4	19.5	11.8	13.0	17.6	7.6	3.2	1.2	0.8	1.8	4.43	1.747	4.29	1.13	0.240	2.04

[1] Swineford and Frye, 1951

grains coarser than about 44 microns, because their mineralogy could be more readily identified.

X-ray diffraction patterns were made from <2 micron samples oriented on glass slides. Patterns were made in each case from the untreated clay and from acid-treated (cold 0.1 N HCl) clay. Selected patterns were also made from glycerol-wet preparations and from clay which had been heated to 550°–600° C. for 30 minutes. Diffractometer patterns were made with a GE XRD3 unit operated at a scanning speed of 0.2° 2θ per minute and using nickel-filtered copper radiation. Scanning was begun at 2° 2θ for several samples, but no low-angle superlattice reflections were observed. The observed d-values and estimated intensities are shown in tables 3, 4, 5, and 6. Some of the diffractometer patterns are shown in figure 3.

The <2 micron fraction characteristically consists of a complex mixture of layer silicates, calcite, and quartz. All samples contain significant quantities of illite and/or comminuted muscovite (and other micas) and most have smaller amounts of kaolinite. The most obvious differences among the loesses lie in the relative proportions of these minerals and in the presence or absence of a 14.7 Å mineral and of a mineral which expands on treatment with glycerol. As will be noted in descriptions of individual samples, some of these minerals have mixed-layer characteristics. The mineralogy of the loesses is so complex, however, that definitive interpretations would have doubtful validity.

In general, the observable effect of treatment with cold 0.1 HCl is as follows. If montmorillonite is present it is converted to hydrogen clay and a broad 13 Å reflection is produced. The 14.7 Å reflection may tend to be sharpened (by shifting of 15 Å band, in part?) and in most instances shifts

TABLE 2.—*Chemical analyses of samples of loess from western Europe and selected samples from Kansas*[1]

No.	SiO₂	Al₂O₃	Fe₂O₃	TiO₂	CaO	MgO	P₂O₅	SO₃	K₂O	Na₂O	S	Loss 1000° C.	CaCO₃, calc.	Total
1	47.43	9.80	4.91	1.42	16.10	3.11	0.13	tr	1.43	0.89	tr	14.59	28.61	99.81
2	64.63	16.03	6.56	0.82	1.72	2.61	0.19	tr	2.19	1.10	nil	4.44	2.62	100.29
3	53.55	8.51	3.24	0.79	16.64	0.51	0.11	tr	1.17	0.57	tr	14.72	29.46	99.81
4	66.92	12.71	4.34	0.90	4.36	0.77	0.13	tr	1.76	1.02	tr	6.28	7.48	99.19
5	58.84	11.74	4.67	0.91	7.99	2.55	0.15	nil	1.70	0.84	0.02	9.87	13.89	99.26
6	43.57	5.78	2.68	1.28	22.87	2.44	0.13	tr	1.08	0.90	tr	19.89	40.69	100.62
7	63.44	8.75	3.64	1.36	8.83	1.71	0.11	nil	1.54	1.17	tr	9.17	15.65	99.72
8	80.62	9.03	2.88	1.18	0.53	0.64	0.07	tr	1.43	1.08	tr	2.34	0.77	99.80
9	54.05	8.17	2.83	1.07	14.60	2.09	0.11	tr	1.41	1.04	nil	14.12	25.81	99.49
10	70.62	10.41	4.19	0.45	3.98	1.16	0.05	nil	1.83	0.89	nil	5.50	6.98	99.08
11	58.88	8.37	3.13	0.73	11.88	2.19	0.11	tr	1.44	0.84	0.02	11.91	20.95	99.48
12	59.36	6.13	2.66	1.89	12.79	1.90	0.10	tr	1.45	1.03	tr	12.31	22.76	99.62
13	59.04	7.44	2.67	1.13	12.96	1.55	0.09	tr	1.45	1.00	nil	12.50	22.92	99.83
14	51.86	6.82	2.40	0.89	17.59	1.69	0.08	tr	1.21	0.92	nil	16.78	31.20	100.24
15	72.77	7.09	2.98	1.79	5.33	1.02	0.10	tr	1.64	1.04	tr	5.76	9.94	99.52
[2]	73.33	12.00[4]	3.02	—	2.67	1.06	0.08	tr	2.55	1.48[5]	—	3.89		
[3]	72.35	13.56[4]	3.46	—	2.04	1.77	0.14	tr	2.78	0.24[5]	—	3.66		

[1] Analyses by Russell Runnels, State Geological Survey of Kansas.
[2] Doniphan County, Kansas (Frye and others, 1949, p. 85).
[3] Norton County, Kansas (Frye and others, 1949, p. 85).
[4] Includes TiO₂.
[5] Determined by difference.

to about 14.5–14.3 Å. Acid treatment also tends to sharpen and accentuate the 4.75 and 3.55 Å reflections in clays which give a strong 14.5 Å reflection. Orientation is improved and basal reflections are enhanced in acid-treated samples by removal of nonflaky calcite particles. Acid treatment also produces a broad unexplained 8 Å band in several samples.

Glycerol saturation produces a weak to moderate 18 Å reflection in some of the clays. This reflection is not as symmetrical as that in bentonites or in Kansas loess, but seems to tail off in the low-angle

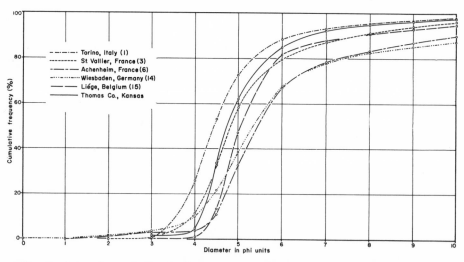

FIG. 2.—Selected cumulative size frequency distribution curves for samples from western Europe (samples 1, 3, 6, 14, and 15) compared with the curve of a typical sample from northwestern Kansas (Swineford and Frye, 1951).

direction. Glycerol improves the symmetry of the 10 Å peak in some samples.

Heat treatment for 30 minutes at 550°–600° C. shifts the 14.7 Å reflection in most of the clays to 14.2–14.0 Å and tends to sharpen it. The 7 Å reflection is destroyed in most clays but not in all of them. All kaolinite reflections are destroyed. Most of the clays show a broad weak reflection at 11.2 to 12.3 Å (accompanied by a reduction in intensity of the 14 Å peak) after heating. The observed *d*-values for heated clays are shown in table 6. Slight warping in a few of the slides (particularly 4, 7, and 14, table 6) has decreased the accuracy of measurement. The component which shrinks to 11–12 Å on heating may be interstratified with a chloritic component, or partly chloritized, as described by MacEwan (1949). The mineral may include vermiculite layers. Whether montmorillonite is primarily interstratified with other layers or is a separate component is not known.

The following criteria were used for identification of minerals in the <2 micron fraction. Normal chlorites were identified by a 14 Å reflection which persists after heating to 550°–600° C., and by a 4.7 Å reflection (003). Antigorite is indicated by 7.3 Å and 3.63 Å reflections which are not destroyed by heat (MacEwan, 1951). Kaolinite is suggested in some samples by particularly strong 7.2 Å reflections which are not completely destroyed by hot HCl, accompanied by reflections at 3.59 Å, 2.39 Å (003), and in some cases 2.28 Å. Micas and illite are shown by basal reflections at about 10 Å and 5 Å. That most of the mica is dioctahedral is suggested by the relatively intense second order basal reflection. Montmorillonoid is identified in glycerol-saturated clay by the presence of an 18 Å reflection. Quartz is identified by a 4.27 Å reflection, and calcite by a 3.04 Å and associated reflections. The behavior of the 3.26 and 3.22 Å reflections under various conditions is not explained.

[*Editor's Note:* Material has been omitted at this point.]

CONCLUSIONS

Judging from the localities studied in western Europe the field appearance of loess in that region is similar to that of the central Great Plains. The buried soils (or paleosols) examined were surprisingly similar to those of the Great Plains region, not only in great soil groups but also in degree of development,

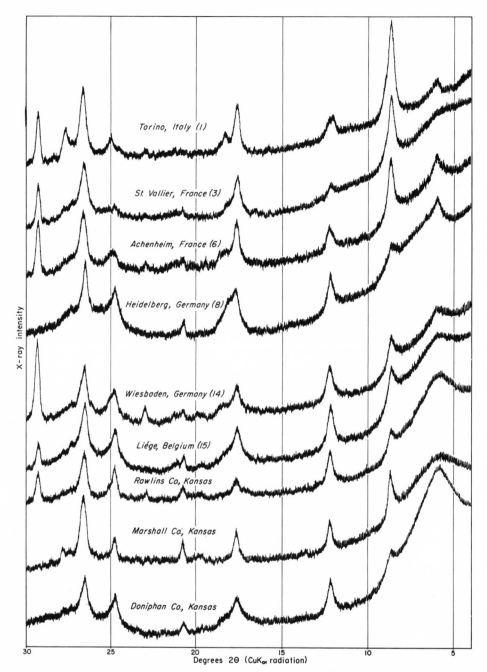

Fɪɢ. 3.—X-ray diffractometer patterns of minus two micron fraction
of loess oriented on glass slides.

color, textural contrast, and structure. It is worthy of note that the term "loess" is used by some workers in western Europe to include deposits of various origins (e.g., Achenheim) that have the appearance of massive silt, but the samples reported here were collected from beds that were judged to be predominantly eolian in origin, either from field relationships, presence of buried soil, etc., or by the statement of a local expert.

In general the European samples studied have the same size and sorting characteristics as Kansas loess (fig. 2). They differ in having more clay in the finer fractions and generally more and coarser sand in the coarser fractions. In the European samples the sorting is less perfect and the skewness is also greater. Sample 1 from near Torino, Italy, is virtually identical with some samples of Kansas loess in texture, even though the mineralogy is quite different.

Chemically, the Kansas loess is generally similar to the European loess analyzed (table 2), although it is relatively high in Al_2O_3 and alkalies and low in CaO.

The size fraction of the Kansas loess coarser than 44 microns is distinctly more uniform in composition geographically than this fraction from the European samples. This difference is judged to be a result of the diversified and local source areas for the European samples, although they lie within a rectangular area substantially smaller than the State of Kansas. Volcanic ash shards invariably occur in this fraction of the Kansas samples but none were found in any of the European samples.

In the size fraction smaller than two microns Kansas loess is notable for its montmorillonite whereas most of the European samples have a 14.7 Å chlorite-vermiculite? mineral which is thought to be derived from altered basic igneous rocks. In many of the European samples the 10 Å reflection is sharp and may be produced by finely divided micas rather

than illite. Kansas loess has somewhat more quartz in the fine fraction than do most of the European samples studied.

In the electron microscope (pls. 1 and 2) the Kansas material has a notably different appearance because of the prevalence of large cottony areas (montmorillonite), which are lacking in most of the European samples. Small rodlike particles are present in both the Kansas and European samples. The Kansas rods are long, straight-sided, and relatively uniform in size, and occur in clusters; whereas the rods observed in the European samples lack these characters and display many varied forms. The European samples commonly show flat, well-crystallized flakes.

The most significant conclusion of the study is that materials in widely separated regions that possess so much similarity in field appearance, topographic relationships, soil profile development, texture, and stratigraphy display such strong differences in mineralogy. Obviously the contrast in mineralogy in the coarse fraction reflects the source, and it is our judgment that source has exerted the controlling influence on the fine fraction as well. As the clay minerals found in the lower A horizon of an interglacial soil (sample 10 near Eltville, Germany) are essentially the same as those found in adjacent fresh loess, it may be argued that the clay minerals did not develop in the loess but were produced by a prior period of weathering or alteration of the source rock or were transported in an unaltered condition from the source rock. The strong contrast in the montmorillonite content seems clearly attributable to the presence of abundant volcanic glass both in the Kansas loess (where it may have altered to clay in place) and in the source rocks to the northwest which contain both volcanic glass and bentonites, as it has been pointed out (Grim, 1953) that montmorillonite is not an abundant mineral in the glacial deposits of North America.

REFERENCES

DAVIDSON, D. T., and HANDY, R. L., 1953, Studies of the clay fraction of southwestern Iowa loess (abstract): Abstracts of Papers Presented at the Second Conf. on Clays and Clay Minerals, Columbia, Mo., p. 4.

FRYE, JOHN C. and LEONARD, A. BYRON, 1951, Stratigraphy of the late Pleistocene loesses of Kansas: Jour. Geology, v. 59, pp. 287–305.

———, 1952, Pleistocene geology of Kansas: Kansas Geol. Survey Bull. 99, pp. 1–230.

FRYE, JOHN C., PLUMMER, NORMAN, RUNNELS, R. T., and HLADIK, W. B., 1949, Ceramic utilization of northern Kansas Pleistocene loesses and fossil soils: Kansas Geol. Survey Bull. 82, pt. 3, pp. 49–124.

GRIM, RALPH E., 1953, Clay mineralogy. McGraw-Hill Book Co., Inc., New York, 384 pp.

HERMANN, F., 1937, Carta geologica delle Alpi Nord-Occidentali; Foglio E: Arti Grafiche Moreschi & C., Milano.

———, 1937a, Carta geologica delle Alpi Nord-Occidentali; Foglio W: Arti Grafiche Moreschi & C., Milano.

MACEWAN, D. M. C., 1949, Some notes on the recording and interpretation of X-ray diagrams of soil clays: Jour. Soil Sci., v. 1, pp. 90–103.

———, 1951, The montmorillonite minerals (montmorillonoids): X-ray identification and crystal structures of clay minerals: Mineralogical Soc., London, pp. 86–137.

———, 1954, "Cardenite," a trioctahedral montmorillonoid derived from biotite. Clay Minerals Bull., v. 2, pp. 120–126.

MIELENZ, R. C., HOLLAND, W. Y., and KING, M. E., 1949, Engineering petrography of loess (abstract). Geol. Soc. America Bull., v. 60, p. 1909.

SWINEFORD, ADA and FRYE, JOHN C., 1951, Petrography of the Peoria loess in Kansas: Jour. Geology, v. 59, pp. 306–322.

ZEUNER, F. E., 1946, Dating the past. Methuen and Co., Ltd., London, 444 pp.

———, 1952, Dating the past, Methuen and Co., Ltd., London, 495 pp.

24

V.V. DOKUCHAEV AND THE LOESS PROBLEM *

M. P. Lysenko
Leningrad University

Dokuchaev considered earths, along with soils, as particular natural history bodies directly related to present-day and earlier modern physical and geographical circumstances. According to Dokuchaev, "soils and *earths* [a difficult word to translate: emphasized by Lysenko. Ed.] are mirrors which are a clear and quite correct reflection, so to speak the direct result of the combined and extremely close interaction, over the centuries, between earth, air and land . . . on the one hand and between vegetable and living organisms and the age of countries on the other . . . ." Obviously, earths are "mirrors" revealing their nature to a lesser extent than soils. It follows from Dokuchaev's argument, however, that for earths a connection should always be established between their look, composition, and properties and the entire range of natural conditions around them (climate, geographical outline, local hydrogeological conditions, etc.).

Owing to the recognized relationship of earths to physical and geographical conditions, products of weathering are distributed zonally in them. Owing to this circumstance and to Dokuchaev's concept of earths as subsoils, it follows that earths must be broken up rocks. Soils, developed to any extent, cannot form directly from rocks. For soils to form, rocks must, as a result of many processes of weathering, go through a marl stage; that is, they must be converted into porous rock.

When Dokuchaev investigated black earth soils, he was principally concerned with different types of loess earth. Dokuchaev developed his opinions on these in stages. In a speech concerning A. V. Stoletov's address "The Black Earth" in 1876, Dokuchaev pointed out that here, as in Western Europe, the loess lies on the slopes of river valleys and in watersheds. In the first case the loess is a river silt deposit; in the second case it is a product of weathering. At that time Dokuchaev had not yet carried out a detailed study of Russian black earths and the loess earths associated with them, and only knew them basically from published data. Later his extensive fieldwork and careful study of published works greatly improved Dokuchaev's theories regarding loess earths.

In 1873, A. P. Karpinskii corrected Lyell's theory regarding typical loess. On the basis of geological research carried out in Volynskii district, he wrote that the absence of lamination is not compulsory sign of loess, and that the absence of well-defined lamination is the only such sign. Closer acquaintance with loesses convinced Dokuchaev that in nature there are not only loess earths as conceived by Lyell, that is, river and glacier deposits in river valleys, but also many other types of loess representing different geneses and having quite different petrographic characteristics. In

*This excerpt was edited by Ian J. Smalley from the complete translation of "V. V. Dokuchaev and the Loess Problem," *Pochvovendenie*, No. 7, 59–67 (1956), which is available from the British Library Lending Division, Boston Spa., Wetherby, West Yorkshire, UK, as RTS 8864, price £4.20.

addition to typical loess (glacial mud deposits), Dokuchaev also isolated variations such as "coarse-grained loess," "rubble loess," "valley loess without rubble," "humus loess," "loess-type clay," and so on. Dokuchaev did not explain the differences among all these types of loess earth. The differences, however, follow from their names. Specifically loess features (carbonate content, silty grain size composition, capacity for vertical structural features, and the presence to porosity visible to the eye) enabled Dokuchaev to relate all these, as it were, petrographically differentiated substances to the category of loess earths. Dokuchaev used the term, still used now, of "loess-type" rocks; he differentiated among typical loesses, these being glacial mud deposits, and loesses subjected to different changes of form, or formed as a result of the weathering of rocks rich in carbonates (limestones, marls, etc.).

Editor's Comments
on Papers 25 and 26

25   TERUGGI
The Nature and Origin of Argentine Loess

26   PÉWÉ
An Observation on Wind-Blown Silt

The South American loess still awaits the full and searching examination that the European and North American loesses have received. Paper 25 is probably the only available paper to devote a careful analysis to the South American material and it suggests that there are significant differences between the Argentinian loess and other loesses, particularly with regard to the presence of significant amounts of volcanic material.

Charlesworth (1957, p. 515) referred to the South American material as a "periglacial loess," and it is certainly tempting to imagine a fairly simple system whereby the Andean Quaternary glaciers provide the silt and the Hobbsean winds emplace it, as shown in Teruggi's map; but the mineralogical evidence suggests that this may be a total misconception. Or does the difference in composition simply reflect the difference in the source rock? Much North European and North American loess material was doubtless initially derived from quartz-rich shield rocks of the high northern latitudes. The South American material comes from the volcanic Andes, possibly by normal glacial grinding plus eolian transportation mechanisms. More petrographic examination (particularly by electron microscope) would help to resolve this problem.

Some interesting papers have appeared since Teruggi's, and the reader is referred in particular to those by Bertoldi de Pomar (1968), Depetris et al. (1970), and Riggi (1968).

Other relatively neglected deposits are those associated with present-day glaciers. Hobbs drew attention to the Greenland loess (see Editor's Comments on Paper 10), but the most significant investigations have been carried out on Alaskan material. Alaska, like Greenland, has a close association with glaciation and therefore loess deposits are to be expected. Péwé has described them in a series of papers. He has also been concerned about the rate of loess deposition and in 1969 stated: "Although loess has been studied for more than 100 years, only few data (Péwé and Holmes, 1964; Reger, Péwé, West, and Skarland, 1965; Péwé, Hopkins, and Giddings, 1965, p. 362) are available for detailed evaluation of rates of accumulation or even of the environment in which modern loess forms."

Péwé's remarks suggest that loess investigators might usefully investigate various aspects of contemporary loess deposits, in particular in Alaska, Greenland, and Antarctica. Some Antarctic loess has been described by Bardin and Sudakova (1969).

REFERENCES

Bardin, V. I., and Sudakova, N. G. 1969. "Loess" of Antarctica. Antarktika: doklady komissii 1969 (1971), p. 113–121 (in Russian).

Bertoldi de Pomar, H. 1968. Petrologia de algunos sedimentos superficiales de la provincia de Santa Fe. Journados Geol. Arg. 3rd Actas·3, 25–48.

Charlesworth, J. K. 1957. The Quaternary era, 2 vols. Arnold, London, 1,700 p.

Depetris, P. J., Vassallo, M. C., and Scherma, G. L. 1970. Arcillas en sedimentos loessoidos de Canals, provincia de Cordoba, Republica Argentina. Assoc. Geol. Arg. Rev. 25, 467–474.

Péwé, T. L. 1969. The periglacial environment, past and present. Proc. 7th INQUA Congr., vol. 11. McGill-Queens University Press, Montreal, 487p.

Riggi, J. C. 1968. El loess de Rio Tercero y el probable origen de los mallines (Cordoba). Jornados Geol. Arg. 3rd Actas 2, 67–77.

THE NATURE AND ORIGIN OF ARGENTINE LOESS[1]

MARIO E. TERUGGI

Instituto Nacional de Investigación de las Ciencias Naturales y Museo, Buenos Aires

ABSTRACT

Though the loess deposits of Argentina (Pampean Formation) are geologically well known, their granulometric and mineralogical composition has hitherto not been analyzed. The data obtained from a study of more than fifty samples are here summarized. It is shown that Argentine loess is similar in field appearance and texture to the North American and European loess. However, the mineralogical composition is completely different, especially in the coarse fractions where an assemblage of volcanic minerals is predominant. The presence of abundant montmorillonite and glass shards in the fine fraction is a point of resemblance between the Argentine and the American sediments. All this material, doubtless of volcanic-pyroclastic origin, has been transported by winds to the place of deposition.

INTRODUCTION

In a recent paper, Swineford and Frye (1955) have made a substantial contribution to our knowledge of the petrography of loess by comparing samples from western Europe with those of Kansas. As is well known, there are two other regions in the world where loess and loess-like deposits cover vast areas: China and Argentina. The Argentine deposits, which are the most important in the southern hemisphere, have been known since the middle of last century from descriptions by D'Orbigny (1842), Darwin (1846), Bravard (1857), Ameghino (1881), and many others. However, very few petrographic data have been available, as the previous work has been mainly stratigraphic, with the exception of the study of some isolated samples by Wright and Fenner (1912) and some partial observations by European authors (Meigen and Werling, 1915; Principi, 1915). In order to fill this gap in our knowledge, a detailed investigation on typical samples were carried out by the author and collaborators (M.E.C. de Di Lorenzo and J. Remiro) at the Instituto Nacional de Investigación de las Ciencias Naturales y Museo of Buenos Aires. The results obtained show that the Argentine loess is entirely different in mineralogical composition from the European and North American loess, and probably from the Chinese as well (Viglino, 1901; Barbour, 1927).

[1] Published with the permission of the Interventor, Instituto Nacional de Investigación de las Ciencias Naturales y Museo, Buenos Aires. Manuscript received September 1956.

CHARACTERISTICS AND STRATIGRAPHY OF ARGENTINE LOESS

The vast distribution of loess deposits in Argentina is shown in figure 1. These Quaternary sediments are grouped under the collective name of Pampean Formation or, in short, Pampean, a name first used by D'Orbigny (*argile pampéene;* 1842) and later by Darwin (*Pampean Mud* or *Pampean Formation;* 1846). The average thickness of the formation is 90 feet; the maximum registered is 210 feet. The Pampean sediments are light yellow or brown in color, sometimes with a reddish or gray tinge; they are devoid of stratification, stand for a long time in vertical walls, possess tubes and rods of calcium carbonate formed around plant roots, contain remains of vertebrate fossils and have all the other characteristics typical of loess deposits the world over. It is not strange then that two European geologists (Heusser and Claraz, 1866) early recognized their resemblance to similar loess deposits of the Rhine valley.

Not all the Pampean sediment is really a loess, although the greater part of it has the appearance of such. It is now generally accepted that reworked or secondary loess is nearly as abundant, or even more abundant, than primary loess. It is extremely difficult to distinguish between these two, and in many cases almost impossible. Argentine geologists have been using, for some 30 years, the adjective *loessoid* to denote material with appearance of loess, regardless of origin. Frenguelli (1925) analyzed all the geological characteristics that may be used, at least in Argentina, to dis-

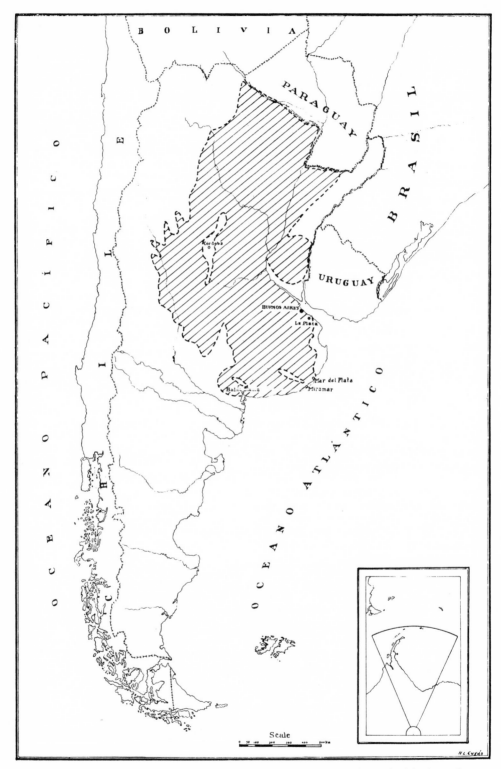

FIG. 1.—Approximate distribution of loessoid sediments (Pampean Formation) in Argentina.

tinguish primary from secondary loess. His contribution is of great value, even now-adays, though further systematic investigation of these deposits is necessary to elucidate this debated question (Smith, 1949).

The stratigraphy of the Pampean deposits was first established by Ameghino (1881). In table 1 the main stratigraphic units are listed, with their approximate equivalents in North American stratigraphy. The names introduced by Kraglievich (1952) are locally applied to the formations which he recognized in the coastal cliffs between Mar del Plata and Miramar, where the best and most complete section of the Pampean Formation is exposed. It has to be borne in mind that no close-fitting equivalence between Argentine and North American (or for this matter European) stratigraphic units is possible, as glaciations in Argentina were either small or non-existent.

SAMPLING

Most of the samples were collected by J. L. Kraglievich from the Mar del Plata-Miramar cliffs. Other samples were collected at the same place by the author in February 1954. The sampling was done on surface exposures, where the stratigraphic relationships were clear.

C. Cortelezzi collected a certain number of samples from La Plata. These were obtained from excavations for the building of new water mains. No auger samples were studied.

The sampling covers practically all the Pampean Formation, i.e., from the lowest Pleistocene up to the present. The results obtained were also checked with samples collected by the author at different points of Buenos Aires Province.

LABORATORY PROCEDURE

Particle-size analysis.—Mechanical analyses were made by a method different from the one used by Swineford and Frye (1951; 1955). The samples were first treated with cold diluted (N/10) hydrochloric acid until effervescence ceased. The dispersion of the filtered material was made with distilled water, to which lithium carbonate in 0.2% solution was added until the pH of the suspension was 8.3. The choice of the peptizer was made after numerous trials, since all the common substances used in mechanical analyses (ammonia, sodium carbonate, sodium silicate, sodium oxalate, etc.) failed to produce a good dispersion of

TABLE 1.—*Stratigraphic divisions of the Pampean Formation of Eastern Argentina*

EASTERN ARGENTINA (Frenguelli, 1950)		MAR DEL PLATA-MIRAMAR AREA (Kraglievich, 1952; Groeber, 1952)	
Cordobense / Platense	Epipluviar		Post-Wisconsin (Post-würm)
Querandinense / Lujanense	Fourth Pluviar	Lobería	Wisconsin (Würm)
Bonaerense	Third Interpluviar	Cobo	Peorian (Riss-Würm)
	Third Pluviar	Santa Isabel	Illinoian (Riss)
Ensenadense	Second Interpluviar	Arroyo Seco	Yarmouth (Mindel-Riss)
	Second Pluviar	Miramar	Kansan (Mindel)
Chapalmalense	First Interpluviar	San Andrés	Aftonian (Günz-Mindel)
	First Pluviar	Vorohué	Nebraskan (Günz)
		Barranca Lobos	Danube Glaciation
Pliocene		Chapadmalal	Upper Pliocene

the samples. Therefore, our analyses are not entirely comparable with those of Swineford and Frye; however, it is thought that the differences in results ought not to be very great. The agitation of the samples was done by hand during a period of 24 hours; the use of laboratory shakers was precluded as in many cases they produced a strong coagulation. Some of the samples were so compact that they had to be disaggregated with a rubber pestle in a mortar prior to the chemical treatment. All the mechanical analyses were made by the pipette method.

Microscopic study.—The sand and silt fractions were examined with a petrographic microscope. The separation of heavy minerals was made with bromoform. The clay fraction was also observed under the microscope, after the pipette samples were evaporated on a glass slide. The mineralogical composition was determined by counting.

Determination of the clay fraction.—Besides the microscopic study, X-ray diffraction patterns of the less-than-1μ fraction of some samples were obtained with a "Debyeflex" (Rich, Seifert & Co.) unit. The patterns were interpreted by Engineer M. Butschowskyi, Instituto Nacional de Investigación de las Ciencias Naturales.

TEXTURE

Of a total of more than 50 analyzed samples, 12 were selected to illustrate the textural composition; the remaining unillustrated ones do not differ in appreciable degree from the ones here selected. With the exception of two samples from La Plata (134 and 20, analyzed by C. Cortelezzi), all the samples selected are from the cliffs between Mar del Plata and Miramar. They range from the lowest Pleistocene to the Recent.

The results of the mechanical analyses show a similarity in texture between the Argentine loess and the North American and European samples, as can be seen by comparing the values of table 2 and the cumulative curves of figure 2 with the data published by Swineford and Frye (1955).

In Argentine loessoid samples the value of Mϕ varies between 4.72 and 6.04, i.e., within the same range as in European and American loesses, or even smaller. Equally

TABLE 2.—*Mechanical analyses of loessoid samples from Eastern Argentina*

No.	Formation	0-1	1-2	2-3	3-4	4-4.5	4.5-5	5-6	6-7	7-8	8-9	9-10	10	Mϕ	$\sigma\phi$	Mdϕ	QDϕ	Skϕ
				Size distribution in phi units (percent by weight)														
62	Lobería		1.0	2.0	32.2	17.3	12.3	9.1	6.9	6.7	3.5	3.5	5.5	5.12	2.10	4.41	1.19	0.585
15	Santa Isabel			0.5	43.3	35.7	6.2	0.6	0.6	0.2	3.8	5.5	3.6	4.72	1.93	4.07	0.32	1.028
58	Miramar			2.9	34.0	22.5	12.2	10.2	7.2	7.9	1.7	0.6	0.8	4.73	1.49	4.26	0.77	0.495
8	Arroyo Seco			1.0	48.6	24.1	7.1	2.7	1.6	0.3	3.1	5.0	6.5	4.83	2.15	4.18	0.40	0.874
11	Barranca Lobos			0.2	41.1	19.6	8.1	5.5	3.3	2.8	4.9	6.6	7.9	5.29	2.25	4.18	1.19	0.600
50	Cobo			2.7	47.4	15.0	11.2	1.9	2.9	3.6	0.7	4.6	10.0	5.02	2.37	4.00	0.69	0.728
2	Vorohué			0.8	25.5	35.7	6.1	6.8	1.4	0.6	3.8	8.5	8.0	5.48	2.32	4.28	1.06	0.577
24	San Andrés			0.5	42.3	8.9	12.5	4.8	0.6	2.1	5.0	7.7	11.1	5.40	2.55	4.43	1.18	0.558
60	Vorohué	9.9	2.0		16.3	24.5	10.1	11.1	3.9	1.8	5.3	8.6	5.1	5.06	2.64	4.40	1.03	0.193
38	Chapadmalal			1.1	16.8	29.1	10.3	10.7	3.2	6.5	6.2	7.9	14.0	6.04	2.49	4.70	2.21	0.404
134	Ensenadense				23.5	12.5	11.3	29.7		6.5	0.5	0.5	13.0	5.43	2.19	5.10	0.92	0.774
20	Ensenadense				35.9	12.5	15.5	10.5	8.3	5.5	2.5	1.2	8.1	5.24	2.10	4.55	1.15	0.674

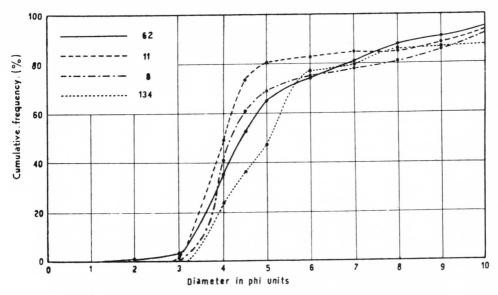

FIG. 2.—Selected cumulative curves for loess samples from eastern Argentina.

similar are the values for Mdφ, though it must be pointed out that in Argentine samples there is a tendency for Mdφ to be smaller (4.00 to 4.40 are the commonest values, as against values larger than 4.40 and up to 6.55 in samples from the Northern hemisphere). The close agreement in Mφ and Mdφ is even more remarkable when compared with Peorian loess from Nebraska and Kansas (Swineford and Frye, 1951).

The phi quartile deviation (QDφ) is also comparable, though it tends to be slightly higher in Argentine samples; it ranges from 0.32 to 2.21. No significant differences exist in the values of phi skewness (Skφ) and phi standard deviation (σφ), though the latter is generally larger in Argentine samples.

COMPOSITION

Sand Fraction

The composition of the "light minerals" fraction is fairly constant, the main constituents being plagioclases, quartz, orthoclase, volcanic glass, altered feldspars, fragments of volcanic rocks and organic opal.

Plagioclases are the most abundant minerals, in all the samples, varying in proportion from as much as 65% down to 20%; normally, however, they make up from one

third to one half of the light minerals. They appear as irregular or cleavage fragments, somewhat rounded and usually showing polysynthetic twinning, Carlsbad-albite and albite being the most frequent laws. They are always remarkably fresh, a fact that did not escape recognition by Wright and Fenner (1912). U-stage determinations showed that the most common species is acid labradorite, followed by andesine, oligoclase (rare), and a variable quantity of albite (between 1 and 10%). Many of the plagioclase grains are zoned, and the presence of solid inclusions and "negative crystals" is to be mentioned.

Quartz is not very abundant; the maximum found was 30%, but more frequently it does not exceed 20% and can even be as low as 2%. The grains are subangular to subrounded, some with liquid inclusions which occasionally are arranged in rows. Grains without inclusions are very common, but those with solid ones are extremely rare. Wavy extinction is practically absent, as are secondary growths and ferric coatings.

Orthoclase appears in rounded grains, usually altered to some clay mineral (kaolin, allophane?). Microcline is extremely scarce, but when found is always unaltered. On the other hand, there is in all the samples a variable proportion of grains so greatly altered that, optical determination is extremely difficult; the greater part, however, seem to be derived from some potash feldspar.

Volcanic glass is always present in the form of angular shards; its percentage varies between 1 and 25% (not counting the thin intercalations of volcanic ashes found in most sections of Pampean loess). It is a colorless variety (rarely light brown), usually with flow structure and with refractive index between 1.49 and 1.51. These acid shards are always fresh and show no signs of alteration. Basic green or brown glass is very rare and is confined to one single level within the whole thickness of the Pampean Formation.

Rounded fragments of volcanic rocks, especially groundmass, are very characteristic in the Argentine loessoid sediments. Some of them are basaltic or andesitic in nature, while others are rhyolitic. Recognition is sometimes difficult because of alteration and impregnation with iron oxides and hydroxides. Their proportion varies between 1 and 25%.

A very typical and frequent constituent of Pampean loess is organic opal, found as minute saw-edged tablets which are really siliceous epidermic cells of grasses. When their form is lost by comminution, the small particles resemble irregular fragments of volcanic glass and only by the determination of their refractive index can they be distinguished (Teruggi, 1955). The percentage of opal is normally low (less than 1%), but in some cases it is as high as 5%.

Some other constituents are occasionally found, such as chalcedony, chert, sanidine, gypsum, fragments of argillaceous rocks, etc., but never more than one or two grains per sample.

The "heavy minerals" make up between 0.6 and 6% of the total sand fraction; in most samples, however, they vary between 0.7 and 1.5%. The principal components are iron ores, amphiboles, and pyroxenes, with a minor amount of other accesory minerals.

Opaque iron ores constitute the greater part of the heavy minerals. Magnetite is by far the predominant species and appears as well-rounded grains which under reflected light show signs of alteration into hematite; hematite is also found as independent grains of similar form and size. The magnetite is a titaniferous variety, as can readily be shown by chemical analysis; its presence in the Pampean sediments has been known since the middle of the last century (Bravard, 1857; Heusser and Claraz, 1866), when the names of "fer titané" and iserite were given to this species. Ilmenite is also common, but its percentage is small in comparison with that of magnetite. Limonite and leucoxene make up a small part of the opaque ores; pyrite is extremely rare and found in few samples. Many of the opaque grains, especially those of hematite and limonite, could very well be fragments of basic igneous rocks deeply penetrated by these secondary iron products

The amphiboles are represented by a green or brown green hornblende, which appears as subrounded prismatic grains; it is always very fresh, in spite of showing signs of having endured long transportation. After the ores, it is the most abundant heavy mineral; it is always accompanied by smaller amounts of dark brown lamprobolite. Fibrous fragments of actinolite are occasionally seen.

The pyroxenes, nearly as abundant as the amphiboles, are hypersthene and augite. Hypersthene is found in long or short prisms, which sometimes are perfectly rounded and other times are fractured across or have the hacksaw terminations typical of intrastratal solution. Oriented rutile or opaque inclusions are frequent; pleochroism is very weak, from colorless to extremely pale brown. Augite appears as rounded grains, ovoidal in shape and dark green in color; in most samples a pale green diopsidic variety is also present in irregular fragments. Both clino- and orthopyroxenes are extremely fresh and clear, though they may have suffered strong attrition. Colorless enstatite is confined to samples from the northern part of Buenos Aires Province; its percentage is always very low.

The remaining minerals are very scarce, with the exception of zoisite and epidote, which in certain levels of the Pampean Formation may be common; they usually appear as rounded grains and prisms. Garnet is found in practically all the samples, but in small quantities; it is a colorless or light red variety, which sometimes shows signs of attrition. Micas, both muscovite and biotite, may be found as occasional flakes; the same happens with the chlorites Tourmaline, zircon, titanite, and rutile are occasionally present, but rarely exceed one or two grains in a sample. The systematic absence of apatite is noteworthy.

Silt Fraction

The composition of the silt fraction is, on the whole, similar to that of the sand fraction, but with one important difference: in all the samples there is a sharp increase in the contents of fresh and altered volcanic glass shards.

Coarse silt (0.062–0.031 mm): The percentage of glass shards varies between 15 and 60%, amounts of 30 and 40% being the most common. The other main minerals are plagioclase, quartz, and orthoclase, in the order mentioned. In some samples, montmorillonite replaces fresh volcanic glass; it appears under the form of altered and more or less fragmented shards. Gypsum in fibrous crystals is occasionally found, sometimes in amounts as great as 20%; it is an authigenic mineral.

Medium silt (0.031–0.011 mm): The composition is comparable to that of coarse silt, but here the percentage of glass shards is even greater (usually more than 60%). Montmorillonite, in aggregates, is also common and abundant. The amount of quartz is slightly increased; it can be as

abundant as plagioclase, or more so Orthoclase is scarce, but opal is present in small quantities.

Fine silt (0.011–0.0039 mm): Glass shards and montmorillonite aggregates make up more than three fourths of this fraction. Usually glass is greatly predominant over montmorillonite, but sometimes the reverse is true. The remaining minerals are, as usual, quartz, feldspars, rarely gypsum and opal.

Clay Fraction

In all the samples the clay fraction is composed of montmorillonite. This mineral appears in minute flakes; optical properties were studied with comparative ease on a film obtained by drying the suspension in a watch-glass. The individual flakes show parallel extinction, are length-slow and the optical figure is biaxial with small 2V. Measurements of refractive indexes gave the following results: α:1.512–1.25; γ:1.532–1.550.

X-ray diffraction data of the less-than-1μ fraction of six samples indicated the presence of a montmorillonite mineral as the main component; there were also small quantities of quartz, feldspar, and probably illite.

The remaining minerals of this fraction, as determined optically, are: volcanic glass, quartz, feldspars, chlorite, illite, kaolinite, etc. All these together very seldom constitute more than 10% of the fraction.

TABLE 3.—*Chemical analyses of Argentine Loess*

Location	SiO$_2$	Al$_2$O$_2$	TiO$_2$	Fe$_2$O$_2$	CaO	MgO	K$_2$O	Na$_2$O	Ign. Loss	H$_2$O (120°)	Total
Alvear[1]	66.81	15.04	0.65	3.11	1.65	1.03	2.31	1.79	4.07	3.34	99.80
Alvear[2]	71.7	12.0	0.9	5.0	4.0	1.0	1.9	1.4	nd	0.5	100.5
Rosario[2]	65.5	15.6	1.1	6.7	2.1	1.4	1.6	3.1	nd	3.4	100.5
Baradero[3]	59.86	17.40	nd	4.80	3.08	1.71	1.71	1.97	3.22	6.04	99.79
La Plata[3]	62.70	15.00	nd	6.00	2.80	1.90	1.88	1.40†	3.38	4.32	99.38
Salto[3]	61.46	14.72	nd	4.91	2.46	1.85	1.56	1.74	4.56	7.11	100.37
Miramar[3]	57.16	17.28	nd	5.43	2.83	1.67	3.68†		3.60	8.35	100.00
La Plata[1]	63.34	16.13	0.74	5.10	3.46	1.64	5.79†		3.80	nd	100.00
La Plata[4]	63.34	17.72	0.80	4.95	2.35	1.84	4.33†		4.67	nd	100.00

† Determined by difference.
[1] Wright, F. E. and Fenner, C. N., 1912.
[2] Meigen, W. and Werling, P., 1915.
[3] Bade, P., 1920.
[4] Laboratorio de Química Analítica, LEMIT, 1955.

Chemical Composition

No chemical analyses were made for the present study. For purposes of comparison, some of the analyses found in the literature have been represented in table 3. The analyses of two samples from La Plata are new and were made at the Laboratorio de Ensayos de Materiales e Investigaciones Tecnológicas (LEMIT), of Buenos Aires Province.

CaCO₃ Content

Calcium carbonate is found in Pampean sediments as calcite in two forms: (1) as concretions, veins and hard-pan layers (caliche), which appear at different levels of the Formation. This is what locally is known as *tosca*, a name already registered by Darwin (1846); and (2) as comminuted particles distributed in the mass of the sediments.

The determination of calcium carbonate refers to the second type, as the first one is an epigenic deposit probably related to climatic variations. It is found that Argentine loessoid sediments have a lower content of calcite than the North American and European samples. This content varies from practically nothing up to a maximum of 8%. Most samples, however, usually contain less than 2% of calcium carbonate, and a few as much as 4%. This fact had been noted by previous workers (Scheidig, 1934).

SOURCE AND ORIGIN OF ARGENTINE LOESS

None of the important minerals composing the Argentine loess deposits are of local origin. The two ranges of the Province of Buenos Aires, the Sierrras de la Ventana (mostly quartzites) and the Sierras de Tandil (migmatites, gneisses, limestones, orthoquartzites), have not contributed in appreciable degree to the loess sedimentation. No data are available for similar deposits in the central and northern parts of Argentina, but there are indications that the Sierras de Córdoba have had some effect on the mineralogical composition.

The Argentine loessoid sediments are mainly formed by minerals of volcanic origin, especially those pertaining to andesitic and basaltic rocks. Characteristically metamorphic minerals are either scarce or absent. The really abundant constituents are plagio-clases (labradorite and andesine, with smaller quantities of oligoclase and albite), which undoubtedly are derived from volcanic (and pyroclastic) rocks, as shown by their common zonal structure and the nature of the inclusions. On the other hand, quartz is rather scarce and, judging by the nature of its inclusions, mainly of igneous origin. Moreover, the constant occurrence of rounded fragments of volcanic rocks is further evidence in support of the petrographic nature of the predominant minerals. Some contribution has also been made by acid volcanic rocks, as is shown by the presence of felsitic groundmass fragments and altered orthoclase; it is supposed that these constituents are derived from rhyolites.

The heavy minerals are essentially represented by hornblende, pyroxenes (hypersthene and augite), and ores. It is difficult to establish which has been the parent rock of the amphiboles, but as they are common in modern volcanic ashes (Larsson, 1937) and in Tertiary and Quaternary tuffs (both amply distributed in Argentina), their derivation from these rocks is suspected. Hypersthene and augite (or diopside) are undoubtedly derived from vulcanites, since plutonic pyroxenic rocks are very scarce in Argentina. The same applies to the ores, which probably are also of volcanic origin,

In the finer grades the abundance of glass shards, either fresh or altered, is striking. This fact points to the important participation of pyroclastic rocks in the make-up of Argentine loessoid sediments. Admittedly, Argentina is a country in which volcanic activity has been very intense during the Tertiary and Quaternary, and as a result vast accumulations of pyroclastic materials cover a large part of its territory. But, in addition to this fact, it has to be remembered that apparently glass shards tend to concentrate in eolian sediments; this phenomenon may be due to their peculiar shape, which makes them more easily lifted by air currents, as pointed out by Russell (1936). The presence of montmorillonite as practically the only clay mineral adds still another proof of the importance of the pyroclastic contribution.

The establishment of the nature of the mineralogical constituents of Argentine loess allows us to determine, in a rough way, the source areas of these minerals. Vast stretches of Patagonia and the Cordillera are covered with andesites and basalts, i.e., the rocks which may have provided most of the loess minerals. There are also in Patagonia large areas in which the predominant rocks are acid tuffs and rhyolites, fragments of which are found in the sand fraction of the Pampean Formation. Therefore, it is reasonably safe to assume that the source areas of the loess deposits were situated far to the west and southwest of the Pampas region.

The transport of the allochthonous material to the Pampas areas must have been effected by winds, since no important streams exist in the region. At present, the strongest and most frequent winds blow from the west and southwest. In 1932 a cordilleran volcano, the Quiza-pu, threw into the atmosphere an enormous volume of volcanic ashes, which were transported to the eastern part of Argentina by the westerly winds, as has been shown by the study of the dispersion of this pyroclastic material (Frenguelli, 1933; Larsson, 1937). It would seem, then, that the same atmospheric conditions prevailed during the Quaternary.

The wind-transported particles and grains which make up the loessoid deposits must have settled down slowly on the surface of the Pampas, where they were trapped by a thick grass cover; the existence of this vegetation is shown in the sediments by the numerous siliceous cells found in all the levels of the Pampean Formation. It is impossible to establish, on the basis of our present knowledge, whether or not the loess deposits were formed in the places where they are now found. However, since in certain levels of the Pampean Formation there are loess phenoclasts imbedded in a matrix of the same nature, it is to be assumed that superficial waters (rivulets, swamps, marshes, etc.) have played a certain role, either important or not, in the reworking and redistribution of the originally eolian sediment.

The majority of the loess minerals, with the exception of the potash feldspars, are completely fresh and devoid of alteration, in spite of the fact that many of them are relatively unstable. Most of the heavy minerals, such as hornblende, hypersthene, and augite, have endured intense wear and are, accordingly, very well rounded, but there are never signs of alteration. The conclusion to be drawn, then, is that they have undergone a long transport under arid conditions, and at the same time that conditions during and after deposition were not adequate for the alteration of the unstable species. The problem that still has to be solved is whether it was a hot or a cold arid climate that predominated during the formation of Argentine loess.

The lack of alteration of the loess minerals is even more striking in comparison with the abundant montmorillonized shards which make up the larger bulk of the fine silt and clay fractions. This fact must be explained without forgetting that, in the same sediments, there is always a considerable proportion of perfectly fresh glass shards and also that intercalations of unaltered volcanic ashes are frequent in the thickness of the Pampean Formation. The only seemingly acceptable explanation of this phenomenon is that the alteration of the shards into montmorillonite did not take place *in situ*. The incorporation of montmorillonite to the Pampean sediments, therefore, must have occurred as such, i.e., as material already altered, most probably derived from the volcanic tuffs so widely distributed in the western and southwestern parts of the Argentine territory. Montmorillonite, then, would be a "second hand" constituent, which has undergone at least two cycles of sedimentation: one, in Pliocene or early Pleistocene times, when the tuff deposits were formed and then altered; and another in the Pleistocene, when the formation of the loess deposits took place. As for the unaltered glass shards found in these sediments, they must be viewed as products of the volcanic explosions that intermittently punctuated the Quaternary and still occur.

Another point to be stressed is that no significant mineralogical or textural variations are found in the whole thickness of the Pampean Formation, i.e., during practically all the Pleistocene. It is then obvious that the same source areas, through similar sedimentary processes, have contributed during all that time to the genesis of the loessoid sediments.

CONCLUSIONS

The results of this study allow us to draw some conclusions by comparison with North American and European loess.

1.—All loess samples seem to have similar granulometric composition; this applies equally to Argentine, North American, and European sediments, whether primary or secondary. Argentine loessoid sediments, like the others, are silts or sandy silts. The sand fraction is somewhat larger in Argentine samples, but it is mostly very fine sand, which lies practically on the border line separating it from coarse silt. The Mdϕ range is noticeably smaller in Argentine samples than in those of the northern hemisphere.

2.—The mineral assemblage composing the sand and coarse silt fractions of Argentine loesses is strikingly different from that found in North American or European samples. The abundance of plagioclases (mainly basic or intermediate), the scarcity of quartz, and the presence of a considerable amount of glass shards and fragments of volcanic rocks all point to a volcanic-pyroclastic origin. This conclusion is confirmed by the nature of the heavy minerals (hornblende, augite, hypersthene, and ores).

3.—The presence of abundant montmorillonite in Argentine loess is a point of similarity to the loess of North America, but of difference from the European. In our case, the origin of montmorillonite is evident, since in many samples it is possible to observe volcanic shards altered into that mineral. In this respect, the study of the Argentine loess has yielded clearer results than the North American samples. The observation of the montmorillonized shards, however, is only possible with direct microscopic investigation of the untreated samples, as dispersion of the samples destroys the aggregates of the clay mineral. This effect has to be taken into account in considering the granulometric composition, because the presence of these aggregates makes Argentine loess in the natural state coarser than what is shown by mechanical analysis.

It is evident, then, that loess is a sedimentary material which may vary considerably in mineralogical composition, but not so much in texture. The Argentine sediment, at least the one composing the Pampean Formation of Buenos Aires Province, is akin to what may be called a sedimentary tuff (Wentworth and Williams, 1932) or a tuffaceous silt (Hay, 1952), and constitutes another example of the importance of the pyroclastic contribution to terrestrial sedimentation, a question which has been recently stressed by Ross (1955). More research is needed to determine, in every particular case, the nature and origin of loess and its relation to climatic conditions (Bryan, 1945).

REFERENCES

AMEGHINO, F., 1881, La formación pampeana o estudio de los terrenos de transporte de la cuenca del Plata. Buenos Aires-Paris, 371 p.

BADE, F., 1920, Investigaciones petroquímicas del loess pampeano: Mus. La Plata Rev., v. 25, p. 213–251.

BARBOUR, G. B., 1927, The loess of China: 1926 Smithsonian Inst., Ann. Report p. 279–296.

BRAVARD, A., 1857, Observaciones geológicas sobre diferentes terrenos de transporte de la hoya del Plata. Buenos Aires, 107 p.

BRYAN, K. 1945, Glacial versus desert origin of loess: Am. Jour. Sci., v. 243, p. 245–248.

DARWIN, C., 1846, Geological observations in South America. Smith, Elder and Co., London, 279 p.

FRENGUELLI, J., 1925, Loess y limos pampeanos: An. Soc. Arg. Est. Geogr. Gaea, v. 1, p. 1–88.

———, 1933, I vulcani delle Ande e l'eruzione del Quiza-Pú: Le vie d'Italia e del mondo, Anno I, Milano, 95–112 p.

———, 1950, Rasgos generales de la morfología y la geología de la Provincia de Buenos Aires: Lab. Ens. Mat. Inv. Tecn. (LEMIT), II ser., N°. 33, p. 1–72.

GROEBER, P., 1952, Glacial, tardío y post-glacial en Patagonia: Mus. Mar del Plata Rev., v. 1, Entr. 1, p. 79–103.

HAY, R. L., 1952, The terminology of fine-grained detrital volcanic rocks: Jour. Sedimentary Petrology, v. 22, p. 119–120.

HEUSSER, J. C., AND CLARAZ, G., 1866, Essai pour servir à une description physique et géognostique de la province argentine de Buenos-Ayres: Mém. Soc. Helvetique Sci. Natur., v. 21, 139 p.

204

KRAGLIEVICH, J. L., 1952, El perfil geológico de Chapadmalal y Miramar, Provincia de Buenos Aires: Mus. Mar del Plata Rev., v. 1, Entr. 1, p. 8–37.

LARSSON, W., 1937, Vulkanische Asche vom Ausbruch des chilenischen Vulkans Quizapú (1932) in Argentina gesamelt: Geol. Inst. Upsala Bull., v. 26, p. 27–52.

MEIGEN, W., AND WERLING, P., 1915, Über den Löss der Pampasformation Argentiniens: Ber. Naturf. Gesell. Freiburg i. Br., Band 21, p. 159–184.

ORBIGNY, A. D., 1942, Voyage dans l'Amérique Meridional, v. 3, part 3, Paris. 298 p.

PRINCIPI, P., 1915, Alcune osservazioni sul loess del territorio argentino: Soc. geol. italiana Boll., v. 34, p. 219–224.

ROSS, C. S., 1955, Provenance of pyroclastic material: Geol. Soc. America Bull., v. 66, p. 427–434.

RUSSELL, R. D., 1936, The mineralogical composition of atmospheric dust collected at Baton Rouge, Louisiana: Am. Jour. Sci., v. 31, p. 50–66.

SCHEIDIG, A., 1934, Der Löss und seine geotechnischen Eigenschaften. Dresden and Leipzig, 1230 p.

SMITH, H. T. U., 1949, Physical effects of Pleistocene climatic changes in nonglaciated areas, eolian phenomena, frost action and stream terracing: Geol. Soc. America Bull., v. 60, p. 1485–1516.

SWINEFORD, A., AND FRYE, J. C., 1951, Petrography of the Peoria loess in Kansas: Jour. Geology, 59, p. 306–322.

———, 1955, Petrographic comparison of some loess samples from western Europe with Kansas loess: Jour. Sedimentary Petrology, v. 25, p. 3–23.

TERUGGI, M. E., 1955, Algunas observaciones microscópicas sobre vidrio volcánico y ópalo organógeno en sedimentos pampianos: La Plata Mus Notas, v. 18, 66, p. 17–26.

VIGLINO, A., 1901, Il loess del Shan-Si settentrionale: Soc. geol. italiana Boll., v. 20, p. 311–338.

WENTWORTH, C. K., AND WILLIAMS, H., 1932, The classification and terminology of pyroclastic rocks: Nat. Research Council Bull. 89, p. 19–53.

WRIGHT, F. E., AND FENNER, C. N., 1912, Petrographic study of the specimens of loess, tierra cocida and scoria collected by the Hrdlicka-Willis expedition: Bur. Am. Ethnology, Smithsonian Inst. Bull. 52, p. 55–98.

26

Reprinted from *Jour. Geol.*, 59(4), 399–401 (1951)

AN OBSERVATION ON WIND-BLOWN SILT[1]

TROY L. PÉWÉ

U.S. Geological Survey, Washington, D.C.

A paper by Russell (1944) has again stimulated discussions on the origin of loess. After thoroughly reviewing the numerous and varied hypotheses proposed, he suggested a fluvial-colluvial origin for loess along the lower Mississippi Valley and rejected the more commonly accepted hypothesis of eolian origin. Holmes (1944), Leighton and Willman (1949), and Thwaites (1944) have supported the hypothesis of eolian origin for the lower Mississippi Valley loess, and Swineford and Frye (1945) showed that, contrary to Russell's beliefs (p. 24), wind could sort material to the degree represented by some loess deposits.

The hypothesis of eolian origin, which has been generally accepted for the last thirty to forty years, is that fine glacial debris, spread over the relatively broad vegetation-free glacial outwash plains and floodplains, is transported by winds and deposited on adjacent uplands. This phenomenon can be observed today on some floodplains of glacial streams (Tarr and Martin, 1913, pp. 299–300; Tuck, 1938, pp. 651–653; Rockie, 1946, pp. 5, 10) but is restricted largely to the summer months. In winter, ice and snow cover much of the silt, decreasing the possibility of wind transportation.

[1] Published by permission of the director, U.S. Geological Survey. Manuscript received July 20, 1950.

In central Alaska, bluffs along part of the Tanana and Yukon rivers are blanketed with a well-sorted tan silt, whose origin has been discussed for many years. In connection with studies by the U.S. Geological Survey dealing with the character, distribution, and origin of this silt and other unconsolidated sediments in the Tanana Valley, the writer was able to obtain, during the summer of 1949, some quantitative observations of silt transported by wind from the floodplain of the Delta River.

The Delta River, approximately 85 miles long, is one of many glacial streams that flow northward from the Alaska Range to the Tanana River. The upper part of the river flows in a canyon across the Alaska Range; the lower part flows across the broad Tanana lowland. During the Pleistocene the valley served as one of the main channels for ice moving northward from the Alaska Range across the lowland almost to the Tanana River. At present the river still receives meltwater from many glaciers, which are about 30–50 miles upstream from its mouth, and consequently is heavily laden with silt. Where the river flows across the Tanana lowland, the gradient is relatively low; the floodplain is 1–2 miles wide, with the intricately braided channel and numerous silt-covered bars that are characteristic of glacial-drainage streams.

Moderate to high-velocity northward-blowing winds from the Delta River canyon in the Alaska Range commonly sweep across the Tanana lowland in the vicinity of Big Delta. Winds up to 25 and 35 miles per hour are common, and velocities up to 97 miles per hour have been recorded (oral communication, Mr. K. Kulm, August 16, 1949) by the Civil Aeronautics Administration at Big Delta airfield, which is adjacent to the river near the junction of the Alaska and Richardson highways.

At 1:00 P.M. on August 20, 1949, when a 14-mile-per-hour wind from the southeast was recorded at the airfield, a cloud of floodplain silt was raised about 50–100 feet above the ground. One hour later the wind had increased to 30 miles per hour and was transporting great clouds of grayish dust from the

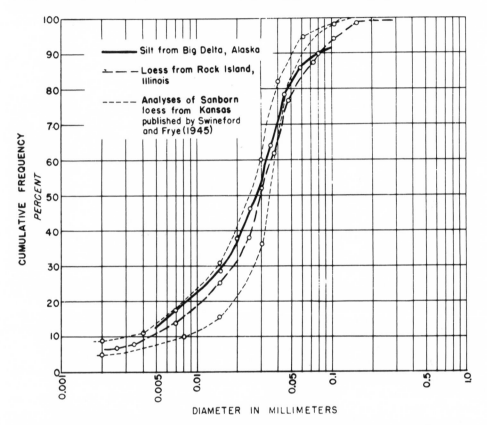

[*Editor's Note:* Plate 1 has been omitted owing to limitations of space.]

Fig. 1.—Cumulative curves of mechanical analyses of silt adjacent to Delta River floodplain near Big Delta, Alaska; loess from Rock Island, Illinois (17th Street and 27th Avenue); and Sanborn loess from Kansas. Alaskan and Illinoian samples collected by T. L. Péwé and analyzed by the Corps of Engineers, U.S. Army, Rock Island, Illinois. Sanborn loess analyses of samples *1* and *10* from Swineford and Frye (1945, p. 252).

floodplain. The writer noted from an airplane that the silt was carried to a relatively even height of 5,000 feet above sea level, or 4,000 feet above the ground, and was suspended over an area of about 300 square miles. The dust was so thick over this area that, from an elevation of 6,000 feet, it almost canceled the ground below. For the most part, the clouds of dust covered the area west of the Delta River to Delta Creek and south of the Tanana River; however, a few square miles north of the Tanana were blanketed, and some of the silt was deposited on the upland, where silt is at least 10 feet thick.

Unfortunately, none of the dust could be collected in the air; however, silt adjacent to the floodplain on the east was analyzed for size-grade distribution. Comparison of the cumulative-frequency curve of this silt with curves from the analyses of loess from Illinois and Kansas (fig. 1) reveal close similarity in shape. The cumulative-frequency curve of the Alaska silt is also similar to cumulative-frequency curves of wind-blown dust published by Swineford and Frye (1945, p. 252) and Zeuner (1949, p. 27).

Excellent aerial photographs of a similar phenomenon were taken by the U.S. Navy in the summer of 1948 (see pl. 1, A and B). These photographs illustrate the type of dust clouds observed in 1949; however, the photographed clouds are less extensive than those reported by the writer.

This observation supports the assumption that large quantities of silt are transported from floodplains of glacial-drainage streams and deposited on adjacent flats and uplands. The velocity of 30 miles per hour compares favorably with the velocities of dust-bearing winds cited by Bryan (1927, pp. 39–40). Mechanical analysis of deposited silt compares well with analyses of wind-blown dust and loess from Illinois and Kansas. The size-grade distribution fulfils Russell's requirement for loess, that at least 50 per cent of the distribution must fall between 0.01 and 0.05 mm. (Russell, 1944, pp. 4, 5).

REFERENCES CITED

BRYAN, K. (1927) The "Palouse soil" problem: U.S. Geol. Survey Bull. 790-B, pp. 21–46.

HOLMES, C. D. (1944) Origin of loess—a criticism: Am. Jour. Sci., vol. 242, pp. 442–446.

LEIGHTON, M. M., and WILLMAN, H. B. (1949) Loess formations of Mississippi Valley: Geol. Soc. America Bull. 60, pp. 1904–1905.

ROCKIE, W. A. (1946) Physical land conditions in the Matanuska Valley, Alaska; U.S. Dept. of Agriculture, Physical Land Survey No. 41, 32 pp.

RUSSELL, R. J. (1944) Lower Mississippi Valley loess: Geol. Soc. America Bull. 55, pp. 1–40.

SWINEFORD, A., and FRYE, J. C. (1945) A mechanical analysis of wind-blown dust compared with analyses of loess; Am. Jour. Sci., vol. 243, pp. 249–255.

TARR, R. S., and MARTIN, L. (1913) Glacial deposits of the continental type in Alaska: Jour. Geology, vol. 21, pp. 289–300.

THWAITES, F. T. (1944) Review of: Lower Mississippi Valley loess, by Richard Joel Russell, 1944: Jour. Sedimentary Petrology, vol. 14, pp. 146–148.

TUCK, R.(1938) The loess of the Matanuska Valley, Alaska: Jour. Geology, vol. 46, pp. 647–653.

ZEUNER, F. E. (1949) Frost soils on Mount Kenya, and the relation of frost soils to aeolian deposits: Jour. Soil Science, vol. 1, pp. 20–30.

Editor's Comments
on Papers 27, 28, and 29

27   SIMONSON and HUTTON
      Distribution Curves for Loess

28   WAGGONER and BINGHAM
      Depth of Loess and Distance from Source

29   FRAZEE et al.
      Loess Distribution from a Source

These papers deal with the problem of describing the variation of the thickness of a loess sheet with distance from its source. The only detailed studies of this type seem to have been done in the lands to the east of the Mississippi–Missouri river system, and the papers presented here deal with this region. It is surprising that similar undertakings have not been made in other parts of the world; or does the Mississippi provide a special and uniquely definable loess source?

The other obvious place to investigate loess thinning would seem to be in the lower Danube valley where the river flows between Romania and Bulgaria. Here is a well known and demarcated loess deposit that could be related to the Danube in the same way that deposits in Illinois are related to the Mississippi, although Maruszczak (Paper 39) has indicated the possibility that loess material was delivered into this part of the Danube valley by eolian transportation from the west.

Frazee et al. decide in favor of an additive exponential model to describe thickness variation, which suggests two physical factors operating in the transportation system. This model may be considered to have superseded the Waggoner and Bingham model, but the earlier paper presents a very thorough and systematic consideration of problems involved in the study of the thickness of loess deposits, in particular the theory of atmospheric turbulence.

Simonson and Hutton were concerned, as were Waggoner and Bingham, to develop the studies of loess distribution and particularly of variations in thickness undertaken by Krumbein (1937) and Smith (1942). The improvement on the earlier work that they proposed and achieved was to use much longer traverses over which variation in thickness could be measured. Their longer traverse was effectively 260 miles, compared with lengths of 13 and 70 miles used by Krumbein and Smith, respectively. They suggest that loess may be deposited much farther from its source than was previously expected.

The particle size of the loess as well as the deposit thickness changes with distance from source: the particles as expected getting smaller. Schönhals (1955) has considered this problem and proposed a *Kennzahl*, or "characteristic figure," to represent the fineness parameter. He has presented data showing variations in *Kennzahl* with distance from source for loess in Illinois and Saxony (Sachsen) but his work has not received the attention it deserves. It will be observed that neither Waggoner and Bingham nor Frazee et al. cite the Schönhals paper.

REFERENCES

Krumbein, W. C. 1937. Sediments and exponential curves. *J. Geol.* 45, 577–601.

Schönhals, E. 1955. Kennzahlen für den Feinheitsgrad des Lösses. Eiszeitalter Gegenwart 6, 133–147.

Smith, G. D. 1942. Illinois loess— variations in its properties and distribution. *Illinois Agr. Exptl. Station Bull. 490.*

27

Reprinted from *Amer. Jour. Sci.,* **252**, 99–105 (Feb. 1954)

DISTRIBUTION CURVES FOR LOESS*

ROY W. SIMONSON and CURTIS E. HUTTON

ABSTRACT. Loess distribution was studied in southwestern Iowa and northern Missouri by measuring maximum thickness at a number of points along two traverses which begin at or near the bluffs beside the flood plain of the Missouri River. The greatest measured thickness of Wisconsin loess is 1380 inches near Council Bluffs, Iowa. The loess thins to the southeast according to an exponential function, and the maximum thickness gradually drops off to as little as 50 inches on an uneroded divide in Boone County, Missouri. The slope of the curve showing loess thickness decreases with increasing distance from the source. The data obtained indicate that the equations already developed for loess distribution curves are valid within restricted geographic ranges but not over the full distance of loess deposition from a single source. The distribution pattern of the loess also suggests the need for further study of silty surface mantles in areas where loess has not been identified in the past.

The nature and distribution of loess formations in the Mississippi Valley, exclusive of the Missouri Valley, have been discussed recently by Leighton and Willman (1950), who cite curves for loess distribution in Illinois. The principal systematic measurements of loess thickness over sizable areas have been made in the Mississippi Valley, especially in Illinois (Smith, 1942). Systematic measurements have also been made in Kentucky, Tennessee and Mississippi (Wascher, Humbert and Cady, 1947), although no effort was made to develop curves for loess distribution patterns in those states. The curves and equations worked out by Krumbein (1937) and Smith (1942) were both based on data obtained in Illinois where possible traverses from single loess sources have maximum lengths of less than 100 miles. Traverses can be extended for approximately 3 times that distance across southwestern Iowa and northern Missouri without encountering important secondary loess sources. This permits the testing of equations for loess distribution curves over greater distances.

NATURE AND SCOPE OF INVESTIGATIONS

The observations of loess distribution in southwestern Iowa were part of an effort to improve the classification of soils for the soil survey of Taylor County (Scholtes, Smith and Riecken, in press). Earlier studies (Kay and Graham, 1943) had shown that the loess in Iowa was not uniform in thickness or composition. Moreover, studies by Bray (1937) and Smith (1942) had demonstrated that the nature of soils could be related to loess distribution patterns in Illinois. Consequently, it was believed that study of the distribution pattern of Wisconsin (Peorian) loess in southwestern Iowa, when coupled with investigations of soil morphology and composition, would improve the understanding and classification of the soils. Inasmuch as the data obtained in studies of soil morphology and composition in southwestern Iowa have been reported elsewhere by Hutton (1950) and Ulrich (1950), they will not be repeated here.

It now seems clear that the Peorian formation, as generally identified in the past, consists of more than one sheet of loess, as was suggested by Leigh-

* Contribution from the Iowa Agricultural Experiment Station and the Soil Conservation Service, U. S. Department of Agriculture. Journal Paper No. J-2319 of the Iowa Agricultural Experiment Station, Ames, Iowa, Project No. 1151.

ton and Willman (1950). Ruhe (1952) has concluded that a major twofold division of Wisconsin drift and its loess equivalents exists in western Iowa. Mickelson (1950) has interpreted an exposure in western Iowa as indicating the presence of a post-Loveland pre-Iowan loess which might correlate with the Farmdale loess of Illinois. In the loess depth measurements reported in this paper, it was not possible to detect consistent breaks in the sections of loess above the Loveland formation. Moreover, the loess above the Loveland formation seems to have originated from the flood plain of the Missouri River along the western edge of the state.

Evidence for the Missouri River flood plain as the principal source of loess in southwestern Iowa and north-central Missouri seems overwhelming. The thickest deposits of loess, or the highest bluffs, lie immediately southeast of the widest segments of the flood plain. Where the flood plain becomes narrow the adjacent loess deposits are much thinner. Over much of western Iowa the loess was apparently laid down upon a dissected land surface, and the deposits are thickest on the southeast sides of the ridges. These are the lee sides to winds blowing from the northwest. The loess mantle is often deep on the southwest slope and very shallow or absent on the comparable northwest slope of a single ridge. With increasing distance to the southeast the loess becomes thinner according to an exponential function as was found earlier by Smith (1942) in Illinois. On the other hand, the thickness of loess in south-central Iowa is essentially constant along a line running from northeast to southwest or perpendicular to the direction chosen for the two traverses. The distribution pattern of loess in southwestern Iowa, together with changes in the nature of that loess with increasing distance from the Missouri River, is the basis for the conclusion that the Wisconsin (Peorian) loess was blown almost entirely from the flood plain of the Missouri River and that the sediments were carried mainly to the southeast. This is in accord with observations in Illinois (Smith, 1942), other observations in Iowa (Hunter, 1950; Simonson, Riecken, and Smith, 1952, p. 13-20, 58-60, 84), and studies of the lower Mississippi Valley (Wascher, Humbert and Cady, 1947). This interpretation forms the basis for laying out traverses in a southeasterly direction from the Missouri River.

Measurements were made along two traverses extending southeast from the flood plain of the Missouri River which forms the western boundary of Iowa. Traverse no. 1 begins 10 miles from the bluff in Monona County and runs southeast to the Iowa-Missouri state line in southwestern Wayne County. No reliable measurements of maximum loess depths were possible nearer the bluff in Monona County. Traverse no. 2 begins at the bluff in the northern edge of Council Bluffs and runs southeast into Ringgold County. Locations of the two traverses are shown in figure 1. Traverse no. 1 extended for a distance of approximately 170 miles and Traverse no. 2 about 90 miles within Iowa. Traverse no. 2 has been extended for a total distance of approximately 260 miles by the addition of measurements at three sites in Missouri.

Measurements of thickness were made by boring with a soil auger and extensions through the Wisconsin (Peorian) loess into the underlying material. three principal kinds of which occur in southwestern Iowa. In most of the

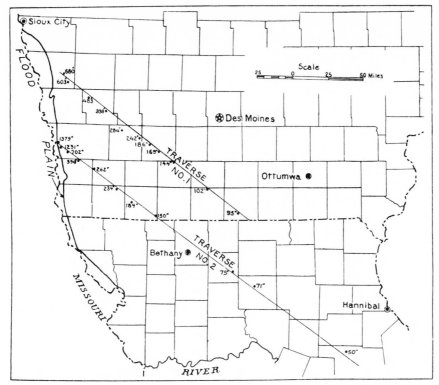

Fig. 1. Map showing approximate locations and loess depths along Traverses no. 1 and 2 in Iowa and Missouri.

area, the loess is underlain by fine-textured glacial till, and the contact between the loess and this till is commonly sharp. On the broad divides in south-central Iowa, the loess is commonly underlain by a soil profile formed from the till (Simonson, 1941), and the contact is seldom as sharp. Apparently, some mixing of the loess and the A horizon of the older soil took place during the early stages of loess deposition. Near the traverse origins, a fine-textured reddish-brown loess (Loveland) underlies the thicker Wisconsin loess in places. Measurements of thickness of Loveland loess were not made because the formation does not occur uniformly and is of limited importance as soil parent material in Iowa.

The actual sites for measurement were chosen to provide data on maximum loess thickness and to space the borings along the traverses. In the hilly to rolling landscapes near the loess bluffs which border the flood plain most of the borings were made in deep road cuts. The loess deposits commonly reach maximum thicknesses of more than 300 inches in a belt ranging from 3 to 30 miles in width east of the Missouri River flood plain. A hand auger does not function effectively to depths much greater than 300 inches, and borings were therefore not feasible on the crests in higher divides unless road cuts were available. In the dominantly hilly to rolling uplands the estimates of loess thickness are certain to be conservative and may be appreciably low. Some loess has doubtless been removed by erosion during and after the period of deposition. Beyond a distance of approximately 40 miles southeast from the bluff, the uplands include some high-lying tabular divides. These flat upland divides, especially common in south-central Iowa, were chosen as sites for measuring loess thickness. On these divides there has been little opportunity for removal of loess by erosion and it is believed that the measurements of maximum thickness have a high degree of accuracy.

DESCRIBING LOESS THICKNESS

In southwestern Iowa the maximum loess thickness on uneroded sites ranges from as much as 1380 inches near Council Bluffs in Pottawattamie County to 95 inches in Wayne County. The data on loess depths together with the locations at which borings were made are given in table 1. Included in table 1 are three measurements of loess thickness at points in

TABLE 1

Depths of Loess in Southwestern Iowa and Adjacent Missouri

Distance from Source in Miles	Depth of Loess in Inches	Location by Legal Subdivision					
		Subdivision*	Township North	Range West	County	State	
		Traverse no. 1					
10.3	680	NW NE NE 16	83	43	Monona	Iowa	
20.5	603	SW SE SW 14	82	42	Monona	Iowa	
33.0	483	NW NE NE 27	81	40	Shelby	Iowa	
55.0	338	NE NE 21	79	37	Shelby	Iowa	
65.5	284	NE NW NE 12	77	37	Cass	Iowa	
83.5	242	SE SW SE 14	76	34	Cass	Iowa	
88.0	186	NW NE 18	75	33	Adair	Iowa	
98.5	169	SE SE NE 3	74	32	Adair	Iowa	
117.5	144	NW 14	73	30	Union	Iowa	
155.0	102	SE SE 8	69	25	Decatur	Iowa	
167.5	95	NW NW SE 1	67	23	Wayne	Iowa	
		Traverse no. 2					
0.1	1379	SW 11	75	44	Pottawattamie	Iowa	
1.6	1231	Council Bluffs	75	44	Pottawattamie	Iowa	
6.9	682	SW SE NE 2	74	43	Pottawattamie	Iowa	
16.4	390	SW NW NW 4	73	41	Mills	Iowa	
30.6	262	NE SE NW 1	71	40	Montgomery	Iowa	
46.8	234	SE SE SW 4	70	37	Page	Iowa	
66.4	184	Middle 20	68	35	Taylor	Iowa	
86.7	150	NW SW SW 33	67	32	Taylor	Iowa	
157.0	75	Sec. 26	61	24	Grundy	Mo.**	
175.0	71	Sec. 5	57	22	Livingstone	Mo.**	
257.0	50	Sec. 10	51	11	Boone	Mo.**	

* Legal subdivisions within sections are abbreviated by omission of the figure "¼." For example, a full description would read NW¼ NE¼ NE¼ Sec. 16.
** Personal communication from G. D. Smith and W. D. Shrader.

Missouri which represent an extension of Traverse no. 2. These three sites lie 150 to 260 miles from the point of origin of the traverse. Some question may be raised about the inclusion of the three measurements in Missouri. The

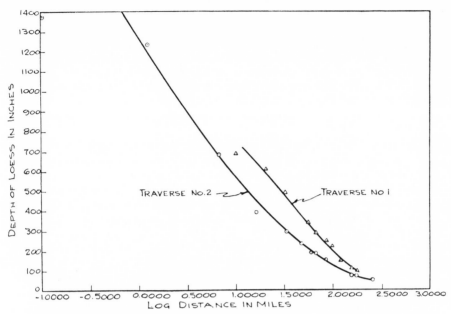

Fig. 2. Graph showing relationship of maximum depth of loess to the logarithm (to base 10) of distance from the source.

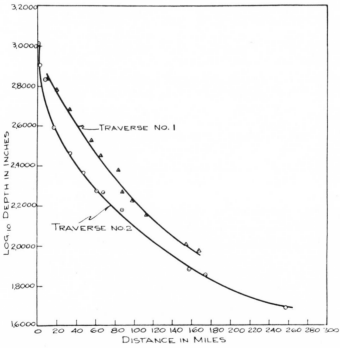

Fig. 3. Graph showing relationship of the logarithm (to base 10) of the maximum depth of loess to the distance from source.

validity of extending the curve on the basis of those measurements cannot be fully tested until more measurements are available. Examinations of loess thicknesses in south-central Iowa and northern Missouri in the course of field trips, however, indicate that maximum loess thickness can be predicted with a high degree of accuracy from the curves along the full extent of Traverses no. 1 and 2. Data for both traverses indicate the gradual thinning of the loess with increasing distance from the Missouri River.

The data given in table 1 are shown graphically in figures 2 and 3. In figure 2, the depth of loess in inches is represented on the "y" axis, and the logarithm to the base 10 of the distance in miles from the bluff is given along the "x" axis. This method for plotting the data follows that of Smith (1942). In figure 3, the logarithm to the base 10 of loess depth in inches is plotted along the "y" axis, whereas the distance in miles from the bluff is plotted on the "x" axis. This follows the method used by Krumbein (1937). The relationships between loess thickness and distance from source, as indicated in figures 2 and 3, are essentially similar to those demonstrated earlier by Krumbein and Smith. The loess becomes thinner with increasing distance from source according to an exponential function. The curve for Traverse no. 2 in figure 2 also shows a discontinuity or break comparable to that observed earlier by Smith (1942). A similar break is not evident in the curve for Traverse no. 1, but the observations nearest the bluff are 10 miles from the edge of the flood plain. A discontinuity in the curve could occur between that point and the bluff itself.

The traverses made in Iowa were longer than those possible in Illinois (Smith, 1942; Krumbein, 1937), and they could be extended into Missouri without danger of encountering important secondary sources of loess. The traverses made by Krumbein (1937) and Smith (1942) have maximum lengths of 13 and 70 miles respectively. Traverse no. 1 was 170 miles long in Iowa, whereas traverse no. 2 was 260 miles long when extended to cover the observations in Missouri. These observations over longer distances indicate that the slope of the loess distribution curve changes as the distance increases.

When the distribution curve for loess is extended beyond points that are 100 miles from the source, the slope begins to change and the curve becomes asymptotic to the "x" axis. This trend is evident in both figures but is more prominent in figure 3. The rate of thinning decreases sharply 170 to 200 miles from the loess source. Thus, the curves in figures 2 and 3 indicate that equations already developed to describe loess distribution are valid within certain geographic limits. They cannot be extended with confidence to cover the entire distance over which loess may be deposited from a given source. Although this study indicates that previous equations are not fully adequate to describe loess distribution, the data obtained do not seem adequate in themselves for the development of new equations. The curves presented in either figure 2 or figure 3 can be used to estimate maximum loess thickness with a high degree of accuracy in southwestern Iowa and north-central Missouri. Moreover, the studies of soils (Hutton, 1950; Ulrich, 1950) have indicated that there is a close relationship between the nature of the soils formed from loess and the maximum thicknesses of the deposits in southwestern and south-central Iowa.

The shape of the distribution curve for loess in south-central Iowa and north-central Missouri is also of interest because it suggests that loess may be deposited in quantity much further from the source than has been commonly supposed. This emphasizes the need for further investigation of silty surface mantles which have not been considered loess in the past. It does not demonstrate that all silty mantles are loess. It does, however, indicate that the study of silty mantles ought to include as one working hypothesis the possible origin of the sediments as loess. For example, occasional areas of Dewey and Etowah soils in Franklin County, Tennessee are far more silty in the upper horizons than is common. The Dewey soils are normally derived from limestone which is relatively low in chert and other impurities, whereas Etowah soils have been formed from alluvial material in stream terraces. The topographical position of the soils in the present landscape would permit them to have been covered by a shallow mantle of loess in the past, but it is hard to imagine conditions which would allow deposition of silty sediments by other means. Yet the area is far removed from important loess sources such as the Mississippi and Ohio Rivers. The fact that loess has been carried great distances across southern Iowa and north-central Missouri suggests re-examination of the thin silty mantles in areas such as south-central Tenessee and northern Alabama. Similar problems may exist in other areas. These surface mantles are of importance to the formation and classification of soils even when they comprise but part of the soil profile. More effective consideration can be given to such mantles when their nature and origin is understood.

REFERENCES

Bray, R. H., 1937, Chemical and physical changes in soil colloids with advancing development in Illinois soils: Soil Sci., v. 43, p. 1-14.

Hunter, R., 1950, Physical properties of some loess-derived Prairie soils of southeastern Iowa: Unpublished thesis, Iowa State College Library.

Hutton, C. E., 1950, Studies of the chemical and physical characteristics a chrono-litho-sequence of loess-derived Prairie soils of southwestern Iowa: Soil Sci. Soc. America Proc., v. 15, p. 318-324.

Kay, G. F., and Graham, J. B., 1943, Illinoian and post-Illinois Pleistocene geology of Iowa: Iowa Geol. Survey Ann. Repts., v. 38, p. 155-203.

Krumbein, W. C., 1937, Sediments and exponential curves: Jour. Geology, v. 45, p. 577-601.

Leighton, M. M., and Willman, H. B., 1950, Loess formations of the Mississippi Valley: Jour. Geology, v. 58, p. 599-623.

Mickelson, J. C., 1950, A post-Loveland pre-Iowan loess in western Iowa: Iowa Acad. Sci. Proc., v. 57, p. 267-269.

Ruhe, R. V., 1952, Classification of the Wisconsin glacial stage: Jour. Geology, v. 60, p. 398-401.

Scholtes, W. H., Smith, Guy D., and Riecken, F. F., Soil survey of Taylor County, Iowa: U. S. Dept. Agr., in press.

Simonson, R. W., 1941, Studies of buried soils formed from till in Iowa: Soil Sci. Soc. America Proc., v. 6, p. 373-381.

Simonson, R. W., Riecken, F. F., and Smith, Guy D., 1952, Understanding Iowa soils: Dubuque, Wm. C. Brown Publishing Co.

Smith, G. D., 1942, Illinois loess—variations in its properties and distribution: A pedologic interpretation: Illinois Agr. Exper. Sta. Bull. 490.

Ulrich, R., 1950, Some chemical changes accompanying profile formation of the nearly level soils developed from Peorian loess in southwestern Iowa: Soil Sci. Soc. America Proc., v. 15, p. 324-329.

Wascher, H. L., Humbert, R. P., and Cady, J. G., 1947, Loess in the southern Mississippi Valley: Identification and distribution of the loess sheets: Soil Sci. Soc. America Proc., v. 12, p. 389-399.

U. S. DEPARTMENT OF AGRICULTURE
SOIL CONSERVATION SERVICE
WASHINGTON, D. C.

28

Reprinted by permission from *Soil Sci.*, **92**(6), 396–401 (1961)

DEPTH OF LOESS AND DISTANCE FROM SOURCE

PAUL E. WAGGONER AND CHRISTOPHER BINGHAM

The Connecticut Agricultural Experiment Station

Received for publication April 13, 1961

As one travels away from the source of loess, the thickness of the blanket at first decreases precipitously, and then, after a few miles, the decrease becomes a slow attenuation. Accepting the aeolian nature of loess, from a theory of atmospheric turbulence we derived a relation between distance and depth that provides an explicit basis for the discussion of these changes in mantle depth.

Along the borders of the glaciers of the Ice Age, the westerly winds lifted clouds of dust from the barren bottomlands of the glacial rivers, and carried and deposited them eastward. Deposits can be found west of rivers and near many rivers, but the vast deposits that stretch eastward from the rivers of Illinois (7) and Iowa (3) provide ideal examples for our study. The blankets of loess along the eastern edge of the bottomlands form bluffs many hundreds of inches thick. Within 20 miles, the blankets thin to a few 100 or even less than 100 inches and spread more evenly over at least another 50 miles.

Despite the great extent and variable depth of the blanket carried eastward from outwash along the rivers, its particles are remarkably similar in size. Near the source, coarse particles of sand are found where, according to Stoke's Law, they were sorted out. But after this short and rapid decrease in particle size, a diameter of about 30 μ is reached, which from then on shows surprisingly little change, encouraging us to believe that a single process deposited the loess and a simple hypothesis will explain it. The conclusions we deduce from the observations of depth will depend upon the hypothesis we adopt.

Deciding whether the blanket depth falls away eastward more quickly or slowly than normally expected, and whether it is a homogeneous gradient and a single process of transport and deposit, will depend upon what is expected. Krumbein (5) expected that the logarithm of blanket depth B would decrease linearly with distance r from the source. This hypothesis is equivalent to $dB/dr = a B$, where a is a constant of proportionality. Thus the decrease in depth

is steepest where the deposit is deepest. Perhaps this is only restating mathematically that the depth decreases rapidly near the deep bluffs. Or it may be stating that great deposition causes rapid depletion of the moving cloud of dust. The hypothesis is not supported by Nature: near the bluffs the change in the logarithm of depth (fig. 1) is steeper than it is afar (7).

Perceiving the failure, Smith (7) proposed that the depth would decrease linearly with the logarithm of distance. This is equivalent to $dB/dr = a'/r$, or the decrease is steeper near the source. Conceivably this mechanism could be a rate of dilution of the air-borne cloud of slit that is inversely proportional to the distance that the cloud has traveled. But this hypothesis also fails: the change in depth per increment in the logarithm of distance (fig. 2) is greater near the bluffs that it is further eastward (7). Smith attributed this failure to varying wind directions and the shape of the bottomlands; he inferred that the mechanisms for deposition nearby were not the same as those for deposition afar. Alternatively, the mathematical specification may not have corresponded to the natural process, and we should seek another explanation.

It has been said [(1), p. 175] "The Pleistocene epoch provided conditions ideal for loess making, but it was not uniquely the workshop for eolian activity that it was once thought to be." If it is not unique, a sensible hypothesis would be one constructed upon a theory of turbulence that describes the atmosphere of the present.

The statistical theory of turbulence adequately explains gradients of silt-size fungal spores disseminated by the wind (2). It provides a particularly good explanation for gradients of known history (10, 11). Having explicit and realistic bases, a hypothesis based upon this theory may be a more useful standard to which the loessial gradients can be compared.

This meteorological theory, developed by G. I. Taylor and O. G. Sutton, gives concentrations of air-borne material. If the concentrations are then simply to be transformed into depositions,

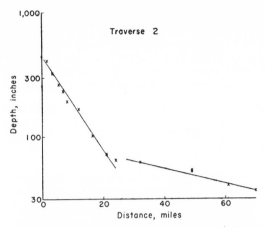

Fig. 1. The logarithm of the depth of all loess as a function of distance from the bluffs [traverse 2, (7)].

the particles must be practically homogeneous in their rates of deposit. At first glance, one might think otherwise, expecting that the air-borne slit would be sorted according to Stoke's Law. But Gregory (2), citing botanical examples, has shown that differences in size within the range of silt cause no differences in deposition. Tamura et al. (9) found that, in fact, little sorting of particles occurs below diameters of 50 μ. This is understandable, because the terminal settling velocities of these small objects are less than the vertical velocities found in turbulent air [(8), table 26]. We shall thus assume that loess particles have effectively uniform rates of deposition, a prerequisite to use of the theory. Later the effect of changing proportions of fast-settling sand and slow-settling clay will be examined.

The physical meaning of the parameters of the statistical theory of turbulence is best shown by an abbreviated derivation. Three assumptions will be made: turbulence in the three dimensions is not correlated; the concentration is twice that specified by a three-dimensional distribution symmetrical about the origin, because all material from the source is in the upper hemisphere; and the traps we consider are so far from the source that the distribution of concentration will not change while the cloud drifts over a single trap.

Two relations are well known and will be incorporated. Concentration is distributed in all three dimensions according to the Gaussian or normal distribution [(8), p. 277]. The variance σ^2 down- and cross-wind is $\frac{1}{2} C^2 r^m$ [(8), p. 286],

while the variance in the vertical direction is $\frac{4}{9}$ of this (6). The constant C is near 1. The exponent m varies between 1 and 2; it describes the degree of turbulence and lies nearer 2 if transfer persists over a long time.

These three assumptions and two relations specify the function relating concentration χ g./cm.^{3} to the quantity of silt particles in a cloud, to the variances and to the distances x, y, and z down-wind, cross-wind, and vertically from the center of the cloud. The quantity in the cloud, an instantaneous emission from 1 cm. of line source extending crosswind, is $q(r)$ g. of material remaining after a drift of r cm.

$$\chi = \frac{2q(r)}{(2\pi)^{3/2}\sigma^2(2/3\sigma)} \exp - \tfrac{1}{2} \frac{x^2}{\sigma^2} + \frac{y^2}{\sigma^2} + \frac{9z^2}{4\sigma^2}$$

The deposition d in g. per cm.^{2} of collecting surface can be derived from the concentration at $z = 0$. First we introduce p, the proportion of the particles passing through a vertical 1-cm.^{2} area or wicket, which then settles upon the cm.^{2} that lies down-wind. Next we sum over an infinite distance cross-wind, because the deposit comes from an infinitely long line source that emitted q g./cm. Finally we sum an infinite distance in the direction of the wind because ample time has certainly elapsed for the cloud to drift through the wicket.

$$d = \int_{-\infty}^{\infty}\int_{-\infty}^{\infty} p\chi \; dy \; dx$$

The third assumption permits the number of particles in the cloud and the variance to be treated as constants with respect to x and y.

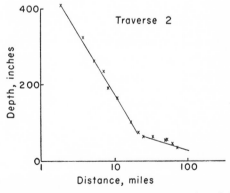

Fig. 2. The depth of all loess as a function of the logarithm of distance from the bluffs [traverse 2, (7)].

Therefore,

$$d = \frac{2pq(r)}{(2\pi)^{\frac{1}{2}}(2/3\sigma)} \int_{-\infty}^{\infty}\int_{-\infty}^{\infty} \frac{e^{-\frac{x^2}{\sigma^2}}}{(2\pi)^{\frac{1}{2}}\sigma} \cdot \frac{e^{-\frac{y^2}{\sigma^2}}}{(2\pi)^{\frac{1}{2}}\sigma} \, dy \, dx$$

$$d = \frac{3pq(r)}{(2\pi)^{\frac{1}{2}}\sigma} = \frac{3pq(r)}{\pi^{\frac{1}{2}}Cr^{m/2}}$$

Because many puffs arose from the line source during the centuries of the Pleistocene, the depth b cm. of the blanket is obtained from the deposit by summing over the many clouds that drifted eastward during the Pleistocene. Considering our cloud as the average, the summation can be replaced by multiplication by a constant of proportionality, k cm.^{3}/g.

$$b = kd \tag{1}$$

This expression for blanket depth as a function of distance from the source is the gist of our hypothesis. It can be refined, allowing for depletion of the cloud through settling and for a wide-source region.

The depletion is equal to the settling proportion times the quantity of silt in the 1-cm. slice of the cloud nearest the ground.

$$\frac{dq(r)}{dr} = -p \int_{-\infty}^{\infty}\int_{-\infty}^{\infty} \chi \, dy \, dx;$$

$$q(r) = q(0) \exp\left\{ -\frac{3pr^{(1-(m/2))}}{C\pi^{\frac{1}{2}}(1-(m/2))} \right\} \tag{2}$$

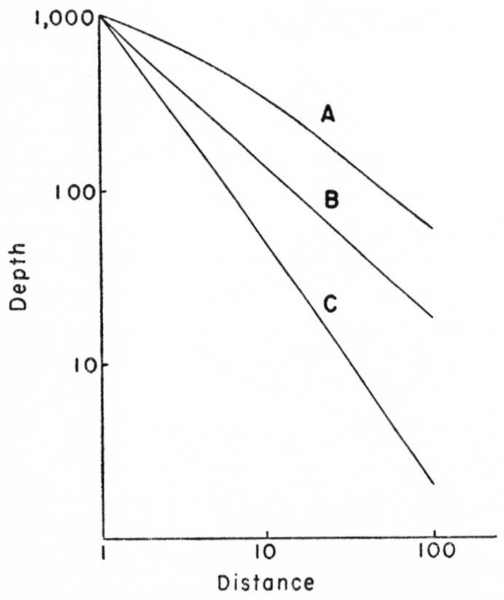

FIG. 3. Effect upon blanket depth of a source 10 km. wide (A) and of depletion of cloud by p = 0.05 (C). The middle curve (B) is the prediction for no depletion, and a line source where m is 1.75.

The effect of a wide-source region can be analyzed by redefining q (0) as the g. emitted in a puff from a source 1 cm. long and 1 cm. wide. Then the depth of blanket B at distance r_1 from the lee and r_w from the windward side of the source is the sum of the deposits b from each of the 1-cm. strips of source.

$$B = \int_{r_1}^{r_w} b \, dr = \frac{3pq(r)}{C\pi^{\frac{1}{2}}(1-(m/2))} \tag{3}$$

$$\cdot \left(r_w^{(1-(m/2))} - r_1^{(1-(m/2))} \right)$$

The consequences of our hypothesis (equation (1)) and the refinements (equations (2) and (3)) can be seen graphically. Because equation (1) is of the form $b = (\text{constant}) \, r^{-m/2}$, our hypothesis predicts that on double logarithmic graph paper the curve relating blanket depth to distance will be linear with slope $-m/2$. The center curve B in figure 3 is drawn from this hypothesis with the frequently observed m of 1.75 [(8), p. 277]. If the settling coefficient p is 0.05, depth will decrease more rapidly with distance, as in the lower curve C in figure 3. Finally, if the source is 10 km. wide, depth will at first decrease more slowly with distance, the graphical line attaining a slope of $-m/2$ only after the cloud has drifted a distance much greater than the width of the source (upper curve A in figure 3). Thus we expect that a graph of log (depth) versus log (distance) will be nearly a straight line. We expect that rapid settling will steepen the slope while broad sources will flatten it. Finally, we expect that any curvature will be a steepening with distance.

The effect of variable particle size and settling rates can be seen intuitively. A relatively large contribution of sand near the source will add more to the logarithm of depth near than far; hence the m estimated from the graph will be greater than the m that represents the transfer of a homogeneous material. If clay from far sources were carried down in proportion to the silt deposition, it would not disturb our analysis. In fact, however, clay is relatively less plentiful in the deposits near the source, adding less to the logarithm of depth near than far; hence, the m estimated from the graph will be less than the m that represents the transfer of a homogeneous material. Since the proportion of silt itself is rather constant with distance, and since the disappearance of sand and appearance of clay with distance affect m in opposite directions, variable particle size has little effect upon our graphs.

The erosion of a constant depth of mantle would decrease the observed m.

The many observations (7) east of the Mississippi near St. Louis and east of the Illinois provide a test for these expectations. The traverse number 2, east of St. Louis, extends from 1.9 to 70 miles, with few particles of diameter greater than 50 μ or less than 5 μ that settled out and confused the application of our simple hypothesis. Although the river bends near the beginning of the traverse, the source is not as wide as we shall encounter later [(1), fig. 46]. Figure 4 shows that depth of the mantle falling away eastward as expected. The estimate of m is 1.43 ± a standard error of 0.07, well within the theoretical bounds for this characteristic. Both the smallness of the estimate of m, which is usually about 1.75, and the slight curvature are undoubtedly manifestations of this source width that is at least a few miles.

The traverse number 1 and Cass County, southeast of the broad outwash of the Illinois, is shown in figure 5. The depths are of Peorian loess, and within 2 miles of the bluffs we have reduced them according to their considerable sand content. Particles less than 1 μ in diameter were few at any distance. The traverse was likely not in the direction of the wind, but this

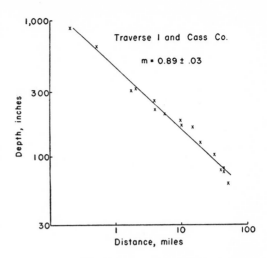

FIG. 5. The logarithm of the depth of Peorian loess as a function of the logarithm of distance from the source [traverse 1, and Cass Co., (7)].

will not affect the value of m. The estimate of m is significantly smaller than either the theoretical lower limit 1 or the 1.43 observed east of St. Louis.

The most obvious cause of the low m is the great width of the source; not only do the bottomlands spread wide where the Sangamon flows into the Illinois, but the Mississippi lies less than 60 miles to the west and undoubtedly contributed dust to the clouds that drifted over this traverse. Our hypothesis not only predicted that depth would decrease slowly near a broad source, it also predicted that the graphical line would attain a slope of $-m/2$ only after the cloud had drifted a distance much greater than the width of the source. In fact, the change in depth between 14.7 and 57 miles corresponds to an m of 1.5, well within the theoretical limits.

Deep drifts of loess are also found east of the Missouri, spreading into Iowa. The contours of loess depth drawn by Kay and Graham (4) in their figure 70 provide an approximation. If the contours are crossed eastward, beginning on the boundary between Monona and Harrison counties, the change in depth is steep and concave downward on double logarithmic coordinates. The estimate of m is about 2.2, exceeding both the theoretical upper limit and the observations in Illinois. If we ignore any contributions from the sand hills of Nebraska, the source can be called thinner than the others. Still, the esti-

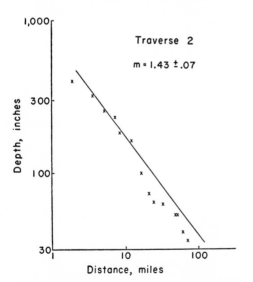

FIG. 4. The logarithm of the depth of all loess as a function of the logarithm of distance from the bluffs [traverse 2, (7)].

mated m is larger than expected, and we turn to the precise depths published by Hutton (3).

Hutton's traverse 1, which proceeds southeast from Monona county, has an estimated m of 1.49 ± 0.15, significantly smaller than the m estimated from the contours of Kay and Graham and well within the theoretical limits. Hutton's traverse 2, which proceeds southeast from Council Bluffs, lies down-wind from the broad source where the Platte joins the Missouri and should display a low m. In fact, the m for the entire traverse is only 0.70 ± 0.04. The change in depth from 47 to 87 miles corresponds, however, to an m of 1.5, well within the theoretical limits.

The curvature of the graphical curves conforms to expectation. Both Illinois and the two Iowa traverses produced curves that are nearly linear, and in all instances the departure from linearity is the anticipated steepening with distance that might be caused either by a rapid depletion of the cloud by settling or by a wide source region. Since the observed m's fall below the commonly observed 1.75, and since the m's vary with width of source region, we conclude that this width is more influential in determining the shape of the loessial gradient than is the settling proportion. This general correspondence between the hypothesis and the observations of the slopes and shapes of 4 traverses of loess leads us to accept the validity of the hypothesis for the aeolian deposition and encourages us to draw conclusions from the estimates of m.

Although within the theoretical limits, the magnitude of the observed m, about 1.5, is surprisingly small. Meteorologists generally observe about 1.75 for short periods of dispersal. They observe near 2 for long periods of dissemination—and the millenia of the Pleistocene were certainly long. Since m characterizes turbulence, the difference between Pleistocene and present ostensibly signifies a calmer Pleistocene and gustier present. But we have no reason to believe this, and so we look for other implications in the smallness of m. First, the m-increasing depletion of the cloud by settling was clearly less important than other processes. First among these m-decreasing processes was undoubtedly the breadth of the source. This was not, however, the only reason: the m attained even after many miles of travel from the source was still less than 1.75, which indicates that other processes were at work. The addition of a constant amount of clay both near and far from the source many have decreased the observed m, especially in the finer-textured soils of the traverses east of the Missouri. Finally, erosion undoubtedly changed the m: wind and water surely carried away about the same depth of mantle both near and far from the source, decreasing the m we observe today.

SUMMARY

From the statistical theory of atmospheric turbulence, a hypothesis is derived, predicting the change in depth of loess blankets observed as one travels away from the source of the windblown deposit. It predicts that the log (depth) will decrease linearly with the log (distance from source) at a rate $m/2$, dependent upon turbulence. Where the source is broad, the decrease will be less at first and reach $m/2$ only at great distance from the source. Where the silt is trapped efficiently by the surface, the decrease will be greater and increase with distance.

When the prediction is compared to loess east of the Mississippi, the form and slope of the line conform nicely to expectation. When it is compared to loess southeast of the Illinois, the slope shows the effect of the broad source. Finally, when it is compared to loess east of the Missouri, the line conforms to expectation, being flatter where the source is broader. In all four graphs the curves are nearly linear, curving slightly downward as expected.

REFERENCES

(1) FLINT, R. F. 1947 *Glacial Geology and the Pleistocene Epoch.* John Wiley and Sons, New York.

(2) GREGORY, P. H. 1945 Dispersion of airborne spores. *Trans. Brit. Myco. Soc.* 28: 26–72.

(3) HUTTON, C. E. 1947 Studies of loessderived soils in southwestern Iowa. *Soil Sci. Soc. Amer. Proc.* 12: 424–431.

(4) KAY, G. F., AND GRAHAM, J. B. 1943 The Illinoian and post-Illinoian geology of Iowa. *Iowa Geol. Survey* 38: 1–262.

(5) KRUMBEIN, W. C. 1937 Sediments and exponential curves. *J. Geol.* 45: 577–601.

(6) SCRASE, F. J. 1930 Some characteristics of eddy motion in the atmosphere. *Gt. Brit. Met. Off. Geophys. Mem.* 52: 3–16.

(7) SMITH, G. D. 1942 Illinois loess. *Illinois Agr. Expt. Sta. Bull. 490*, pp. 137–184.

(8) SUTTON, O. G. 1953 *Micrometeorology.* McGraw-Hill Book Co., New York.

(9) TAMURA, T., *et al.* 1957 Characteristics of eolian-influenced soils in Connecticut: II. *Soil Sci. Soc. Amer. Proc.* 21: 536–539.

(10) WAGGONER, P. E. 1952 Distribution of potato late blight around inoculum sources. *Phytopathology* 42: 323–328.

(11) WAGGONER, P. E., AND TAYLOR, G. S. 1955 Tobacco blue mold epiphytotics in the field. *Plant Disease Rep.* 39: 79–85.

29

Reprinted from *Soil Sci. Soc. America Proc.*, **34**(2), 296–301 (1970)

Loess Distribution from a Source[1]

C. J. FRAZEE, J. B. FEHRENBACHER, AND W. C. KRUMBEIN[2]

ABSTRACT

Loess thickness and particle size changes along six Peoria loess traverses in Illinois and southwestern Indiana were investigated by fitting the data to various mathematical models by computer. An exponential model, a logarithmic model, a power function, and an additive exponential model developed in this study were tested. The decrease in mean particle size, which is due to changes in the coarse silt content, both windward and leeward from the source is explained best by the additive exponential model. Mean particle size changes much more rapidly near the source than afar.

Leeward of the source, two changes in the rate of loess thinning occurred. The change closest to the bluff is related to particle size differences with distance from the source. The second change is attributed to greater loess deposition in a wider zone near the source by winds from different directions. Windward of the source, no change in the rate of loess thinning due to particle size differences was observed.

Based upon the particle size and thickness data presented, the decrease in loess thickness from its source is explained best by an additive exponential model. Leeward of the source where two changes in the rate of loess thinning occur, three exponential terms are needed to accurately characterize loess thickness. The decrease in loess thickness with distance from the source which is due to the more rapid decrease in particle size near the source is defined by the first exponential term. The decrease in loess thickness associated with deposition by winds of different directions is described by the second exponential term. The third exponential term depicts the regional loess thinning due to a decrease in the number of loess particles with distance from the source. Windward of the source only two exponential terms are needed to accurately describe loess thickness because no change in the rate of loess thinning due to particle size differences was observed.

[1] Contribution from the Dept. of Agronomy, Univ. of Illinois, Urbana. Part of a thesis, submitted by the senior author in partial fulfillment of the requirements of the Ph.D. degree at the University of Illinois. Presented before Div. S-5, Soil Science Society of America, Nov. 12, 1968 at New Orleans, La. Received Aug. 8, 1969. Approved Oct. 8, 1969.

[2] Formerly Graduate Research Assistant, Univ. of Illinois, now Assistant Professor, South Dakota State Univ.; Professor, Univ. of Illinois; and Professor, Northwestern Univ., Evanston, Ill.

Additional Key Words for Indexing: loess thickness, mean particle size, additive exponential, mathematical model.

ALTHOUGH loess distribution from a source has been studied in various regions (Krumbein, 1937; Smith, 1942; Hutton, 1947; Simonson and Hutton, 1954; Ruhe, 1954; Hanna and Bidwell, 1955; Caldwell and White, 1956; Sitler and Baker, 1960; Waggoner and Bingham, 1961; Foss and Rust, 1962; Fehrenbacher et al., 1965), the exact nature and quantitative description of the loess thinning pattern are still debatable. Once suitable equations are developed, loess thickness at a specific location may be calculated and if more than one source area is involved, the loess increments from each indicated.

Because of deviating assumptions and varying sampling conditions previous investigators proposed various mathematical models to specify the decrease in loess thickness away from a source. In the first quantitative attempt by Krumbein (1937) an exponential equation was used to depict the loess depths for an area between 3.2 and 20.9 km (2 and 13 miles) southeastward from the Mississippi River bluff in Henderson County, Ill. In this equation, $y = ae^{-bx}$, y is the loess depth at distance x from the bluff, a is loess thickness at the bluff, and b is the coefficient of loess thinning. This model predicts that the logarithm of the thickness will decrease linearly with distance from the source. The rate of depletion of the loess cloud is proportional to the quantity of particles which the cloud contains as well as to distance traveled. Krumbein's data as well as that of Smith (1942), Hutton (1947), and Fehrenbacher et al. (1965) show the change in the logarithm of the thickness, or the rate of thinning of the loess, to be greater near the source than afar. Krumbein (1937) suggested the change in the rate of loess thinning with distance from the source may be due to particle size differences.

In an attempt to solve this problem, Smith (1942) proposed a logarithmic equation, $y = a - b \log x$, where y is loess thickness, a and b are constants for a given source, and x is distance from the river bluff to account for the decrease in particle size and loess thickness for two traverses in Illinois. According to this model, loess thickness will decrease linearly with the logarithm of the distance. The depletion of the loess cloud is inversely proportional to distance traveled. The decrease in mean particle size of the loess also was linear with respect to the log of the distance.

Similar to the exponential model, the logarithmic model fails to adequately describe the decrease in loess thickness from the source as evidenced by the data of Smith (1942), Hutton (1947), and Fehrenbacher et al. (1965). Smith (1942) and others found it necessary to use two of the log models to fit their loess thickness data. Smith attributed this change in the constants of the equation for loess thickness to greater loess deposition near the source by winds from variable directions rather than to a change in rate of particle size decrease.

Not accepting Smith's explanation, Waggoner and Bing-

ham (1961) derived a power function to characterize the decrease in loess thickness away from a source. They assumed little sorting of loess particles below 50 μ and based their power function, $y = ax^{-b}$ (where y is loess thickness, a is a constant, x is distance from the source, and b is the rate of loess thinning which is related to air turbulence) on a statistical theory of turbulence. On the basis of this model the logarithm of the thickness is expected to decrease linearly with the logarithm of the distance.

In 1965 Wright and Ruhe introduced yet another mathematical model to represent loess thinning from a source. They used a hyperbolic function $y = 1/a + bx$ (where y is loess thickness, a and b are constants, and x is distance from the source) to explain the decrease in loess thickness along Rock Island Railroad cuts in Southwestern Iowa. This equation is similar to the logarithmic model in that the rate of loess thinning decreases progressively with distance from the source. However, the rate of the decrease in loess thickness is greater for the hyperbolic function.

Because the above equations do not appear to adequately describe the distribution of loess from a source, and also because we disagree with the assumption that loess particles have uniform rates of deposition, an additive exponential model will be proposed to define loess distribution. The loessial particles were swept from the glacial outwash deposits in the Pleistocene drainageways and deposited upon the landscape systematically according to the physical conditions of the wind. The decrease in loess thickness away from the source is the result of two factors. First and most important is the decrease in the number of particles which occurs as the loess cloud is being depleted. This decrease probably is the result of lower wind velocities away from the source as well as distance traveled. Secondly, there is a decrease in the size of the loessial particles with distance from the source. Most of the size decrease is near the source so that the effect of decreasing particle size is more local to the source while the decrease in the number of particles is regional.

Mathematically the result of these two factors operating simultaneously can be expressed by an additive exponential model, $y = ae^{-bx} + ce^{-dx}$. The first term, which describes the more rapid change in loess thickness near the source due mainly to particle size differences, decreases asymptotically to the second, which describes the regional decrease in loess thickness that is due to differences in the number of loessial particles. The loess thickness at the bluff equals $a + c$. The loess thinning coefficient b is expected to be much larger than the loess thinning coefficient d. This model has been used by Krumbein (1941) to describe size, shape, and roundness data of rock fragments.

The distribution of loess from a source is of pedologic significance because many chemical and physical properties of loess soils have been related not only to source of the loess but also to its thickness and mean particle size.

The purpose of this investigation is to test the proposed additive exponential model as well as some of the previous equations in order to determine which model will best describe the changes in loess thickness and particle size with distance from the source.

EXPERIMENTAL PROCEDURE

A survey of the literature did not reveal any satisfactory data illustrating changes in loess thickness and particle size in which the parameters for these two characteristics were obtained from the same site. Although Smith (1942) presented information on loess thickness and particle size, in most cases, the data are not from the same site. Because our proposed additive exponential model is based upon the changes in particle size it was necessary to obtain this kind of data.

Six Peoria loess traverses were laid out from major loess source areas in Illinois and southwestern Indiana (Fig. 1). The thickness of Peoria loess was measured by using a soil probe truck, boring with a bucket auger, or observing road cuts. Measurements were made at sites which were as nearly uniform as possible in slope position on stable, upland divides where accumulation and erosion of soil material were at a minimum. At each site thickness and depth of leaching of the Peoria loess and nature of the underlying material were recorded. The Peoria loess was treated as one bed because the individual increments cannot be traced visually or texturally.

In order to compare particle size at the same relative position in thick and thin loess sections, the samples were obtained as follows. First, the Peoria loess section was divided into quarters; next, each quarter was divided into thirds; then for the calcareous portion of the loess section the middle of each third was sampled. If textural stratigraphic changes occurred in the loess section, additional samples were taken.

Particle size distribution was determined on both calcareous and carbonate-removed samples by the pipette method as described by Kilmer and Alexander (1949) using calgon as a dispersant. The particle size classes determined were 2—1 mm, 1—.5 mm, 500—250μ, 250—177μ, 177—125μ, 125—88μ, 88—62μ, 62—44μ, 44—31μ, 31—15μ, 15—2μ, and $< 2\mu$. In samples where carbonates were removed, they were destroyed with hydrochloric acid. $CaCO_3$ equivalent was determined gravimetrically as described by Richards (1954).

The different models which have been proposed to explain loess distribution from the source were fitted by computer to the loess thickness and particle size data using a least squares basis. When two logarithmic models are used to explain the loess thickness distribution the region of the traverse which each equation will describe must be chosen by observation before fitting with the computer.

RESULTS AND DISCUSSION

Particle Size

Variation of Particle Size—The greatest variation of mean particle size within a vertical loess section is near the source, probably because of greater differences in wind velocities.

The mean particle size for only the calcareous samples of each site in five of the six traverses studied is shown in Table 1. Samples were not collected along Traverse 6. Only calcareous samples were used to avoid the effects of weathering on particle size. Other than distance from the source, the mean particle size of loess at a location is dependent upon texture of the parent sediment, wind velocity, and topography of the bluff. The finer mean particle sizes at sites along Traverse 4 as compared to Traverses 1 and 3 can probably be explained by the fact that Traverse 4 is located much farther down the glacial drainageway (Table 1, Fig. 1). The mean particle size of the parent sediment would be expected to decrease with distance along the Pleistocene drainageway. The turbulent velocities of the wind determines not only the size of particles transported but also the distances from the source to which the coarser particles may be carried.

The topography of the bluff also has an important influence upon particle size changes away from the source. The bluff undoubtedly sets up turbulent eddies and affects wind velocity. It is the coarser particles, those which move mainly by saltation but also to some extent by suspension, which are most affected. These coarser particles, fine sand to coarse silt, can move farther downwind or leeward from the source as the slope of the bluff decreases. If a high vertical or near vertical bluff is encountered by wind-blown particles, very few sand and very coarse silt particles would be expected on top of the bluff. The slope of the bluff for all traverses studied appeared to be between 45–60%.

The coarser mean particle sizes at sites 21 and 22 of Traverse 3 and site 48 of Traverse 4 (Table 1) are due to the presence of preloessial side valleys cut through the bluff of the larger loess source valley. The sand and coarse silt

Fig. 1—Traverses along which loess was studied.

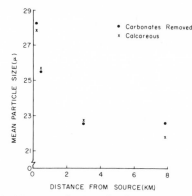

Fig. 2—Mean particle size of calcareous and carbonate-removed samples from middle third of third quarter of sites along Traverse 4.

Table 1—Mean particle size of calcareous samples and loess thickness changes along traverses studied*

Site	Distance from bluff	Thickness	Mean particle size	Coarse silt 62 - 31μ	Fine silt & clay < 31μ
	km	cm	μ	%	%
			Traverse 1 - Leeward		
1	0.08	1,565	35.4	56.4	38.8
2	0.64	945	28.9	47.4	51.1
3*	1.9	671	27.0	40.1	58.7
4	2.8	500	25.8	37.2	61.7
5	5.1	442	24.2	34.8	64.4
6	7.8	376			
7	13.2	305			
8	22.0	226			
9	27.2	198			
			Traverse 2 - Windward		
10	51.2	198			
11	43.1	203			
12	29.3	211			
13	21.7	231	16.3	11.6	87.8
14	11.9	269	18.2	14.6	84.6
15	8.9	279	19.4	17.0	82.4
16	6.0	376	19.7	17.5	82.6
17	3.7	432	20.4	19.2	80.2
18	1.3	605	21.2	21.6	78.4
19	0.16	813	22.1	23.7	76.1
			Traverse 3 - Leeward		
20	0.40	1,981	33.6	41.5	53.5
21	0.97	1,194	46.4	35.2	46.9
22	1.4	1,003	38.4	36.9	49.9
23	2.7	828	25.4	31.7	66.9
24	5.1	610	24.6	28.8	69.9
25	10.1	396	22.3	22.4	75.7
26	19.0	358	22.0	21.7	77.0
27	25.4	318			
28	29.8	244	20.6	20.3	76.3
29	47.3	211	20.7	21.4	76.4
30	58.7	185			
31	67.9	180			
32	77.9	178			
33	88.4	180			
34	102.2	150			
35	112.2	157			
36	121.7	140			
37	145.5	127			
38	152.9	127			
39	162.5	127			
40	178.2	124			
41	183.8	109			
42	209.2	102			

(continued)

Table 1-continued

Site	Distance from bluff	Thickness	Mean particle size	Coarse silt 62 - 31μ	Fine silt & clay < 31μ
			Traverse 4 - Leeward		
43	0.24	1,257	27.3	37.3	60.7
44	0.48	1,046	25.9	37.6	61.5
45	0.72	912	25.8	36.3	62.9
46	1.6	702	23.8	31.9	67.6
47	2.8	589	22.9	27.0	71.3
48	5.5	437	25.4	28.2	68.9
49	7.9	417	22.5	28.2	71.0
50	10.6	371			
51	13.2	330			
52	16.0	315			
53	20.6	229			
54	21.7	229			
55	22.5	216			
56	25.7	226			
57	29.0	165			
58	29.3	195			
59	33.8	203			
60	35.4	201			
61	47.0	140			
62	48.3	157			
63	49.9	173			
64	61.2	165			
65	74.8	140			
66	84.5	122			
67	97.4	99			
68	121.5	94			
69	131.6	91			
			Traverse 5 - Windward		
70	38.0	99			
71	22.7	112			
72	19.8	124			
73	14.6	147			
74	12.4	160			
75	9.5	165			
76	6.4	180			
77	5.0	208			
78	4.0	198			
79	2.9	262			
80	2.3	290	20.4	19.2	79.6
81	1.1	442	22.0	21.5	77.6
82	0.72	508	22.6	23.4	76.1
83	0.16	640	23.5	25.9	73.4
			Traverse 6 - Leeward		
84	0.08	1,006			
85	0.32	699			
86	0.48	635			
87	1.0	533			
88	1.1	533			
89	1.6	445			
90	3.5	279			
91	6.0	279			
92	12.9	241			
93	17.1	203			
94	24.1	183			
95	25.7	140			
96	33.0	140			
97	41.8	145			
98	44.3	114			
99	52.3	114			
100	57.9	117			
101	62.8	102			
102	67.6	91			

* Sites along Traverses 1, 2, 3, 4, and 5 for which mean particle size is not shown were not sampled because all of the loess was leached.

particles were able to move farther from the source than expected because the slope of these valleys is more gradual than that of the bluff. Because of the above situation sites 21 and 22 of Traverse 3 and site 48 of Traverse 4 were not used in analyzing the decrease in mean particle size.

Tamura et al. (1957) attributed the decrease in particle size along Smith's Traverse 1 (1942) after 6.4 km (4 miles) to increased weathering of the easily weatherable calcium carbonate of the loess as it thinned. In order to study the effect of differences in calcium carbonate content on the size decrease, a mechanical analysis was performed on several samples along Traverses 1, 2, 3, 4, and 5 from which the carbonates were removed. Data from Traverse 4 which are representative are shown in Fig. 2 to illustrate this relationship. Neither the mean particle size at a site nor the decrease in mean particle size with distance from the source are related to the differences in calcium carbonate content of the loess for the traverses studied.

The decrease in mean particle size is the result of sorting of silt size particles (Table 1). The coarse silt (62–31 μ) content diminishes away from the source while the content of the finer fractions (fine silt and clay — < 31 μ) increases (Table 1). The coarse silt fraction decreases more rapidly near the source (Table 1). This decrease is in the same region of the traverses as explained by the first term of the additive exponential equation. Waggoner and Bingham's (1961) assumption that loess particles have uniform rates of deposition appears not to be supported, since the decrease in mean particle size is due to changes in the silt particle size range.

Decrease of the Mean Particle Size with Distance from Source—Mean particle size is coarser leeward than windward from the source (Table 1). On the leeward side the mean particle size changes more rapidly near the source and is best explained by the additive exponential model (Fig. 3). On the windward side the decrease is fairly constant with distance and is also described quite well by the single exponential model. (Fig. 4).

Based upon the data presented, the particle size decrease with distance leeward from the source may be illustrated as in Fig. 5. The greater decrease near the source, from the bluff to distance x which is explained by the first term of the additive exponential model, is postulated to be the result of coarse silt particles settling out according to Stokes's Law (Fig. 5). This sorting occurs only for short distances from the source and this distance lengthens or

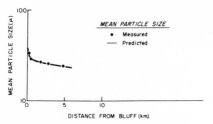

Fig. 3—Mean particle size decrease along Traverse 1 as predicted by additive exponential model.

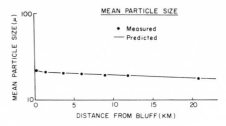

Fig. 4—Mean particle size decrease along Traverse 2 as predicted by additive exponential model.

diminishes according to wind velocity and the topographic conditions at the bluff. The coefficient c of the additive exponential model is indicative of the mean particle size above which sorting occurs according to Stokes's Law. The c coefficients of the additive exponential model suggest a fairly rapid decrease in particle size until a mean particle size of approximately 25 μ is reached. Therefore, unless the loess has mean particle sizes greater than approximately 25 μ, no change in the rate of particle size decrease would be expected and the decrease in particle size with distance from the source would be defined by a single exponential model. This appears to be the situation windward of the source. The decrease in particle size at distances farther away from the source, beyond distance x (Fig. 5), which is explained by the second term of the additive exponential model, is much more gradual because the terminal velocities of the small silt and clay size particles are similar to or less than the velocity of the turbulent air in which they were transported. This decrease is due to a depletion of the loess cloud as it traverses the landscape and is proportional to the particle size composition of the loess cloud as well as distance traveled.

Thickness

Change in the Rate of Loess Thinning—In this study as well as others (Krumbein, 1937; Smith, 1942; Hutton, 1947; Fehrenbacher et al., 1965), the rate at which the loess thinned changes with distance from the source. On the basis of the particle size analysis this change should be

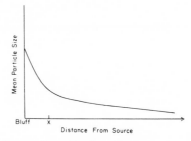

Fig. 5—Decrease in particle size with distance from source.

related to particle size differences. The data suggest the possibility of two changes, one due primarily to particle size differences, the other due to greater deposition of loess near the source by winds of variable directions. Along Traverse 3 there appears to be a break at about 6 km and another at 35 km (C. J. Frazee, 1969. Distribution of loess from a source. Ph.D. thesis, University of Illinois, Urbana). This first change occurs in the same area as the more rapid change in particle size. The loess thickness distribution along Traverse 4 also shows two changes in the rate of thinning (Fig. 6). Again the first change in thinning rate is at about the same distance from the bluff as the change in rate of particle size decrease. This change occurs at about 4 km (Fig. 6). The second change occurs at a similar distance from the bluff as the change in the rate of thinning found by Smith (1942) at about 32 km (20 miles) along his Traverse 2 which is in the same area. Similar changes in the rate of loess thinning are not apparent for Traverse 1 because most of the particle size decrease occurs between sites 1 and 2. Also, only one change in rate of loess thinning occurs in Traverse 6, probably because this area has a blend of loess deposits from several sources, the Wabash, Ohio, and White Rivers. Two changes in the rate of loess thinning are not evident for the windward Traverses 2 and 5, because as pointed out in the particle size analysis, no change in the rate of particle size decrease is expected when the loess has mean particle sizes less than 25 μ (Table 1).

Decrease of Loess Thickness with Distance from Source—The loess thinning equations for Traverses 1 and 4 were computed first. Then, the thickness data of the other traverses were adjusted for loess increments on the basis

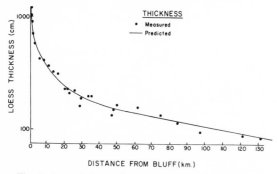

Fig. 6—Loess thickness along Traverse 4 as predicted by a three term additive exponential model.

of the equations for Traverses 1 and 4 before the loess thinning equations were developed for them. As expected, from the analysis of the changes which occurred in the rate of loess thinning with distance from the source, the additive exponential model or two of the logarithmic equations adequately defines the loess thickness decrease along the traverses which have one change in the rate of thinning. These traverses are 1, 2, 5, and 6. However, for Traverses 3 and 4, which have two changes in rate of loess thinning, the additive exponential model tends to underestimate the loess depths beyond 110 km along Traverse 3 and 60 km along Traverse 4. This is because no provision has been made for the change in the rate of loess thinning due to greater deposition of loess near the source by winds from different directions. This problem may be solved by adding an additional exponential term giving an equation, $y = ae^{-bx} + ce^{-dx} + fe^{-gx}$. Figure 6 shows the fit of the three term exponential equation, $y = 328\,e^{-2.6} + 190e^{-.17} + 87e^{-.01}$, to the loess thicknesses along Traverse 4. Although the power function explains the rapid decrease of loess thickness near the source, it does not adequately depict the more gradual decrease farther away from the source. The loess depths in this region are overestimated.

The different models were fitted to the loess traverses reported in literature. The traverses may be divided into two groups. First, those which do not start at or near the bluff, Krumbein (1937); Hutton's Traverse 1 (1947); and Smith's Traverse 1 (1942). These traverses do not show a change in the rate of loess thinning and the loess thicknesses are explained adequately by the single exponential or logarithmic model. The other traverses, Hutton's Traverse 2 (1947); Smith's Traverse 2 (1942); and Smith's Traverse 1 and Cass County (1942), start near the bluff and provide a good test for the various models. These traverses show a change in the rate of loess thinning with distance from the source. Both the additive exponential and two logarithmic models explain the loess thicknesses along these traverses well. As stated previously for Traverses 1, 2, 3, 4, 5, and 6 of this study, the power function cannot characterize the more gradual regional decrease farther away from the source.

Based upon the thickness and particle size data presented in this study, the decrease in loess thickness from a source and the decrease in particle size with distance from the source, which has an associated change in the

rate of loess thinning, are best described by the additive exponential model. The single exponential, logarithmic, and power functions do not adequately explain the distribution of loess from a source because they do not account for simultaneous particle size decrease and loess thinning with distance from the source.

LITERATURE CITED

1. Caldwell, R. E., and J. L. White. 1956. A study of the origin and distribution of loess in southern Indiana. Soil Sci. Soc. Amer. Proc. 20:258–263.
2. Fehrenbacher, J. B., J. L. White, H. P. Ulrich, and R. T. Odell. 1965. Loess distribution in southeastern Illinois and southwestern Indiana. Soil Sci. Soc. Amer. Proc. 29:566–572.
3. Foss, J. E., and R. H. Rust. 1962. Soil development in relation to loessial deposition in southeastern Minnesota. Soil Sci. Soc. Amer. Proc. 26:270–274.
4. Hanna, R. M., and O. W. Bidwell. 1955. The relation of certain loessial soils of northwestern Kansas to the texture of the underlying loess. Soil Sci. Soc. Amer. Proc. 19:354–359.
5. Hutton, C. E. 1947. Studies of loess-derived soils in southwestern Iowa. Soil Sci. Soc. Amer. Proc. 12:424–431.
6. Kilmer, V. J., and L. T. Alexander. 1949. Methods of making mechanical analysis of soils. Soil Sci. 68:15–24.
7. Krumbein, W. C. 1937. Sediments and exponential curves. J. Geol. 45:577–601.
8. Krumbein, W. C. 1941. The effects of abrasion on the size, shape, and roundness of rock fragments. J. Geol. 49:482–520.
9. Richards, L. A., (ed.). 1954. Diagnosis and Improvement of Saline and Alkali Soils. USDA Handbook 60.
10. Ruhe, R. V. 1954. Relations of the properties of Wisconsin loess to topography in western Iowa. Amer. J. Sci. 252:663–672.
11. Simonson, R. W., and C. E. Hutton. 1954. Distribution curves for loess. Amer. J. Sci. 252:99–105.
12. Sitler, R. F., and J. Baker. 1960. Thickness of loess in Clark County, Ill. Ohio J. Sci. 60:73–77.
13. Smith, G. D. 1942. Illinois loess: variations in its properties and distribution. U. of Ill. Agr. Exp. Sta. Bull. 490.
14. Tamura, T., A. Ritchie, Jr., C. L. W. Swanson, and R. M. Hanna. 1957. Characteristics of eolian influenced soils in Connecticut: II. Chemical and mineralogical properties as keys to profile mixing. Soil Sci. Soc. Amer. Proc. 21:536–539.
15. Waggoner, P. E., and C. Bingham. 1961. Depth of loess and distance from source. Soil Sci. 92:396–401.
16. Wright, H. E., Jr., and R. V. Ruhe. 1965. Glaciation of Minnesota and Iowa, p. 29–41. In H. E. Wright, Jr. and D. G. Frey (ed.) The Quaternary of the United States. Princeton University Press, Princeton.

Editor's Comments
on Papers 30, 31, and 32

30   GINZBOURG and YAALON
Petrography and Origin of the Loess in the Be'er Sheva Basin

31   YAALON
Origin of Desert Loess (abstract)

32   BRUNNACKER and LOŽEK
The Presence of Loess in Southeastern Spain (summary)

Papers 30–31 offer a look at a specific deposit of "desert" loess, that in the Be'er Sheva basin in Israel, and also a brief consideration of Yaalon's views on loess formation. There are desert deposits in other parts of the Middle East, and interesting observations on loess in Iraq have been made by Fookes and Knill (1969) and by Kukal and Saadallah (1970).

Material in Israel and Iraq is obviously not glacial; it cannot be the "glacigene eolian steppe deposit" described by Fairbridge (see Editor's Comments on Paper 44), and yet it is usually referred to as loess. It presumably falls within Ruhe's definition of loess: "Loess is a wind-deposited sediment that is commonly unstratified and unconsolidated and is composed dominantly of silt-size particles" (Ruhe, 1969, p. 29). It differs from the larger, obviously glacially related deposits in one very interesting way: it can contain a large proportion of carbonate particles. Kukal and Saadallah report a carbonate percentage of 60+. Perhaps a special carbonate loess should be specified and recognized as a definable material. Desert conditions may not allow the efficient production of silt-sized quartz particles, but silt-sized carbonates could be a different matter. Brunnacker and various coworkers operating in regions near the Mediterranean have reported several high-carbonate deposits. The extract printed here is from a paper by Brunnacker and Ložek in which they report the Granada loess with a carbonate content of up to 70 percent. Although totally different in mineralogy, these carbonate loesses appear to show definite stratigraphic similarities to the normal European loesses. (For more discussion of the desert loess problem, see Editor's Comments on Papers 39, 40, and 41.)

REFERENCES

Fookes, P. G., and Knill, J. L. 1969. The application of engineering geology in the regional development of northern and central Iran. Eng. Geol. 3, 81–120.

Kukal, Z., and Saadallah, A. 1970. Composition and rate of deposition of dust storm sediments in central Iraq. Cas. Mineral. Geol. (Prague) 15, 227–234.

Ruhe, R. V. 1969. Quaternary landscapes in Iowa. Iowa State University Press, Ames, Iowa, 255p.

30

Reprinted from *Israel Jour. Earth Sci.*, **12,** 68–70 (1963)

Petrography and origin of the loess in the Be'er Sheva basin

D. GINZBOURG AND D.H. YAALON, *Geological Survey of Israel, Jerusalem and Department of Geology, The Hebrew University of Jerusalem*

The Pleistocene loess deposits in the Be'er Sheva basin are of variable thickness, covering an area of about 1,600 km². Greatest concentrations of loess are found on the banks of the larger wadis in a tongue extending towards Tel-Arad, and on the slopes of the Hebron anticline. Exposed sections up to 12 m have been measured.

Particle size determinations indicate a textural grouping into: clay silt, sandy clay silt and sandy silt (Figure 1). The calcareous fraction is generally finer-grained. The material is well sorted, without stones, but in vertical sections the

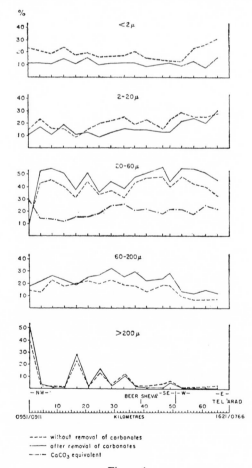

Figure 1
Particle size distribution of the loess along a NW-E traverse.

loess is sometimes seen to be intercalated by one or two layers of gravel. The older, deeper loess is usually finer-grained and poorer in carbonates.

The heavy mineral suite of the loess is composed of, in decreasing order, opaques (mainly magnetite), hornblende, epidote, and zircon, together comprising 80% to 90% of the heavy minerals. Rutile, staurolite, hypersthene, tourmaline, augite,

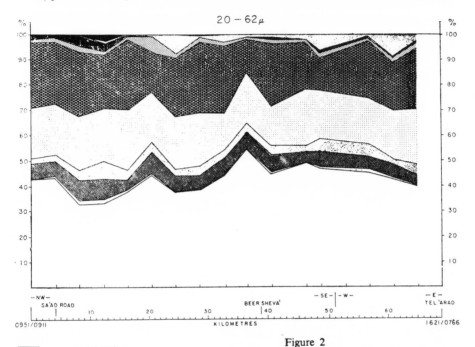

Figure 2
Heavy mineral composition in the silt fraction along a NW-E traverse.

garnet and others are accessory (Figure 2). The light minerals comprise quartz, micas and feldspars, with small amounts of opal phytoliths. In the clay fraction illite and montmorillonite are found in variable amounts, supplemented by 25–35% kaolinite (Analyses by L. Heller).

The content of $CaCO_3$ varies from 5% to 30%. Concentrations of pedogenic calcareous concentrations are found in the subsoil at the depth of rain penetration, and are especially evident in the western part of the basin.

Microfaunal determinations (by Z. Reiss) in various samples show a rich assemblage of Senonian, Eocene and Neogene foraminifera. Macrofauna is mainly composed of *Helicella* and *Spinderochila*. The identification of this fauna indicates that the Be'er Sheva basin serves as an ecological border and transition zone for a number of species. In some sections various artefacts including potsherds and human bones, were found. Evidence of fluviatile erosion and redeposition of the loess, both past and present were noted in numerous places.

The mineralogical and faunal evidence indicate that the loess of Be'er Sheva is mainly derived from the weathering residues of calcareous rocks of the southwestern desert.

31

Reprinted from *Etude Quaternaire monde, Proc. 8th INQUA Congr. Paris 1969*, M. Ters, ed., Vol. 2, 1969, p. 755

Origin of desert loess

by DAN H. YAALON
Department of Geology, Hebrew University
Jerusalem, Israel

ABSTRACT

Pleistocene and Holocene loess of desertic origin is found in the Be'er Sheva basin of the Northern Negev, and eolian material of silt and clay size has been added to and imbedded into the soils of central and northern Israel. The gradient of decreasing particle size and thickness from the loessial desert fringe region is one of the factors which point to the Sinai desert as the major provenance of the eolian material. Additional dust could be brought in from the more distant North African deserts.

Studies in the Sinai desert show that silt and clay size material is obtained in the desert by the disintegration and weathering of preexisting sedimentary rocks. Infrequent floods wash down this loose material into the larger wadis, onto fans and plains. The granulometry and mineralogy of the Wadi-el-Arish and its major tributaries especially in its middle and lower reaches is equivalent to that of the Be'er Sheva loess. In one recent large flood in the western Sinai a sedimentary layer of silt and clay 4 cm thick was formed on the alluvial fan. When dry the crust cracked and curled extensively. The absence of such a fine grained sedimentary layer before and several months after the flood is taken as evidence that the material is being blown out of the desert.

The total area of such dry washes, receiving infrequent deposition of fine grained sediments is less than 4 % of the desert surface. It is postulated that it is these wadi floors and fans which are the ultimate source area of desert loess. Wind will in general not be able to blow out material directly from the rocky desert or from desert pavement surfaces. Only after it has been washed out onto the wadi floor or fan surfaces is wind able to detrain it and transport the dust out to the fringes of the desert.

It is suggested that desert loess of the Central Asian and the Gobi deserts originated in a similar sequence of processes.

234

32

Reprinted from *Zeit. Geomorph.*, **13**(3), 297–298, 316 (1969)

THE PRESENCE OF LOESS IN SOUTHEASTERN SPAIN

Karl Brunnacker, Köln, und Vojen Ložek, Praha

Summary. In certain areas of South East Spain some of the loess exhibits deluvial facies. On high ground, a more humid layer (above 700 m, e. g. near Granada), is found on top of argillacious loess (gley-loess) and dust loam, whereas in the lowland a dryer facies predominates (e. g. west of Murcia). Some of the dust was blown direct from a frost detrital layer (above 1000 m, e. g. Velez Rubio – Granada) and some originates from valleys (e. g. Tarragona – Lerida). These different origins have an effect on the analysis of the loess (e. g. Granada loess up to 70 % $CaCO_3$, Tarragona loess up to 35 % $CaCO_3$). However, heavy Mediterranean rainfalls caused extensive displacements during the depositing phase.

In general, the mollusc fauna of the loess is sparse, and exhibits approximately the same relationships as the existing fauna of Central Europe, taking into consideration the warmer latitudes. A special feature of the area near Velez Rubio (800 metres) is a marsh loess fauna with particularly pronounced similarity to the periglacial fauna communities of Central Europe. The fauna of Altafulla, approximately 300 metres below the Wurm periglacial stage, is very similar to the loess fauna of the Neretva valley – which also correlates as regards location – from the ecological aspect. In addition, striking similarity with the Mediterranean loess of southern France is evident.

Overall, the mollusc fauna of Altafulla indicates a steppe landscape interspersed with isolated small woods. Conclusions drawn from the state of the loess indicating a dry climate have been reinforced by analyses of pollen taken from a lower level, e. g. Murcia, approx. 700 metres below the periglacial strata (Padul – J. Menéndez Amor & F. Florschütz, 1962).

The stratigraphic column can easily be compared with the Central European Wurm strata. It would appear that the strata in arid areas is always of relatively simple composition (soil formation at the commencement of Wurm and in the Stillfried B-interstadial). In contrast, the strata in less arid areas is disrupted by intercalated soils. In the latter, "threshold values" analogous to those of Central Europe can be detected.

Even in southern Spain, the Wurm glacial period as such produced a marked cold region with temperature drops of some 10° C (estimated from the position of the Wurm periglacial strata). The stadials of this period were obviously very dry. Conversely, the interstadials were more of a pluvial nature; this is particularly true of the initial period of the Wurm, less pronounced in the Stillfried B-interstadial, and with limited distribution to more humid areas of a further one or two interstadials which (in Estepona) cannot be accurately defined.

BIBLIOGRAPHY

BRUNNACKER, K. (1957): Die Geschichte der Böden im jüngeren Pleistozän in Bayern. — Geol. Bavarica, H. 34, München.

— (1959): Zur Parallelisierung des Jungpleistozäns in den Periglazialgebieten Bayerns und seiner östlichen Nachbarländer. — Geol. Jb., 76: 129–150, Hannover.

BRUNNACKER, K., DJ. BASLER, V. LOŽEK, H. J. BEUG & H. J. ALTEMÜLLER (1969): Zur Kenntnis der Löße im Neretva-Tal. — N. Jb. Geol. Paläont. Abh. 132: 127–154, Stuttgart.

BUTZER, K. W. (1964): Pleistocene cold-climate phenomena of the Island of Mallorca. — Z. Geomorph., N. F. 8: 7–31, Berlin.

FRENZEL, B. (1967): Die Klimaschwankungen des Eiszeitalters. — Braunschweig.

HEMPEL, L. (1966): Klimamorphologische Taltypen und die Frage einer humiden Höhenstufe in europäischen Mittelmeerländern. — Petermanns geograph. Mitt., 1966: 81–96, Gotha.

KUBIENA, W. L. (1953): Bestimmungsbuch und Systematik der Böden Europas. — Stuttgart.

MENÉNDEZ AMOR, J., & F. FLORSCHÜTZ (1962): Un aspect de la végétation en Espagne méridionale durant la dernière glaciation et l'holocène. — Geol. en Mijnbouw, 41: 131–134, 'S-Gravenhage.

PÉSCI, M. (1965): Zur Frage der Typen der Lösse und lößartigen Sedimente im Karpatenbecken und ihrer lithostratigraphischen Einteilung. — Geograph. Mitt., 13: 305–332, Budapest.

RUTTE, E. (1958): Kalkkrusten in Spanien. — N. Jb. Geol. Paläontol., Abh. 106: 52–138, Stuttgart.

WALTER, H., & H. LIETH (1960): Klimadiagramm-Weltatlas. — Jena.

WICHE, K. (1964): Formen der pleistozänen Erosion und Akkumulation in Südostspanien. — Report VIth. Internat. Congress on Quaternary, Warsaw 1961, IV: 187–197, Lodz.

Editor's Comments
on Paper 33

33   GERASIMOV
Excerpts from *Loess Genesis and Soil Formation*

Paper 33 is one of the most significant of recent Russian publications on loess and is much cited by Russian authors. It is particularly significant because Gerasimov makes a tentative attempt to reconcile the classic Russian, Berg-based *in situ* view of loess formation with the accepted Western eolian view of deposit formation. Western geoscientists sometimes find it hard to believe that Berg-based theories can still be held, but the current status can be illustrated by two short quotations:

> The loose packing of the loesslike parent material supports in some measure the explanation of the origin of these deposits, which became loesslike as a result of drying and dehydration of colloids, weathering and diagenetic transformations in a specific geochemical environment (Dobrovolskiy and Shoba, 1972).

> Periods of intense soil formation during interglacial and interstadial epochs alternated with epochs of loess formation. Loesses are typical continental deposits formed, according to Gerasimov's investigations, directly under the effect of weathering and soil formation processes (Morozova, 1972).

Gerasimov accepts Berg's basic contention about the *in situ* formation of loess but allows that Berg overstated his position when he declared that the eolian emplacement of loess material did not occur. Gerasimov appears able to accept the eolian emplacement of fine quartz material; if this can be achieved, the remaining problems are those of definition. The total *in situ* theory is absolutely untenable because it allows no mechanism for the formation of the predominant silt-sized quartz material; if this constraint can be avoided, a certain reconciliation is possible (see Paper 45 for further discussion).

Several papers on loess were presented at the INQUA meeting at which Gerasimov offered his paper. One of the most interesting was on the Chinese loess, by Liu and Chang. This long paper could not be included here but merits some discussion.

In 1965 Liu Tung-sheng and his coworkers published a book in Peking on the Chinese loess. This work has been described by Ho Ping-ti (1969b) as "probably the most systematic study of the loess in any language." The paper presented at the 1961 INQUA meeting was a precursor to the major endeavor and is closely related to a very similar paper published soon afterward in Chinese (1962); it is believed to give the best, relatively recent, account in English of loess investigations in China. According to Ho (1969b, p. 4), Liu et al. (1965) reach the conclusion that the exceptional textural homogeneity of the soil can be explained only by the high probability that it was the wind, rather than any other natural agent, that transported the loess material from far and near and deposited it during the long periods of desiccation that characterized the Pleistocene climate of North China.

An earlier paper (Liu et al., 1959) showed signs of possible influence from the Eastern European–Russian pedological school, but the Chinese workers appear to have come down firmly on the side of eolian deposition. The problem of loess particle production does not seem to have been discussed at any great length and there appears to be a residual acceptance of a desert source. Liu et al. (1959) report that, in the younger loess, there is a particle size gradation from larger in the northwest to smaller in the southeast. "Pedological" interpretations of the Chinese loess have been made, in particular by Gellert (1962).

A most interesting recent book on the Chinese loess was published in Chinese in Hong Kong by Ho Ping-ti (1969a). It deals in particular with agricultural aspects but includes much information about the loess. Ho has published a detailed English summary in the *American Historical Review* (1969b). He quotes some interesting palynological results, which appear to resolve the long controversy about the existence of trees on the loess. Ho writes, "the over-all meager forest resources and the likely special habitats of the two numerically significant groups of trees would indicate that the level areas of the semiarid loess highlands were little, if at all, forested" (Ho, 1969b, p. 9).

The Chinese loess has been exposed to many extraneous influences and these have wrought great changes in North China. Since the deposition of the primary loess, there has been a vast uplift in the region of the loess area, causing climatic changes and much movement of loess material by the Yellow River. The fluvial action has formed the North China plain, and the changing climate appears to have stimulated population movement. An attempt to review geological events and their consequences has been made by Smalley (1968), and this paper discusses the theory that the Chinese loess is glacial rather than desert material.

REFERENCES

Dobrovolskiy, G. V., and Shoba, S. A. 1972. Micromorphological study of a secondary podzolic soil with a scanning electron microscope. *Pochvovedeniye* 1972, no. 7, 105–110. (English translation in *Soviet Soil Sci.* 1972, no. 4, 468–473.)

Gellert, J. F. 1962. Das Lössproblem in China. Petermanns Geog. Mitt. 106, 81–94.

Ho Ping-ti. 1969a. The loess and the origin of Chinese agriculture. Hong Kong (in Chinese).

——. 1969b. The loess and the origin of Chinese agriculture. Amer. Hist. Rev. 75, 1–36.

Liu Tung-sheng, Wang Ting-mai, Wang Keh-loo, and Wen Chi-chung. 1959. Die Verbreitung des Löss in den Provinzen Shansi and Shensi (Gebiete des mittleren Hoangho, China). Geologie 8, 123–130.

—— and Chang Tsung-yu. 1962. The loess of China. Acta Geol. Sinica 42, 1–14 (in Chinese).

—— et al. 1965. The loess deposits of China. Peking (in Chinese).

Morozova, T. D. 1972. Evolution of soil formation processes of the Russian Plain during the Quaternary. *Pochvovedeniye* 1972, no. 7, 3–10. (English translation in *Soviet Soil Sci.* 1972, no. 4, 385–390.)

Smalley, I. J. 1968. The loess deposits and Neolithic culture of northern China. Man 3, 224–241.

33

Reprinted from *Rept. 6th INQUA Congr., Warsaw 1961,* Vol. 4, 1964
pp. 463–466, 467–468 (1964)

LOESS GENESIS AND SOIL FORMATION

I. P. Gerasimov

Half a century ago in 1912, a young Russian scholar L. S. Berg introduced the following idea in a short preliminary report on the natural geographical observations in the Chernigov Government (Region): „....South Russian loess must be considered as an eluvial soil that originated by way of normal formation from the bedrock under the desert climatic conditions which prevailed in the Post Glacial times (Berg 1960, p. 365).

In his subsequent works from 1916, 1922, 1925, 1927, 1928, 1929 and finally 1946 the writer, who had become the most prominent Soviet geographer, worked out and proved this idea of foremost importance. It became a foundation for the concept of loess being a product of weathering and soil formation under dry climatic conditions. The essentials and main arguments of this famous theory need not be discussed in this paper. Having been published in Soviet scientific issues, they are well-known to Soviet specialists. Foreign experts (not acquainted with the Russian language) may become familiar with this theory by reading articles published in English (Berg 1927, 1932).

It is worth to know that Berg's ideas on the great significance of weathering and soil-formation in loess formation have been worked out by outstanding investigators. The great Soviet pedologist and geographer S. S. Neustruev essentially developed L. S. Berg's ideas (Neustruev 1915, 1925). The academician B. B. Polynov (1934), pedologist and geochemist, and other workers also have largely contributed to this concept. L. S. Berg's concept of loess formation was widely made use of by the author (Gerasimov and Markov 1939). Having been advanced by L. S. Berg a long ago, the soil theory of loess formation was later misunderstood. Unfortunately, this misunderstanding had not until now been set aside. It lay in the fact that the soil theory of loess formation was set against other theories of its origin (for example eolian, fluvioglacial etc.), and not considered as an important contribution to them. Moreover, it was not understood that weathering and soil formation must be recognized as leading agents in the formation of specific loess properties. It

is the only one possibility or a possible creative synthesis of different opinions on loess formation and so for a clearing up of this problem.

It is impossible to shut eyes at the fact that for such a misunderstanding on the general significance of soil theory, were grounds in the categorial
denial of its founder (L. S. Berg) of any role of eolian factor in loess accumulation. This negative position as to the former dominating concept
of eolian loess origin lead to the fact that the soil theory was understood
as an antipode to the eolian theory. Only one, but the most well-known
Russian adherent of the eolian theory of loess genesis, academician V. A. Obruchev took quite another position. By his last years he understood essentially of the soil theory of loess genesis and even proposed to distinguish a genetic type of loess (eolian) soils. In this connection V. A. Obruchev (1951) wrote the following: „Primary loess has an eolian genesis,
and is a final product of a slow interrupted accumulation of atmospheric
dust on the dry grass steppe under dry climatic conditions. Soil formation
creates a real steppe soil, its depth increasing upwards to a considerable
thickness".

The main advantage of the soil theory lies in the strict identification of two important phenomena for the complicated process of loess
formation: (a) accumulation of mineral mass; (b) acquisition of loess
properties by this mass (loessification).

These phenomena may take place in nature in different ways. It is
possible to distinguish three main schemes: epigenesis, syngenesis and
protogenesis. Epigenesis (emphasized by L. S. Berg) is an initial
accumulation in different manner of a mineral mass without loess properties
(high silt and lime content etc.). Under weathering and soil formation
this mineral mass acquires the loess properties and is transformed into
loess. Syngenesis (emphasized by V. A. Obruchev in his last works)
is that the accumulation of the mineral mass (mainly of eolian origin)
and the acquisition of all loess properties (under the influence of soil
formation) go side by side. Protogenesis (emphasized by B. B. Polynov) is that the accumulated mineral material (in different ways) has
already all the main loess propoerties. It is because it has already undergone a weathering and soil formation and was transported later. These
three possible ways of loess formation are likely to combine or interming-
le with each other in time. In all these cases the ways of the mineral
mass accumulation of the loess may be different (eolian, deluvial, alluvial
etc.). It is because all these deposits acquire similar properties only under
the influence of weathering and soil formation. The ways of accumulation
of the mineral mass are determined in every case by local geomorphological
conditions, so the depth, distribution and stratigraphy of loess deposits
vary from place to place, but their lithology is uniform.

The most difficult point in the soil theory of loess formation lies in the establishment of the processes of weathering and soil formation.

It is indisputable that these processes take place under the dry climatic conditions. This is ascertained by the presence of secondary carbonates in loess, not removed by leaching. Further hypothesis upon the genetic essence of the soil processes in loess formation has a provisory character L. S. Berg described the desert features of these processes to start with. Later for the South Russian loess he mentioned a possibility of dry postglacial and interglacial steppes. V. A. Obruchev's views on the subject had undergone a similar evolution.

All these opinions were based upon too general considerations. Little if any objective evidences of soil formation on loess were taken into account. In part these opinions even contrasted with these evidences. I mean the presence of traces of ancient soil formation as seen in well preserved fossil podsol, brown and gray forest, chernozem and others in the loess strata. In most cases these ancient soils are not sygenetic to loess. Moreover, long ago such soils were recognized as interloess ones and considered as boundary horizons between loess layer of different age which were formed during the intervals in the accumulation of loess material. Therefore, in Europe similar buried soils in constrast to loess layers were used as indicators of interglacial (interstadial) stages of the Quaternary. In Asia (China, Middle Asia) they show pluvial stages.

So it is clear that the evidence of soil formation in the loess strata itself has quite another significance. In contrast to the above mentioned buried soils these fossil soils are very poorly defined. With usual field methods they are often missed or not taken into account. But the situation changes if some modern pedological methods are made use of.

Such researches are undertaken by the Institute of Geography of USSR Academy of Science. On the assumption of a probable considerable significance of weathering and soil formation in the formation of main properties of loesses, a special microlithological study of a present-day soils (podsols, chernozems, serozems etc.), loess samples from layers of different age and buried soils between these layers. The main results are given here.

The present-day soils, loesses and buried soils consist mainly of primary minerals (quartz, feldspar, mica, hornblende, accessory minerals etc.) cemented by a highly dispersed „plasma" sometimes with microcrystalline calcite. „Plasma" may be often aggregated. In some in the soil mass between pores and cracks occur big aggregates of different form and size.

The main part of „plasma" consists of clay and in present-day soils of humic and clay substances. There are two types of clay structure: optically oriented and non oriented ones. The former has two main microscopic forms: (a) matted-fibrous clay (argilization in situ), and clay oriented skins in pores (illuviation).

[*Editor's Note:* Material has been omitted at this point.]

At present our research are extended with special reference to micromorphological properties of cryomorphic origin. These results must not be overemphasized. One general conclusions is doubtless. It lies in the fact that at first sight homogenous loess strata proved to have a complicated heterogenous structure with doubtless evidences of different ancient processes of weathering and soil formation. For a long time these strata seemed short accessible for further lithological study and subdivision. It would be difficult to imagine another situation. Loess is a typical continental formation originated under the influence of natural processes of weathering, soil formation, denudation and accumulation, which acted always before and at present. All these processes might not take part in the formation of so widely distributed deposit as loess. They must have left traces of their activity in the composition and structure of such a remarkable deposit as loess.

The soil theory of loess genesis was primarily proposed by L. S. Berg in most part by intuition, on the base of a general relation of field data and geographical logic. This theory is a basic scientific concept of extremely progressive character which fell a new light to the loess problem. Due to the imperfection of research methods as concerns loess, this concept has not yet so large significance in the working out of the loess problem, but it had rights to pretend to such significance. Some new methods of research has just been proposed possible to change the present situation.

It is one of these methods, i.e. the research method of loess micromorphology that seems to be very perspective. Being put at the service of the soil theory of loess genesis, this method proves its justness.

References cited

Berg, L. S. 1913 — General geographical observations in the Chernigov Government. A short preliminary report on the natural-geographical observations made in the Novozybkov, Krolevets and Konotop districts in 1913. Moscow. (in Russian).

Berg, L. S. 1927 — Loess as a product of weathering and soil formation. *Pochvovede-nie*, nr. 2, (in Russian).

Berg, L. S. 1932 — The origin of loess. *Gerlands Beiträge z. Geophysik*, Bd. 35.

Berg, L. S. 1960 — Loess as product of weathering and soil formation. (in Russian). Selected works by L. S. Berg, t. III, Moscow.

Gerasimov, I. P., Markov, K. K. 1939 — Ice period on the territory of the USSR. *Trudy Inst. Geogr. Akad. Nauk SSSR*, t. 33, (in Russian).

Neustruev, S. S. 1925 — On the study of Turkestan loess. *Geol. Vestnik*, no. 3, (in Russian).

Neustruev, S. S. 1925 b — The soil concept of loess formation. *Priroda*, no. 1—3 (in Russian).

Obruchev, V. A. 1951 — Loess as a specific soil type, its genesis and problems of its investigation. Selected works by V. A. Obruchev on the geography of Asia, (in Russian) Moscow.

Polynov, B. B. 1934 — Crust of weathering. T. I, (in Russian), Moscow.

Velichko, A. A., Morozova, T. D. 1961 — Buried soils in the loesses of central Russian Plain: their palaeogeographical and stratigraphical significance. *Abstracts of papers*. INQUA, VIth Congress, Poland.

Editor's Comments
on Papers 34 Through 38

Papers 34–38 all relate to the loess of central and eastern Europe. The papers by Fink and Pécsi (Papers 34 and 35) were presented at a meeting of the INQUA loess commission held in Budapest in 1971. This was a very significant meeting and represented a broadening of the scope of the commission to include more attention to loess material and applied research in such fields as geotechnology and agriculture. Fink's introduction gives some background on the Loess Commission and discusses new directions of loess endeavor. Pécsi continues this theme and provides a very valuable discussion of the significance of loess research. Seppälä's article (Paper 36) complements Pécsi's and emphasizes the critical role played by Pécsi in the study of the Hungarian loesses. Maruszczak discusses the entire east-central region and Conea concentrates on the eastern part of Romania.

The editor is concious of considerable inadequacy in dealing with this region. Loess is important in east-central Europe; much research has been done and many papers published, and this section might easily be twice as big. Important material would still be omitted. Several factors conspire to prevent the wider dissemination of local information: pa-

pers published in Bulgarian or Hungarian can only have a small reader-
ship, the journals inevitably have small circulation, and thus important
items may not receive the attention they deserve. The most obvious
omission is work from Bulgaria, and this is largely due to the (not un-
reasonable) fact that most of the literature about the Bulgarian loess is
published in Bulgarian. However, some indication of accessible data
must be attempted.

The most important work on the Bulgarian loess is probably the
book *The Loess in North Bulgaria* by M. Minkov (1968); this is totally
in Bulgarian, without even a contents list or abstract in German or English. Perhaps the most accessible and relatively comprehensive description of the Bulgarian loess is that by Fotakieva and Minkov (1966),
published in German. The following details are taken from this paper.

The loess in Bulgaria takes up an area of 9,800 km^2 and is distributed exclusively in the section along the Danube. The thickness of the
loess depends on the distance from the Danube and the age of the geo-
morphologic element that it overlies. On plateaus in immediate proximity to the river this thickness varies from 50 to 60 m, with loess walls
reaching 102 m. The thickness drops to 25 to 30 m about 10 km south
of the Danube, and in the southern peripheral parts it hardly reaches 4
to 5 m. The investigators assume that the loess in Bulgaria is of eolian
origin and terrigenous source—the overflowing of the Danube. A very
short article in English was prepared for the 7th INQUA Congress, and
appended to this is a very useful list of references with the titles translated into English (Fotakieva and Minkov, 1968).

The Romanian loess literature presents fewer problems, and is better
represented in this collection. The excellent review paper by Marosi (Paper 50) is given a rightfully climactic position at the end of the book;
this paper discusses the original loess problem. A more local view of
loess is provided by Conea in Paper 38. The work reported in this paper
has been developed, expanded, and published as a monograph (Conea,
1970) which gives detailed descriptions of the loess and paleosols of the
Dobogea region. This book contains a contents list and summary in English. The English-speaking loess investigator who is interested in the Romanian loess should also consult the INQUA guidebook produced by
Conea, Bally, and Canarache in 1972.

Maruszczak, the author of Paper 37, has edited a symposium on the
lithology and stratigraphy of loesses in Poland, which gives a good over-
view of loess research in Poland with some emphasis on the problems of
classification (Maruszczak, 1972). J. E. Mojski contributed a consideration of the principal trends in the work done by the INQUA Loess Com-
mision. The development of loess research in Poland is illustrated by Pa-
ers 21 and 22, by Dylik and Tokarski, and studies on mechanism of sedimentation are illustrated by the work of Cegla (Papers 47 and 48).

REFERENCES

Conea, Ana. 1970. Formatiuni cuaternare in Dobrogea (Loessuri si paleosoluri). Edit. Acad. Rep. Soc. Romania, Bucharest, p. 234.

——, Bally, R., and Canarache, A. 1972. Guidebook to excursions of the INQUA Loess Symposium in Romania. Geological Institute, Bucharest, 54p.

Fotakiewa, E. and Minkov, M. 1966. Der Löss in Bulgarien. Eiszeitalter Gegenwart 17, 87-96.

——, and Minkov, M. 1968. The loess in Bulgaria. *In* Loess and Related Eolian Deposits of the World, C. B. Schultz and J. C. Frye (eds.), Proc. 7th INQUA Congr. University of Nebraska Press, Lincoln, Neb., p. 297-301.

Maruszczak, H. (ed.). 1972. Proceedings of National Symposium on the Lithology and Stratigraphy of Loesses in Poland. Lublin 25-30 Sept. 1972. Wydawnictwa Geologiczne, Warsaw, p. 214.

Minkov, M. 1968. The loess in North Bulgaria. Bulgarian Acad. Sci. Sofia, p. 198.

34

Reprinted from *Acta Geol. Acad. Sci. Hungar.*, **16**, 313–315 (1972)

PRESIDENTIAL ADDRESS OF INAUGURATION
Symposium of the INQUA Loess Commission Hungary, 1971

By

J. FINK

DEPARTMENT OF GEOGRAPHY, UNIVERSITY OF VIENNA

It is an unusual pleasure and honour for me as President of the INQUA Loess Commission, to have the opportunity to open its meeting 1971 here in Budapest. It is the first time that we have met as an *independent commission* of INQUA. Although our activities have been largely independent, they were harnessed within the frame of the Commission on Stratigraphy and the scope was restricted to Europe.

It is interesting, that the present Symposium is the first to convene under the joint auspices of both INQUA and IGU. For a long time now this tie has been sought for by both international societies, and it is entirely due to this new connection, that a generous sponsorship for the preparation and execution of the Symposium could be enacted within the frame of the IGU Regional Conference. Financial support could not have been granted for any meeting under the auspices of an INQUA commission alone.

Nevertheless, there are technical reasons to account for the particular value of the Symposium:

1. *Hungary* due to her physical background, is a country with rich traditions in the field of Quaternary research. Whereas in the Alpine areas geology was concerned predominantly with solid rocks, here in the Pannonian Basin the unconsolidated sediments have received the greatest attention. For this reason, a pedological surveying was begun at an early date and it was as early as 1909, that an Agrogeological Conference, a predecessor of the International Soil Science Congresses, was organized in Budapest. It was here that discussions about the genesis of alkaline soils were started and many subsequent impetuses to international scientific progress in this special branch of knowledge were provided by the Hungarian pioneering work. A considerable number of important scientists were active in the domain of earth sciences. Just a few of the names such as R. BALLENEGGER, B. BULLA, H. HORUSITZKY, I. MIHÁLTZ, J. SÜMEGHY, P. TREITZ shall be quoted, who can be considered representatives of the older generation of scientists.

At an early date the development of a classification of the eolian sediments in Hungary was prepared. J. SÜMEGHY recognized the zonality of these sediments and their dependence on paleoclimate, and he provided the foundations for modern field investigations. Unfortunately, his works did not become known abroad.

Therefore, it stood to reason, that European loess specialists came here as early as in 1965 to attend their annual meeting, after having been in the ČSSR and the GDR at similar meetings. Just like today, did my dear friend, academician Pécsi, engage himself in the preparation and execution of the meeting. Like on that occasion, he is supported now by hosts of co-workers and colleagues of INQUA. And as in 1965 the program is now rich in both extent and content, with an agenda which includes a demonstration of the problems of the Hungarian realm.

Let us emphasize that, in spite of the efforts made here, the chronology of the Hungarian loess profiles still includes great difficulties to overcome, so that their co-ordination with other countries is *problematic*. This problem is due to the natural environment because the terraces carrying most of the loess profiles exhibit a striking tectonic deformation throughout their occurrences. Also the paleoclimatic situation in the Pannonian Basin was different from the other Central European territories: The strong influence of the groundwater and rivers resulted in special kinds of sediments, called "infusion loesses"; while like solifluction was responsible for the development of the wide-spread slope-deposited loesses. Despite the afore-mentioned difficulties, in 1965, we were totally convinced, that the results arrived at in Hungary relied on many observations in the field and that they have to be considered a *reality*. In other words, they could be incorporated in some way or other in our European concept.

Numerous works have been produced here since the last Loess Symposium and many of them were written by M. Pécsi and his working team of geographers. The problems of stratigraphy, however, are not the only concern of the Subcomission for they also include a great many practical problems belonging to the domains of geotechnics and agriculture. And these concerns bring us to the second technical reason for the significance of the present Symposium.

2. With this meeting begins a reorientation of our Commission's activities towards practical problems. Questions of this kind have already been tackled in projects such as the map of the distribution of loesses in Europe, now in preparation, or the question of the *nomenclature* of the different eolian sediments as connected with uniform laboratory analyses. The widening of the Subcomission on Loess to form an independent commission of INQUA will allow inclusion of other loess-covered regions beyond the *confines of Europe* into the scope of our activities.

An addition, the new organizational structure will allow more thorough discussion of the problem of *applied research*. In this connection a very real ulterior motive of this reorganization should be mentioned: If the INQUA

Loess Commission is able to contribute to the solution of present-day tasks and objectives, it is entitled to expect a financial support from institutions which are interested only in investigations with practical aims.

A reorganization of this kind requires the closest possible co-operation with colleagues engaged in related disciplines. This point has been taken into account by our Hungarian hosts. We are very glad to see that the Geographical Research Institute of the Hungarian Academy of Sciences has been assisted in both preparation and execution of the Symposium by numerous colleagues active in hydrology, geotechnics and soil sciences. Once more, geography, aware of its integrative duties is functioning as a mediator of various earth-science disciplines. In this sense, we expect to attend a meeting worthy of a geographical congress, a meeting to which we extend our wishes for great success.

35

Reprinted from *Acta. Geol. Acad. Sci. Hungar.*, **16**, 317–328 (1972)

SCIENTIFIC AND PRACTICAL SIGNIFICANCE OF LOESS RESEARCH

By

M. Pécsi

GEOGRAPHICAL RESEARCH INSTITUTE OF THE HUNGARIAN ACADEMY OF SCIENCES, BUDAPEST

Loess plays a significant role in agricultural production and in the planning and construction of technical establishments not only in Hungary, but also in many other parts of the Earth. It carries fertile soils traditionally having been cultivated intensively, and its effects in concentrating the population and settlements have always been significant.

The loessic regions provide very favourable natural conditions for agricultural production and in many places the loess is used as building material. As a result of technical activity and agricultural production, the loess is easily eroded and dissected; generally it is compacted under buildings and its durability is being degraded in this way. Therefore, the investigation of the loess and of its soil cover has practical purposes which include maintaining and increasing the agricultural production on the one hand, and settling and ensuring the operation of the economic and technical establishments, on the other.

Although the practical loess research with an engineering point of view has inevitably become more important, the lithological, chemical and physical evaluations of the regional types and genetic varieties of the loess, too, are significant basic research tasks. The stratigraphical classification of the loess strata — on the basis of paleopedological, paleontological, archeological, geomorphological, geochemical and radiocarbon methods —, however, is one of the most significant means for the reconstruction of the Quaternary historic geology and paleoecology.

Harmony and integration of the theoretical and practical trends in the loess research are absent in many cases, but it must be emphasized that new progress in loess research may be expected only by applying interdisciplinary investigations and by means of joint development of field and laboratory research.

Research of Loess Areas in Service of Agriculture

1. The lithological and genetic investigations of the loesses are necessary for agricultural production because of their connection with the fertility or devastation of loess soils.

[*Editor's Note:* Certain figures have been omitted owing to limitations of space.]

The unstratified eolian and the sandy loesses contain 0.2 to 0.3 per cent humus, while in the redeposited slope loesses the humus content is 0.3 to 0.4 per cent. The soil sediments interbedded in the loess strata may contain 0.6 to 0.7 per cent humus, while the fossil soils occurring in the loess in several levels may have a humus content of 0.8 per cent. Also the clay and humus content and the ground-water balance influence the loess structure.

Due to their more compact structure, the fossil soil, loessic semi-pedolite and loess strata mixed with rock detritus covered by the present-day soil in the sections of the loesses in the hilly regions are not so much devastated by slope-wash erosion. The strata mentioned above are rich in nutriments and therefore the investigation and mapping of the genetic and lithological characteristics of the loessoid sediments are closely connected to agrogeological investigations.

2. The replacement of the overwhelming majority of the natural vegetation by cultivated plants has considerably changed the natural run-off and *ground-water balance* of the hilly regions covered by loess. The anthropogenic interventions accelerated the development of soils, but on slopes they increased the erosive-derasive-suffosive soil devastation, too. At the same time there occurs an increased need for water, and therefore *an artificial drying of the soil and subsoil climate occurs*, too. This damage is increased by the fact that minerals, humus and other nutriments of the soil are carried away by the huge quantities of running water (250 to 800 m³/ha). Another factor is that on the barren loess surfaces the strong solar radiation causes *partial photochemical devastation of the soil's humus content.* Due to the worsened ground-water balance, humus and dust in considerable quantities will be carried away by *wind erosion.*

The precipitation on the barren surfaces causes the compaction of the loess soil, decreases its porosity and reduces the circulation of water or air in the soil. From such an area the run-off carries away huge quantities of mud and also carries away the genetic soil, causing *gully erosion.*

On the slopes of the loess regions the running water washes away the soil and devastates it not only by its kinetic energy, but also by infiltrating into the porous rocks, where *solution and mechanical suffosion* processes take place, as well. Sink-hole-like depressions and cavities are caused by these processes. When no systematic protection is taken against these processes on the cultivated slopes and loess cuts, the deepening of gullies at the expense of arable lands is rapid and it may be as much as several metres in a year.

The establishing of large fields of arable land in the agricultural production co-operatives and state farms, the reorganization of the area belonging to them, the marking out of new unmetalled roads in the heavily dissected and sloping regions of loess necessitates to find the relations between the surface run-off and the geomorphological characteristics of the small watershed areas. In such regions the rearrangement of the other area and road network cannot be done but on the basis of farm management and administrative considerations.

3. For the most part empiric observations and a limited number of measurement figures have been used *to explain the relations between soil erosion and the cultivation method*. In the agrarian loess regions of the hilly areas, the natural vegetation — e.g. forest steppe or forest — had been replaced by the cultivated plants only periodically and/or partially covering the slopes. In the spring and early summer periods the young plants are still weak, and after the harvest the soil remains nearly barren and unprotected against the atmospheric influences. Thus the way of cultivation must be fitted to the natural ecological equilibrium of the area.

The extent of erosion of the loess cover depends not only on the atmospheric, geomorphological and lithological conditions, but also on the cultivation method and the size of the sloping area cultivated in the same way. The regrouping of farm fields on the slopes cannot be based exclusively on administrative considerations, neglecting the threatening danger of erosion. On sloping surfaces the mechanized large-scale farming may also increase the danger of erosion.

4. Complex scientific research is needed for the evaluation of the natural environment, for the protection of loesses and their soils, for the increase of their fertility and for the choice of the suitable crops. This research requires the co-operation of a number of branches of science, including pedology, geology, geomorphology, agrochemistry, agrometeorology, agronomy, agricultural geography, etc.

Taking into consideration the facts mentioned above, the Geographical Research Institute of the Hungarian Academy of Sciences co-ordinated the loess research in lithological, geomorphological, and Quaternary geology *with an agricultural economic evaluation of the natural endowments*. These evaluations of the natural environment based on economic considerations and done mainly referring to the loessic areas of the Transdanubian Hilly Regions are only methodological experiments, but they represent a new direction.

This kind of research is named "geopedology" and was initiated by L. GÓCZÁN, S. MAROSI, M. PÉCSI and J. SZILÁRD.

Technically Oriented Loess Research

Till World War II the loess research in Hungary for the most part was of a soil-geographical and agrogeological character. After this period, research was directed towards geological and geographical fields and was mainly concerned with the investigation of the regional and genetic types of loesses, as well as with Quaternary stratigraphical questions.

The number of papers dealing with the investigations of loesses from an engineering geological point of view has been increasing in the last decade. Recently the complex investigation and practical evaluation of the natural characteristics and economic utility of loess has become a major concern in the regional planning and in the construction.

The dynamic changes in the loess forms and loessic areas due to the effects of natural processes and the economic-technical activity mean a manifold research task for *engineering geology*. The financial and intellectual resources assigned to these investigations recently have, in many cases, surpassed those devoted to the analytical activity for agrarian purposes.

Accurate engineering-geological and soil-mechanical investigations of the physico-mechanical and dynamic features of the loess areas and loess-like sediments are indispensable for planning and constructing engineers. It has become necessary to improve the methods and criteria for the evaluation of loess collapse and loess strength. In addition, various standards and directives referring to the properties of loess had to be provided for the contractors. Such investigations are done and used by soil mechanicians, engineering geologists and experts in the planning offices.

It must be emphasized that the loess collapse and the strength and mobility of the loess strata are closely connected to the geological, geomorphological and ecological factors of the environment. The soil-mechanical investigations alone do not give enough information on the slide processes of the slopes and steep loess-bluffs, or on the periodicity and frequency of these movements. The more detailed knowledge of the dynamic changes of the loessic rocks is supported by engineering-geomorphological observations and by the three-dimensional geomorphological-geological analysis of the loess strata introduced by Hungarian scientists in order to do the stratigraphical classification of the loesses (PÉCSI, 1965). The regularities of the dynamic movements of the loesses and solid predictions may be given only by really co-ordinated interdisciplinary research involving soil-mechanical, engineering-geological and geomorphological investigations.

The present-day loess research does not deal only with the genesis and stratigraphical classification of the loess and loess forms. It can be stated, however, that the results and the rules found by the classical loess research are indispensable for making the estimations and calculations of a technical and economic point of view, also in the future. The problems of the engineering-geological and soil-mechanical research of the loess and the classical geoscientific loess research will continue to be connected to each other in an increasing rate. Therefore, the findings of both directions, as well as their method and views should be taken into consideration by each of them.

Lithological and Chronological Subdivision of the Hungarian Loesses[1]

Loess chronology is closely connected with the subdivision of the litho-logical varieties of the formations occurring in the loess series. This provides for the indispensable basic framework used by several interrelated branches of science and by means of their special methods they give loess chronology a positive form. In addition to this, the lithological characterization of the loess varieties is fundamentally significant also for engineering-geology.

1. The loess cover — loess and loess-like sediments together — occurs in three characteristic geomorphological positions in Hungary and generally in the Carpathian Basin.

a. In the largest and most continuous extension the loessy silt covers occur in flood plains and huge alluvial fans of the Great Hungarian Plain. In these cases the loess strata alternate with sandy or silty layers and flood-plain hydromorphous soils (Fig. 1). On these surfaces the loessic sediments are deposited in normal stratigraphical sequence.

b. On flood-free alluvial fans of the Great Hungarian Plain (i.e., where during the Pleistocene a continuous subsidence took place) the loessic forma-tions are deposited together with blown sand and fluviatile sand intercalating both horizontally and vertically. In the bores made in the southern part of the Danube-Tisza-Interfluve, the alternation of the sand and loess strata was recognized even at a subsurface depth of 150 to 200 metres (MIHÁLTZ, 1954).

The thickest and most easily classifiable loess series developed in the older alluvial fans of the foothills and Pliocene-Pannonian tabular surfaces (Mezőföld) where no considerable subsidence took place during the Pleistocene. In this case the lithologically typical loess strata are varied cyclically by various loess-like sediments, fossil soils, interbedded sand strata and erosion hiatus (Figs 2 and 3). In certain phases of the Pleistocene the thick loess cover was dissected repeatedly by erosion valleys, gullies and numerous dells. The major-ity of these loess valleys was buried by the renewed dust accumulations, and most of them was filled up by "slope loesses", sandy-clayey stratified loes-ses. This process took place repeatedly as buried valleys of loess settled on or in each other (Fig. 4).

c. Loess types of lithologically varied compositions and origins occur on the slopes of mountains and foothill surfaces and in the derasion valleys dissect-ing frequently the slopes of the ridges in the hilly regions. In these areas the typical loess layers play a subordinate role in the loess sequences. The inter-

[1] According to the preliminary programme of the loess symposium detailed information will be given on the lithological types and chronological subdivision of the Hungarian loesses. In this paper only a brief summary of the main results and the analyzed profiles of the expo-sures will be published. These profiles are not identical with those published in the "Guide-book for Loess Symposium in Hungary in 1971", but they are identical with the unprinted figures demonstrated in the course of field trips and with the strata demonstrated and num-bered in the exposures.

LEGEND OF THE PROFILES OF LOESS

Eolian

fine sand, blown sand

loessy sand

sandy loess

loess

clayey loess

Colluvial – Deluvial

slope sand

loessy slope sand

sandy slope loess

slope loess

clayey slope loess

Fluvial – Proluvial

sand

silty sand

silt

clay

sandy gravel

Recent and Fossil Soils

weak humus carbonate

humus carbonate

steppe – type soil

chernozem brown forest soil

brown forest soil

grey-brown podsolic soil (lessivé)

red clay

semipedolite

hydromorphic soil, medow soil

alluvial marchy soil

Iluvial Horizons in Soil

∧ Ca ∧　calcium carbonate accumulation

⋀ Ca ⋀　strong calcium carbonate accumulation

× × × ×　volcanic ash

∞　∞　loess doll

krotovinen, animal burrows

+ + + +　charcoal

macrofauna, worm – cast

Miscellaneous

boundaries of packs, definite

boundaires of packs uncertain

∼　cryoturbation, solifluction

▼　dessiccation fissures

discontinuity of profile

⇨　areal denudation, unconformity

➡　linear dissection, unconformity

M_F　M_F Mende upper soil complex

M_B　M_B Mende base soil complex

B_D　B_D Basaharc double soil complex

B_A　B_A Basaharc soil complex

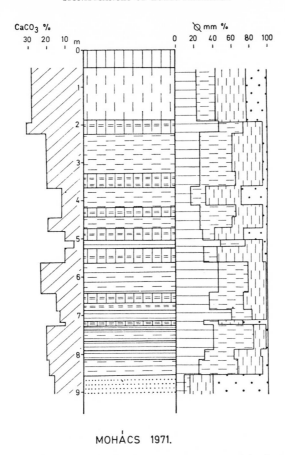

MOHÁCS 1971.

Fig. 1. Profile of "flood-plain loess" at Mohács (according to M. Pécsi). Laboratory analyses evaluated by E. Szebényi, profiling done with the cooperation of F. Schweitzer. There are silt, sandy-silt and sandy-clay layers between the marshy flood-plain soils repeated several times. Due to soil-biological processes, the uppermost part of the profile becomes a loess-like structure. — (see the legend: p.: 322)

calated sandy, clayey loesses, soil sediments usually showing a macroscopically recognizable stratification, are the most prevalent ones. The accumulations of these loess-like sediments are of solifluctional-deluvial origin (Fig. 5).

In these cases the chronological subdivision usually requires a special attention, since the loess series lying on the terraced relief overlie one another partly in reversed, partly in normal stratigraphical sequences. In the heads of the terraces and steep slopes where the loess cover flattens the formerly dissected topography, an "oblique sequence" developed under the loess series with a normal sequence. In these loess areas the frequent relief inversion has also to be taken into consideration. Consequently, the stratigraphical subdivision of the loess cover must be coupled with the three-dimensional investigation of the exposures (Pécsi, 1967).

2. In Hungary the so-called "young and old loesses" can be separated from one another fairly well, from a point of view of both the lithology and stratigraphy. The *young loesses* form a series of 25 to 30 metres thickness and their sequence is quite complete (Figs 6 and 7). The thicker sandy loess and loessic sand strata are characteristic of the upper third of the formation (first loess series). In the thinner typical loess strata two or three pale, humic loess horizons ($>$0.4 metre) also occur (second loess series). In large blown sand and flood-plain areas, only the uppermost part of the young loesses occurs (Fig. 1).

The lower two thirds of the "young loesses" are subdivided by three well developed, dark-coloured soil horizons of a forest-steppe character. In several cases these are doubled soils (Fig. 8).

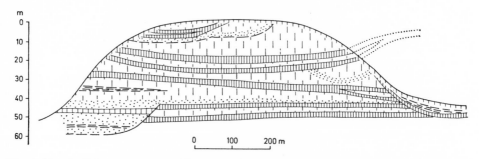

Fig. 4. Sketch of the arrangement of loesses and fossil soils in the undulating regions consisting of loess series (according to M. PÉCSI)

a. In the exposure of the Mende brickyard the first doubled soil complex lies between 10 and 12 metres,[2] the upper part of doubled soil horizon of which dates back to 28 or 29 thousand years according to repeated radiocarbon investigations. This soil complex is repeated prominently in numerous significant loess exposures of Hungary and of the Carpathian Basin, and even the lower part of it has not been found to be older than 32,000 years. On this basis and in the Hungarian loess chronology the *Mende upper soil complex* represents the interstadial period separating the Middle and Upper Würm.

The two fossil steppe soils under this complex were described first from the loess exposure of the Basaharc brickyard (Fig. 7) in the Danube Bend (PÉCSI: Földr. Közl., **4**, 346—351, 1965) where a typically doubled soil series (between about 14 and 16 metres) and a thick humic soil (between about 18 and 20 metres) were designated as *Basaharc D (BD)*, and *Basaharc A (BA)*, respectively. According to our loess chronological classification, the soil complex *Basaharc D* may belong to the interstadial period completing the Lower Würm, while the thick humic soil *Basaharc A* may be assigned to the

[2] Its name is *Mende upper soil complex* (signed by MF). (M. PÉCSI: Földr. Közl., **4**, 332—338, 1965; and Petermanns Geogr. Mitt., 241—252, 1966.)

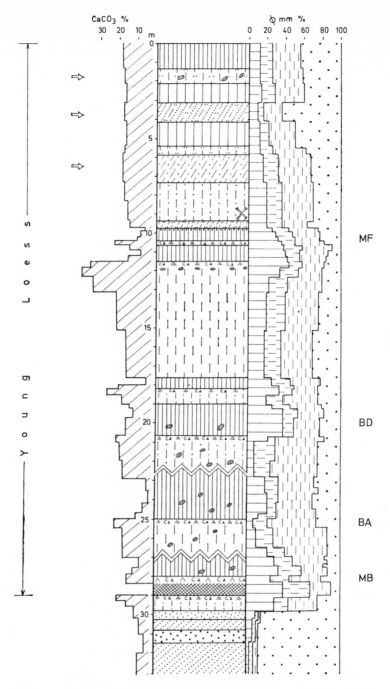

MENDE 1969-72.

Fig. 6. Loess profile in the Mende brickyard (according to M. Pécsi and E. Szebényi). Profiling done with the cooperation of Á. Juhász and M. Di Gléria. X — complete skeleton of Elephas primigenius, ++ — radiocarbon datum: 29.800 years BP

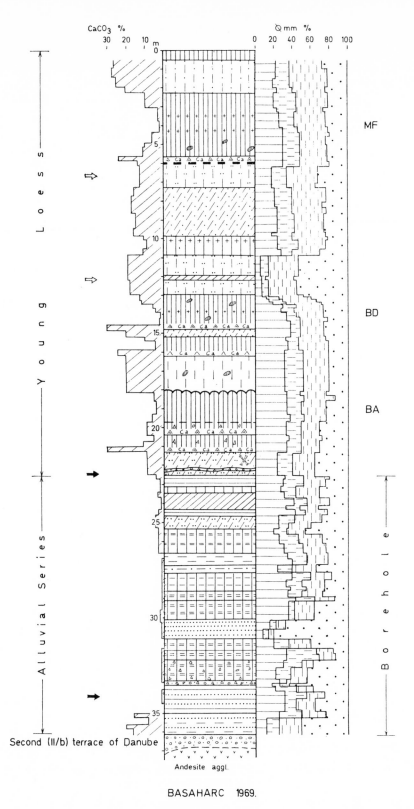

CaCO₃ %

BASAHARC 1969.

Second (II/b) terrace of Danube

Andesite aggl.

MF

BD

BA

Fig. 7. Loess profile in the Basaharc brickyard (according to M. Pécsi and E. Szebényi).
Profiling done with the co-operation of Gy. Scheuer, F. Schweitzer. X — complete skull of
an Ursus speleus Minor, — — radiocarbon datum: 27 045 years BP, 2593 HV

260

late beginning period of the Lower Würm. The underlying loess strata are the lowest member of the younger loess series (fifth loess series).

In larger exposures the "young loesses" are separated from the older ones by the so-called *Mende basal soil complex* and the underlying sharp erosion unconformity. The *Mende B* soil complex (MB) consists of a well-developed brown forest soil and of the directly overlying dark steppe soil (>2 m). On the basis of terrace-morphological, paleontological and other data, the *Mende basal soil complex* was assigned to the second part of the last (Riss-Würm) interglacial period (Fig. 8).

b. On the basis of the results of different investigations, calculations were performed to determine the deposition rate of the "young loesses". It was determined that in Hungary exactly a thousand years were required for the formation of one metre of the typical loess. For the deposition of one metre of stratified slope loess about two thousand years were required, and in the case of buried fossil steppe soils about 3 to 5 thousand years must have passed in the formation of one metre thickness. Taking into account the whole 25 metre thickness of the Upper Pleistocene strata, it is probable that its accumulation may have taken place in a period of 50 to 70 thousand years. If, in addition, it will also be taken into consideration that the accumulation of sediments was periodically interrupted by denudation, the fact can be stated that the length of a glacial stage is not represented by a particular loess layer. On the basis of the results of the investigations also the conclusion may have been drawn that in the lowlands and depressions certain loess-like series are deposited in Holocene period, or of interstadial-interglacial origin. Similar developments occur in the slope loesses of the hilly regions, mainly in the loesses filling up derasion valleys where the soil strata are separated from each other by a layer of several decimetre thickness of colluvial-deluvial loess of a non-glacial age.

3. The chronological subdivision of the young loess series in Hungary can fairly well be made parallel with the youngest glacials of the Pleistocene. In contrast with this, the subdivision of the old loess series can only be given in an outlined way. No satisfactory data are at disposal for the identification of the older loess series and fossil soils on the basis of the older glacial and interglacial stages.

a. The *old loess series* are separated from the younger ones not only by a definite unconformity, but also lithologically. Usually they are more compact, more clayey and they contain numerous horizons of $CaCO_3$ concretions. At the same time the loess stratum itself contains less carbonate.

The old loess series occur in a thickness of 25 to 50 metres. The deeper levels are known only from core samples (Figs 2. and 3).

The thickness of the old loess series as well as the number of the different fossil soil horizons, unconformities and connected sandy and non-loess-like strata are not quite identical in the exposures. Since such profiles of old loess

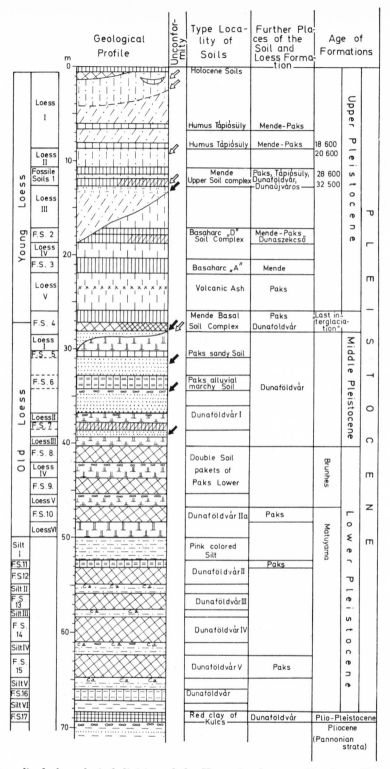

Fig. 8. Generalized chronological division of the Hungarian loess profiles (according to M Pécsi.)

are infrequent in Hungary, the chronological subdivision for each of them has been supplied, in the first place for the Paks profile; ÁDÁM, MAROSI, SZILÁRD, 1954; BACSÁK, 1942; BULLA, 1934; KRIVÁN, 1955; PÉCSI, 1965; SCHERF, 1938; STEFANOVITS, RÓZSAVÖLGYI, 1962. During the last decade in several places the preparation of profiles became possible (Kulcs, Dunaújváros, Dunaföldvár; along the reaches of the Danube in Yugoslavia: Nestin, Stari Slankamen, etc.).

In the exposures mentioned above, 4 to 6 fossil soil horizons occur in the old loess series. At Paks and Dunaföldvár additional 3 to 4 fossil soil horizons were recognized in the bores (Figs 2. and 3). In the old loess series of Paks 7 fossil soil horizons were found, and at Dunaföldvár 9 of them were identified. In addition to these horizons 2 to 3 very sharp unconformities and the same number of fluvial sand and silt intercalations were observed. In certain profiles the oldest soils nearly directly overlie each other and only a thin level with accumulation of $CaCO_3$ can be found between two fossil soils with no loess intercalation at all. This situation makes it difficult to separate or stratigraphically subdivide them. Most of the fossil soils are reddish-brown, reddish-ochre soil and red clay, but three kinds of hydromorphous clay soil (two dark-coloured ones and a pale one) can be found also as key horizons in some places.

Taking into consideration the facts mentioned above, a new attempt was done to perform a general chronological subdivision of the older loess series in Hungary (Fig. 8), based directly on the exposures at Paks and Dunaföldvár.

b. The two rust-coloured, hydromorphous clayey soil horizons and the brown forest soil horizons and the interbedded loess bands (I—III) of the exposures at Paks and Dunaföldvár, as well as the thick fluvial and proluvial sand series occurring conspicuously at Paks were assigned to the *Middle Pleistocene.*

c. The oldest series of the loess profiles in Hungary beginning with the so-called *Paks basal double soil complex* (Fig. 3) or at Dunaföldvár with the so-called "Dunaföldvár complex" (Fig. 2) a thickness of 5 to 6 metres were assigned to the *Lower Pleistocene.* Based on the very recent and detailed paleomagnetic examination of the layers at the exposures of Dunaföldvár and Paks, the Brunhes-Matuyama boundary lies just below these strata mentioned before (according to the analysis of M. A. PEVZNER, 1972).

At Dunaföldvár, below the pink-coloured sandy silt (like stratified "rocky loess") ($>$5 metres) and the dark-grey meadow soil ($>$1 metre), situated in the Danube's level, there are further four red-ochre-coloured fossil soil horizons among sandy, silty loess strata, belonging to the *Lowermost Pleistocene.* At Dunaföldvár and in other places even a red clay stratum occurs in the base of the "old loess". This is generally a product of weathering in the Upper Pliocene and overlies the Pannonian clayey strata.

d. Usually in the stratigraphy of the old loess series lying in a morphologically emerged position, there occur less fossil soils and thinner loess bands due to the surface erosion of the interval period. On the other hand, the loess series deposited in the valleys or in smaller depressions are thicker and more stratified by fossil soils, loessic and non-loessic sediments. In the case of a chronological subdivision also these experiences should be taken into consideration.

Finally it must be emphasized that the thickest and oldest loess series with the greatest number of intercalated fossil soils mostly occur in the Pannonian plateau and terraces along the Danube. The same is true also in the case of the tributaries of the Danube. On the ridges of the hilly regions (when they are non-dissected loess plateaus) and in the foothill surfaces, the loess strata are very incomplete and in the exposures here older loesses are rarely found.

REFERENCES

1. Ádám, L.—Marosi, S.—Szilárd, J.: A paksi löszfeltárás (Loess profile of Paks). Földr. Közl., LXXXII, 239—254, 1954.
2. Bacsák, Gy.: A skandináv eljegesedés hatása a periglaciális övön (Die Auswirkung der skandinavischen Vereisung in der periglazialen Zone). Magyar Orsz. Meteorol. és Földmágn. Int. Kisebb Kiadv. 86, 1942.
3. Bulla, B.: A magyarországi löszök és folyóteraszok problémái (Problems of the Hungarian loesses and river terraces). Földr. Közl., LXII, 7—8, 136—149, 1934.
4. Kriván, P.: A közép-európai pleisztocén éghajlati tagolódása és a paksi alapszelvény (Klimatische Gliederung des mitteleuropäischen Pleistozäns und des Grundprofils von Paks). Magy. Áll. Földt. Int. Évk., 43, 3, 364—512, 1955.
5. Miháltz, I.: La division des sédiments quaternaires de l'Alföld. Acta Geol. Sci. Hung., II, 109—121, 1953.
6. Pécsi, M.: Genetic classification of the deposits constituting the loess profiles of Hungary. Acta Geol. Sci. Hung., IX, 65—85, 1965.
7. Pécsi, M.: Zur Frage der Typen der Löß- und lößartigen Sedimente im Karpatenbecken und ihrer lithostratigraphischen Einleitung. Földr. Közl., 4, 305—323, 332—338, 346—351, 1965.
8. Pécsi, M.: Relationship between slope geomorphology and Quaternary slope sedimentation. Acta Geol. Sci. Hung., XI, 307—321, 1967.
9. Pécsi, M.: Horizontal and vertical distribution of the loess in Hungary. Studia Geomorphologica, 1, Krakow, 1967.
10. Scherf, E.: Versuch einer Einleitung des ungarischen Pleistozäns auf moderner polyglazialistischer Grundlage. (Verhandlungen der III. Internationalen Quartär-Konferenz, 237—247.)   Wien, 1938.
11. Stefanovits, P.—Rózsavölgyi, J.: Újabb paleopedológiai adatok a paksi szelvényről (Weitere paleopedologische Angaben über das Bodenprofil von Paks). Agrokémia és Talajtan, 143—160, 1962.

36

Reprinted from *Bull. Geol. Soc. Finland,* **43,** 109–123 (1972)

STRATIGRAPHY AND MATERIAL OF THE LOESS LAYERS AT MENDE, HUNGARY

MATTI SEPPÄLÄ

SEPPÄLÄ, MATTI 1971: Stratigraphy and material of the loess layers at Mende, Hungary. Appendix by Kalevi Punakivi and Matti Seppälä. *Bull. Geol. Soc. Finland 43,* 109—123.

Various experimental research methods were adopted in the present investigation for the analysis of the samples which were taken from 16 different places from the loess exposures in the brickyard at Mende, Hungary.

The quartz grains in the loess showed a fairly outworn stage of roundness.

The colour of the loess alternates from pale yellow to strong brown (2.5Y, 7.5YR).

The amounts of calcium and magnesium vary according to the stage of weathering. No constant proportion can be noticed between them. Magnesium is preserved proportionally better than calcium in the weathering. Carbonates mainly come from rock minerals.

The weight % of Ca in the carbonate concretion was 21.17, that of Mg only 0.29.

The phosphorus content in fresh loesses is a little larger than in soil horizons.

The pH values in the aqueous suspension of the samples ranges from 8.3 to 8.75.

Electric conductivity values are also high — between 64.1 and 118.7 S.

The C-14 dating was done with material from the third paleosol horizon counting downwards from the surface. The computations yielded a radiocarbon age of $27\ 200\ {}^{+1\ 400}_{-1\ 100}$ yr. B. P., which corresponds to the Paudorf Interstadial between Würm stages II and III.

At the end of the paper, the results obtained are compared with opinions generally held as to the origin of loesses. The present author is in favour of the diagenetic theory of loess origin.

Matti Seppälä, Department of Geography, University of Turku, Turku 2, Finland.

Kalevi Punakivi, Geological Survey of Finland, Otaniemi, Finland.

CONTENTS

Introduction

Hungary, in the middle of the Carpathian Basin, serves as a large sedimentation basin to which weathering-detached loose ingredients from the surrounding mountains have continuously sedimentated since the sub-tropic Middle Pliocene Era (Fink 1964 p. 451). This accounts for the formation of sediment deposits even as thick as 3 000—4 000 metres (Rónai 1965). The surface of the major part of Hungary is covered by loess, which is more than 50 metres thick in some places (Ádám, Marosi & Szilárd 1954).

Palaeogeographically, the stratigraphy of loesses is a very important object of investigation, which concerns conditions during the Pleistocene Era. A loess exposure of only 10 metres often reveals 3—5 buried fossil soil horizons indicating certain stages in the development of the climate. Owing to local erosion

Fig. 1. Part of the loess exposure at Mende. The arrow points to the place from where sample 7C (I—3130) was taken. Photographed by the author. April 14, 1967.

and accumulation during loess formation, it is not easy, however, to connect the fossil soil horizons from different places with each other. Investigations based on both pedological, archeological, palaeontological, and radiocarbon datings have all facilitated the palaeogeographical interpretation of the phenomena observed in loess layers (*e.g.* Haase 1963).

The samples analysed in the laboratory were collected by the author from the pit of a brickyard in the valley of the River Tápio at Mende, 40 km NE of Budapest (47°25′ N Lat, 19°25′ E Long). The characteristic features of the area include many-branched derasion valleys which have been eroded into the loess with ridges between them (Pécsi 1965a p. 332—333) which are about 220 m.a.s.l. Hungarians have investigated the same profiles very intensively. However, the research methods adopted in the present investigation have yielded some new and supplementary data on loess layers. An attempt has been made to apply the observations to opinions generally held as to the origin, age, and ways of formation of loesses.

Stratigraphy of loess layers at Mende

The stratigraphy of the loess layers at Mende has been elucidated thoroughly by several previous articles (*e.g.* Pécsi 1965a, 1965b, 1966, Stefanovits 1965, Pécsi & Hahn 1969). Based on these, some general points concerning stratigraphy are repeated here (see Fig. 2).

On the surface there is the present chernozem soil. Sample 1 was taken from the sandy slope loess beneath it.

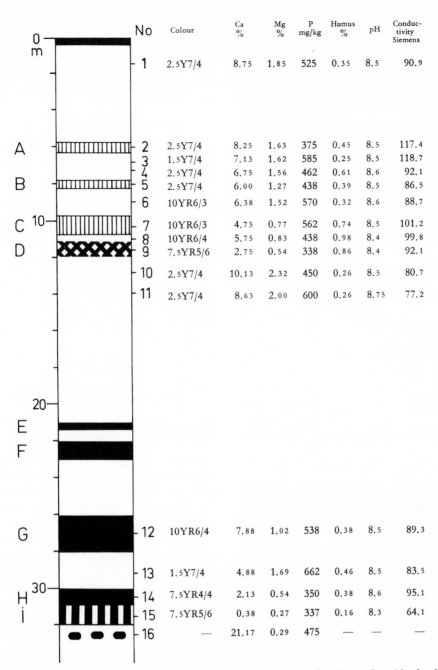

No	Colour	Ca %	Mg %	P mg/kg	Humus %	pH	Conductivity Siemens
1	2.5Y7/4	8.75	1.85	525	0.35	8.5	90.9
2	2.5Y7/4	8.25	1.63	375	0.45	8.5	117.4
3	1.5Y7/4	7.13	1.62	585	0.25	8.5	118.7
4	2.5Y7/4	6.75	1.56	462	0.61	8.6	92.1
5	2.5Y7/4	6.00	1.27	438	0.39	8.5	86.5
6	10YR6/3	6.38	1.52	570	0.32	8.6	88.7
7	10YR6/3	4.75	0.77	562	0.74	8.5	101.2
8	10YR6/4	5.75	0.83	438	0.98	8.4	99.8
9	7.5YR5/6	2.75	0.54	338	0.86	8.4	92.1
10	2.5Y7/4	10.13	2.32	450	0.26	8.5	80.7
11	2.5Y7/4	8.63	2.00	600	0.26	8.75	77.2
12	10YR6/4	7.88	1.02	538	0.38	8.5	89.3
13	1.5Y7/4	4.88	1.69	662	0.46	8.5	83.5
14	7.5YR4/4	2.13	0.54	350	0.38	8.6	95.1
15	7.5YR5/6	0.38	0.27	337	0.16	8.3	64.1
16	—	21.17	0.29	475	—	—	—

Fig. 2. An outline of the loess layers at Mende according to Pécsi 1966 (Tafel 33). A—C weakly developed paleosols, D a chernozem-type soil, E—H chernozem, Ī brown forest soil. The numbers of the samples show the places from where they were taken. Sample 16 is from a carbonate concretion (see Fig. 3).

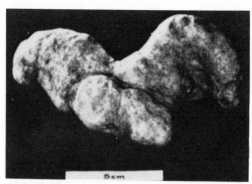

Fig. 3. Small carbonate concretion (»loessdoll»), sample 16 in Fig. 2. Photographed by Martti Valtonen.

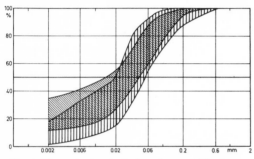

Fig. 4. The grain sizes in the loess layers at Mende as a cumulative graph. Constructed on the basis of the analyses made by Pécsi (1965 a) and Stefanovits (1965). The part with vertical lines shows the grain size curves of fresh loesses. Skew lines show the area of the grain size curves of soil horizons.

Fig. 5. Quartz grains from sample 7. Fraction 0.074—0.125 mm. Photographed by the author.

The youngest fossil soil (A) is at a depth of approximately 6 metres. It is a weakly developed humus carbonate layer (sample 2).

The fresh material beneath this is defined by Pécsi (1965 a p. 335) as slope loess containing a little sand (samples 3 and 4).

Soil horizon B (sample 5) is also a weakly developed humus carbonate layer.

In the loess layer (sample 6) between soils B and C, which is about 1.5 metres thick, there are very many shells of snails and bones of mammoths *(Elephas primigenius)*.

Fossil soil horizon C at a depth of about 10 metres is meadow soil about 1.5 metres thick and containing charcoal and bones of mammoths. Samples 7 and 7 C were taken from this horizon; the latter was dug out of a fire pit in which the loess had partly burned and turned to brick-red. A radiocarbon dating of sample 7 C was made. Sample 7 represents the soil horizon proper.

The material in the following intermediary layer (sample 8) is sandy loess with a fairly large humus content. Rodents and other animals digging holes in the soil have mixed the original stratigraphy by digging soil-filled burrows (crotowinas).

Fossil soil D (sample 9) is a dark brown, roughly 1 metre thick horizon resembling chernozem, and also with crotowinas.

Crotowinas have also mixed the fresh loess beneath horizon D to some extent. Samples 10 and 11 were taken from the upper part of the horizon.

E and F fossil soils are chernozem. The author was not able to collect any samples from the depth between 15—26 metres because there were no fresh exposures in spring 1967.

G and H soils are also chernozem. In the loess between them (sample 13) it is possible to notice a slightly laminated structure.

Fossil soil I (sample 15) is a brown forest soil horizon. It forms together with the chernozem (H) above it a two metre thick soil complex from which plenty of calcium has leached. The lime accumulation has formed potato-like carbonate concretions beneath the soil complex. Calcium, magnesium, and phosphorus analyses were made using the concretion in Fig. 3 (sample 16).

The loess still continues on down at least 10 metres deeper, but the exposures in the spring of 1967 were from no deeper than 33 metres.

The weathered layers in the loess mainly represent the illuvial zones of the soils. Mende is a rewarding object of investigation because all the main soil horizons typical of Hungary since the Riss-Würm Interglacial occur there (Pécsi 1965b, 1968).

There is almost no laminated structure in the loesses at Mende, but material is mainly a relatively unstratified, homogeneous and porous mass.

Granulometric composition

Pécsi (1965a, 1966) has made several analyses of the grain size of the fresh loess layers at Mende. Stefanovits (1965) has investigated the grain sizes in the soil horizons of exposures. The present author has enumerated these data by circumscribing, in a half-logarithmic scale, the areas in which the grain size curves fall (Fig. 4). The medians of the cumulative curves are almost invariably in the grain size area of 0.01—0.05 mm, which is, indeed, the most typical for loesses.

The effect of weathering is clearly shown by the abundance of clay fractions (<0.002 mm) in the samples taken from soil horizons. The finest fractions, usually exceeding 25 %, are dominant in soil horizons representing the chernozem type. The proportion of the clay fraction in a forest soil (I in Fig. 2) even exceeds 30 % (Stefanovits 1965 p. 340).

The proportion of the clay fraction in fresh loesses is less than 10 % in general (Pécsi 1965a p. 336). The coarsest loesses containing sand are situated in the surface part of the layer. Loesses seldom have grains larger than 0.6 mm in diameter.

Roundness of quartz grains

Part of sample 7 was washed in 10 % hydrochlorid acid, rinsed with distillated water, wet-sifted and dried. The quartz grains in the fraction of 0.074—0.125 mm were separated for photographing with a binocular microscope. In the enlargement of the photograph (Fig. 5), the percentages of the plain (P), concave (C), and convex (V) faces in the silhouettes of 100 grains were measured according to the Szadeczky-

Kardóss method (Köster 1964 p. 180—181). The percentual proportions of the faces are to be seen in the triangular diagram (Fig. 6) showing the differences in the roundness of the grains. Most quartz grains are slightly rounded, but there are few very well rounded. The roundness index of sample 7, computed according to Szadeczky-Kardóss, was $V + \dfrac{P}{2} = 5.9$, *i.e.* the material is fairly rounded. The index value of completely rounded material is 10 (Pécsi & Pécsiné 1959).

300 quartz grains from the same sample were examined with the binocular microscope according to the Russell—Taylor method (Köster 1964 p. 183—185). The roundness value obtained was 2.31 (max 5).

These results agree with Charlesworth's (1966 p. 513) general observation that the grains of loesses are sometimes round and subangular but usually sharp and uneffected by weathering.

60 % of the grains had a mat surface probably caused by aeolian activity. Though sample 7 was taken from the illuvial zone of a fossil soil, not all of the grains are mat-faced. This would indicate that chemical weathering is not a partic-

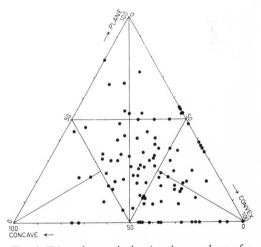

Fig. 6. Triangular graph showing the roundness of a hundred quartz grains from sample 7 (Fig. 5), according to the Szadeczky-Kardóss method.

ularly important factor in the frosting of grains.

It is hardly possible that the material would have worn significantly during the accumulation of loesses, for such fine material moves suspended both in water and in air. It cannot be inferred from the roundness and surface texture of the material whether it was worn during loess accumulation or due to earlier processes. The great dispersion of roundness refers to grains of various kinds of origin. In that case earlier aeolian and fluvial sediments as well as chemical and mechanical weathering products must be considered.

Colour

Limonite and humus are usually mentioned as the agents which cause the characteristic colour of loess (*e.g.* Berg 1964 p. 45). The colours of the samples taken from Mende were identified by comparing them when air-dried with the Munsell Soil Color Charts.

The colours of the soil horizons fell into groups 2.5Y, 10YR and 7.5YR (pale yellow, yellowish or pale brown, and strong brown). Fink (1964 p. 453), for example, has observed that the two latter colours are typical of the B horizon of fossil soils in Austria.

Pale yellow is a common colour of fresh loess at Mende. In the youngest layers, there are no big differences in colour between fossil soil horizons (A—C) and fresh loess (Fig. 2). Colour 1.5Y7/4 in samples 3 and 13 means one between colours 10YR7/4 and 2.5Y7/4, which is alittle browner than pale yellow.

It is evident that the soiling process had been going on at the same time as loess had been accumulating among the vegetation. Thick and dark fossil soil horizons are evidence of long intervals in loess accumulation. The weathering of loess layers alone was hardly responsible for the yellowish colour of the loess. It is most probable that the depositing material was already yellowish owing to earlier weathering. Consequently it is impossible to distin-

guish between the influence of local circumstances and factors that have previously affected the colour.

Calcium and magnesium content

Guenther (1961 p. 11) divides the chalk in loesses into primary and secondary types of occurrence. Primary or syngenetic chalk includes: 1) calcium carbonate grains, which only occur in weakly weathered loesses and 2) organic chalk, which mainly originates from shells of snails which had become buried in the loess. Secondary or epigenetic chalk comprises: 1) the calcium carbonate film covering other mineral grains, 2) the chalk filling the hollows of roots and other places both above and beneath the soils. 3) carbonate concretions or loess dolls which are big lumps beneath soil horizons, and 4) white loess layers accumulated from the ground water containing calcium.

The chalk at Mende occurs in all the other forms except for the latter coherent lime layer. Thus the calcium in loesses is not evenly distributed in the material, but migrates in the loess layers with the ground water and the water filtering down through the layers. The heterogeneousness of the material was taken into consideration by making the analysed loess samples as homogeneous as possible before the material needed for the analyses was separated.

For the Ca and Mg analyses, the 1-gramme samples were boiled in a weak hydrochlorid acid solution to dissolve the carbonates. The analysis was carried out with a Perkin-Elmer 290 AAS (Atomic Absorption Spectrophotometer). The calcium and magnesium contents are given in weight percentages (Fig. 2). The corresponding $CaCO_3$ and $MgCO_3$ content values were obtained by multiplying the Ca values by 2.5 and the Mg values by 3.06. Thus the values of the $CaCO_3$ content in samples 1—15 vary from 0.95—25.32 weight %, and those of the $MgCO_3$ content from 0.83—7.10 weight %. When determining these values it was supposed that all the calcium and magnesium come from carbonates.

Guenther (1961 p. 12) states that the ratio of

CaCO₃:MgCO₃ in loesses is 6:1. On the basis of the observations in the present paper it can be concluded that Mg dissolves relatively less readily than Ca during weathering processes. Thus the proportion of calcium to magnesium alternates in the loesses at Mende, and no ratio of general application can be found (*cf.* Fig. 7).

The values of the Ca and Mg content are lower in soil horizons than in fresh loesses. The scarcity of Ca and Mg in soils H and I is due to a long phase of humid climate and to the strong eluviation caused by it. This also resulted in the formation of the carbonate concretion zone beneath. The calcium content of the concretion (Fig. 3) is 21.17 weight %, while the magnesium content is only 0.29 weight % (Fig. 2).

The analysis of snail shells (*Pedicella* sp.) collected from the surface of the earth at Mende gave 39.89 % (=99.7 % CaCO₃) and 0.07 % magnesium (=0.2 % MgCO₃). It can be inferred from this that all the carbonates in loesses cannot possibly be of organic origin, but that they probably mainly originate from dolomite and dolomitic limestone (see Appendix). The mineragenic origin of calcium and magnesium is also indicated by the fact that well-preserved, unweathered snailshells are regularly found in loesslayers.

Phosphorus content

The phosphorus contents of the samples were examined in the laboratory of Viljavuuspalvelu Oy, Helsinki, by photoelectric colorimeter. The method has been published in the series Kungl. Landbrukstyrelsens Kungörelser M. M. 1965 No. 1 p. 14—15. It uses as a reagent 2.00 M HCl in which 2 g of the sample is boiled for 2 hours in a double boiler. 5 ml of the extract is taken and mixed with 35 ml of distilled water. The measurement is performed with the diluted solution.

On the basis of the results (Fig. 2), it can be concluded that phosphorus compounds eluviate to some extent in connection with weathering. The total phosphorus contents in fossil soils alternated between 338—562 mg/sample kg (=ppm), and in fresh loesses between 438—662 mg/sample kg. The concretion (sample 16) contained 475 mg/kg phosphorus.

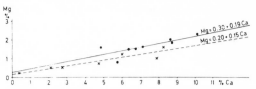

Fig. 7. The linear correlations of the calcium and magnesium contents in samples 1—15. The figures are weight percentages from the whole sample. Points indicate fresh layers (Mg = 0.30 + 0.19 Ca, r = 0.76) and crosses fossil soil samples (Mg = 0.20 + 0.15 Ca, r = 0.92) (see also Fig. 2). r = Bravais-Pearson coefficient of linear correlation.

The value of 750 mg/sample kg in sample 7C is a significant exception to the general phosphorus content values. The abundance of phosphorus may indicate that the sample is probably from a man-made fire pit, for a clear increase of the phosphorus content in the soil has been noticed even in the temporary dwelling places of man (Lorch 1940 p. 633).

Humus content

The humus content was determined from 5 g samples by titration. The reagent consisted of 40 ml concentrated sulphuric acid (H₂SO₄) and 25 ml 2 N potassium bicromate (K₂Gr₂O₇). The solution was boiled in a double boiler for 2 hours. The volumetric flask was filled up to 250 ml with distilled water. 25 ml of the solution was taken for titration and diluted with distilled water up to 200 ml. The unused bicromate was reduced with 5 ml 20 % KJ. The iodine liberated was titrated with 0.1 N sodium tiosulphate solution (N₂S₂O₃ · 5H₂O). In addition to the actual analyses, a »blind test» was made without any soil sample. The amount of sodium tiosulphate used for this test was compared with that needed for the analyses respectively. The amount of oxygen used in the analysis reactions can be calculated by subtraction. The humus content is computed by using the formula:

$$Hu = \frac{0.518 \cdot (B-b)}{a} \%$$

in which Hu = amount of humus in percentages, B = amount of sodium tiosulphate in ml used in the blind test, b = amount of Na₂S₂O₃ solution in ml used in the titrating of the sample, a = amount of the soil sample in grams. The method is an application of the Rauterberg & Kremkus method (Thun 1955 p. 48—49) made by Pentti Alalammi.

271

The humus content in samples 1—15 did not exceed 1 %. Stefanovits (1965 p. 340) got the same result when investigating the fossil soils at Mende. The humus contents of soil horizons G, H and I in the present paper (Fig. 2) are the only ones that diverge significantly from the values 0.33—0.78 % given by Stefanovits. The humus content in sample 8 is markedly high (0.98 %) in comparison with the humus content values in the soil horizons. As also revealed by the colour index, it is very difficult to distinguish this horizon from paleosols C and D. The humus content of sample 7C was 2.74 weight %.

As soiling proceeds and the humus content increases so calcium and magnesium are leaching. They are usually only to be found in small amounts in humus-rich layers (Pécsi 1965a Abb. 16).

The relatively high values of the humus content (0.25—0.98 %) in fresh loesses seem to indicate that soiling was continually taking place during the loess accumulation. On the other hand, the reason may be that humus was driven to the area during the accumulation phase (*cf.* the chapter on colour). The activity of plants' roots and animals are also factors affecting the mixing of humus. The amount of humus in the soils (0.16—0.86 %) is small in comparison with the present soil horizons. Kubiëna (1953 p. 242) points out that the humus content of *e.g.* chernozem is 4—16 %.

pH and electric conductivity

To investigate the pH of the loess samples, 5 grams of air-dried material was silted into 100 ml distilled water and left for 24 hours. The sample was mixed and the measurement of the acidity of the solution was carried out by means of a pH measure model E350B Metrohm Herisau.

The results (Fig. 2) show that loesses are very alkaline. The pH varies in soil horizons from 8.3 to 8.6 and in fresh loesses from 8.5 to 8.75, the reason being the great number of alkaline cations.

Resistance measurements were made with the same water solution using a battery gadget (Norma RW1) to determine the electric conductivity and the extent to which the salt content varies. According to the measurements, loesses contain abundant electrolytes. The electric conductivity in samples 1—15 ranged from 64.1 to 118.7 Siemens (1 S $= 1\Omega^{-1}$). The conductivity values were approximately as great in soils as in fresh loesses. Only sample 15 had a value which was considerably smaller than the others, 64.1 S, while that of sample 7C was 140.8 S.

Conductivity generally increases with increasing humus content, although correlation is not clear.

Radiocarbon age of paleosol horizon C

Pécsi (1965b, 1966, 1968) has given a general chronological scheme as to the age relationships of the fossil soils of loesses in Hungary. According to him, the double soils H and I (Fig. 2) date from the Riss-Würm Interglacial, which is considered to have ended 75 000 years ago (*e.g.* Büdel 1960). The other soils at Mende are younger. Quoting Pécsi further, the Würm glaciation divides into three glacial periods between which there are two interstadial phases. Several soil horizons were formed during these phases.

A radiocarbon dating (Lab.no. I—3130) of sample 7C was made in the Isotopes, Inc., Westwood laboratories (USA). The sample was taken from a buried fire pit approximately in the centre of soil horizon C. Sample I—3130 consisted of soil, small charcoal fragments and other organic material. According to the statement of the laboratory, various components could not be separated, and so the sample was treated and combusted in bulk. The radiocarbon age obtained was

$$27\ 200\ {}^{+1\ 400}_{-1\ 100}\ \text{years B.P.}$$

This dating was also reported in the bulletin, Radiocarbon 1969 (p. 81). It confirms an earlier dating of the same horizon by Pécsi (1966 p. 245), for the difference is not very significant:

29 800 ± 600 yrs. B.P. (Lab.no. MO-422).

The paleosol horizon C in question was formed at the end of Middle Würm during the so-called Paudorf Interstadial, which was between 35 000 and 25 000 years B.P. according to different scientists (*e.g.* Haase 1963, Frenzel 1964 p. 26—27). It has been stated that soil formation also took place in other loess areas at that time. Gross (1962—63 p. 62—63, 65) regards the period between 27 000 and 25 000 as the proper Paudorf Interstadial, which would have been preceded by the second deglaciation phase of the Würm glaciation starting 31 000 yrs. B.P.

For comparison, some other datings of paleosols may be mentioned:

In Austria, Stillfried: 27 900±300 (Gro-2523) and 28 120±200 (Gro-2533), and Unterwisternitz: 28 100±300 and 29 900±300 (Fink 1962 p. 14, 16). In Iowa, USA, the ages of paleosols vary between 16 500±500 (I-1419) and 29 000±3 500 (I-1269) (Ruhe 1968 p. 59). In Zelzate, Belgium, fossil soil was dated as 28 200±270 (GrN-4783) (Paepe 1969 p. 47—49). In Sittard, Holland, the figures were 27 900±670 (Kuyl & Bisschops 1969 p. 101). From Czechoslovakia, a dating in Věstonice is: 28 300±300 (GrN-2092) (Kukla & Lozek 1969 p. 54). The so-called soil of Briansk in the U. S. S. R. is close to the age of paleosol C at Mende. This was dated as 24 920±1 800. Two soils from Molotova are also rather close: 23 000±800 (23 700±320) and 28 100±1 000 (29 650±1 320) (Ivanova 1969 p. 152, 155).

Conclusions

The stage of weathering is of decisive importance in the characteristics of loess layers. There are more clay fractions in soil horizons than in fresh loesses. The colour of the loess varies depending on the stage of weathering. Carbonates and phosphorus compounds are dissolved in weathering processes and precipitated into other layers. The pH values of soil horizons are slightly smaller than those of fresh loesses. However, weathering has not had any significant influence on humus content and electric conductivity. The accumulation of loess has been so slow that weathering has had some influence on all loess. However, the eluvial soil hypothesis of the origin of loesses by Berg (1964) does not

hold good as such, for weathered paleosols of various phases occur in loess exposures. They cannot be explained without consecutive events of accumulation. Loess cannot, however, be regarded as a mere product of weathering because the roundness of its quartz grains is very heterogeneous. Thus the material of loess obviously originates from several different sources. »Chief source of loess was outwash in the valley of the Danube, derived from glaciers in the Alps and in the Carpathians» (Flint 1967 p. 191). This is a very simplified picture, for loesses are derived from varied sources. Fluvial and aeolian processes and solifluction are mentioned as depositing factors (Kádár 1960, Pécsi 1965b, 1968, Rónai 1965). The Tertiary weathered layers in the mountains have obviously been primary sources. The Pannonian deposits which were already in the Carpathian Basin during the glaciations must not be forgotten either. The deposits have provided plenty of material for fluvial and aeolian re-accumulation.

Zeuner's (1959 p. 93) statement that the weathering of soils in younger loesses is comparatively weakly developed also holds good with the loesslayers at Mende, the reason being the lowering in temperature of the climate. Würm III was the coldest period during the last glaciation (Šegota 1967 p. 132). The periods of interstadials became continously shorter.

Acknowledgements — My hearty thanks are due to Professor Márton Pécsi, Ph.D., Hungarian Academy of Sciences, and the staff of his Department for their friendly guidance during my field-trips.

It would not have been possible to do the dating without Professor Birger Ohlson's very positive attitude and kindness and I am grateful to him for this.

Mr. Heikki Papunen, Phil. lic., Department of Geology at the University of Turku, kindly helped me to operate the Atomic Absorption Spectrophotometer during the calcium and magnesium analyses.

This investigation was aided financially by the Finnish-Hungarian Cultural Committee.

My thanks are also due to the Department of Geography at the University of Turku for the permission to use the laboratory and for covering the expenses of the equipment needed in the investigation.

REFERENCES

Áдáм, László; Marosi, Sándor & Szilárd, Jenö (1954) A Paksi löszfeltárás. Zusammenfassung: Der Lössaufschluss von Paks. Földrajzi Közlemények 2, 239—254.

Berg, L. S. (1964) Loess as a product of weathering and soil formation. (Translated from Russian). Israel Program for Scientific Translations. 207 p. Jerusalem.

Büdel, Julius (1960) Die Gliederung der Würmkaltzeit. Würzburger Geographische Arbeiten 8, 1—45.

Charlesworth, J. K. (1966) The Quaternary Era. Vol. 1. 591 p. London.

Fink, J. (1962) Studien zur absoluten und relativen Chronologie der fossilen Böden in Österreich. II. Wetzleinsdorf und Stillfried. Archaelogia Austriaca 31, 1—18.

— (1964) Die Gliederung der Würmeiszeit in Österreich. Report of the VIth INQUA Congress Warsaw 1961. Vol. IV, 451—462.

Flint, Richard Foster (1967) Glacial and Pleistocene Geology. 553 p. New York.

Frenzel, Burkhard (1964) Zur Pollenanalyse von Lössen. Eiszeitalter und Gegenwart 15, 5—39.

Gross, Hugo (1962—63) Der gegenwärtige Stand der Geochronologie des Spätpleistozäns in Mittel- und Westeuropa. Quartär 14, 49—68.

Guenther, Ekke W. (1961) Sedimentpetrographische Untersuchung von Lössen. Zur Gliederung der Eiszeitalters und zur Einordnung paläolithischer Kulturen. Teil I. 77 p. Köln-Graz.

Haase, Günter (1963) Stand und Probleme der Lössforschung in Europa. Geographische Berichte 27, 97—129.

Ivanova, I. K. (1969) Les loess de la partie sud-ouest du territoire européen de l'U. R. S. S. et leur stratigraphie. La Stratigraphie des Loess d'Europe. Supplément au Bulletin de l'Association Francaise pour l'Étude du Quaternaire, 151—159.

Kádár, László (1960) Climatical and other conditions of loess formation. Studies in Hungarian Geographical Sciences (ed. Gyula Miklós), Budapest, 17—24.

Kubiëna, W. L. (1953) Bestimmungsbuch und Systematik der Böden Europas. 392 p. Stuttgart.

Kungl. lantbruksstyrelsens kungörelser M. M. (1965) Nr 1, 1—20. Solna.

Kukla, J. & Lozek, V. (1969) Trois profils caractéristiques de la Bohême Centrale et de la Moravie du Sud. La Stratigraphie des Loess d'Europe. Supplément au Bulletin de l'Association Francaise pour l'Étude du Quaternaire, 53—56.

Kuyl, O. S. & Bisschops, J. H. (1969) Le loess aux Pays-Bas. La Stratigraphie des Loess d'Europe.

Supplément au Bulletin de l'Association Francaise pour l'Étude du Quaternaire, 101—104.

Köster, Erhard (1964) Granulometrische und morphometrische Messmethoden an Mineralkörnern, Steinen und sontigen Stoffen. 336 p. Stuttgart.

Lorch, Walter (1940) Die siedlungsgeographische Phosphatmethode. Die Naturwissenschaften 28, 633—640.

Munsell soil color charts. U. S. Dept. Agriculture Handbook 18. (1954) Baltimore.

Paepe, Roland (1969) Les unités litho-stratigraphiques du Pléistocene supérieur de la Belgique. La Stratigraphie des Loess d'Europe. Supplément au Bulletin de l'Association Francaise pour l'Étude du Quaternaire, 45—51.

Pécsi, Márton (1964) Ten years of physico geographic research in Hungary. 132 p. Budapest.

— (1965 a) Der Lössaufschluss von Mende. Földrajzi Közlemények 13, 332—338.

— (1965 b) Zur Frage der Typen der Lösse und Lössartigen Sedimente im Karpatenbecken und ihrer lithostratigraphischen Einleitung. Földrajzi Közlemények 13, 305—323.

— (1966) Lösse und lössartige Sedimente im Karpatenbecken und ihre lithostratigraphische Gliederung I—II. Petermanns Geographischen Mitteilungen 110, 176—189, 241—252.

— (1968) Loess. The Encyclopedia of Geomorphology (ed. Rhodes W. Fairbridge), New York, 674—678.

— & Hahn, G. (1969) Historique des recherches sur le loess en Hongrie. La Stratigraphie des Loessd'Europe. Supplement au Bulletin de l'Association Francaise pour l'Étude du Quaternaire, 85—91.

— & Pécsiné, Donáth Éva (1959) Elemző módszerek alkalmazá a geomorfológiai kutatásban. Zusammenfassung: Die Anwendung analytischer Forschungsmethoden in der Geomorphologie. Földrajzi Értesitö 8, 165—178.

Radiocarbon 11 : 1 (1969) Published by The American Journal of Science. 244 p.

Rónai, Andre (1965) Some observations concerning the Quaternary sedimentation in Hungary. Acta Geologica Hungarica 9, 17—32.

Ruhe, Robert V. (1968) Identification of paleosols in loess deposits in the United States. Loess and related eolian deposits of the world (ed. C. Bertrand Schultz and John C. Frey), Lincoln, Nebraska, 49—65.

Šegota, Tomislav (1967) Paleotemperature changes in the Upper and Middle Pleistocene. Eiszeitalter und Gegenwart 18, 127—141.

STEFANOVITS, PÁL (1965) Untersuchungsangaben der Begraben Bodenschichten im Lössprofil von Mende. Földrajzi Közlemények 13, 339—344.

THUN, RICHARD (1955) Die Untersuchung von Böden. Handbuch der landwirtschaftlichen Versuchs-' und Untersuchungsmethodik (Methodenbuch) I. Dritte Auflage (ed. Erich Knickmann), 271 p. Berlin.

WOLDSTEDT, PAUL (1961) Das Eiszeitalter I. 374 p. Stuttgart.

ZEUNER, FREDERICK E. (1959) The Pleistocene Period. Its Climate, Chronology and Faunal Successions. 447 p. London.

Manuscript received, May 25, 1970.

APPENDIX

MINERAL ANALYSES OF THE LOESS SAMPLES

by

KALEVI PUNAKIVI and MATTI SEPPÄLÄ

The minerals in loess samples 1—15 (p. 111) in the present paper were examined by means of X-ray diffraction, differential thermal and thermogravimetric analyses.

INTERPRETATION OF X-RAY DIAGRAMS

The apparatuses used were a Philips X-ray diffraction Generator PW—1010 and Goniometer PW—1050 of the Geological Survey of Finland.

CuKα radiation and a Ni filter were used in the analyses. The Goniometer speed was 2°Θ/min, and slides were orientated.

Fresh loesses generally contained abundant calsite and dolomite. Illite, chlorite, quartz and plagioclase occurred in all the samples. Weathered loesses had more clay minerals than fresh loesses.

Sample 10 contained considerably more carbonates than the other samples. Sample 15 had no carbonates according to the analysis (cf. Fig. 2 p. 111). It must be pointed out, however, that even as much as 5 % calsite may remain undetected in X-ray analyses (Webb & Heystek 1957 p. 329).

TABLE 1

Occurrence intensities of minerals estimated on the basis of the X-ray diffractions

No	Chlorite 14.0 Å	Illite 10.0 Å	Chlorite 7.1 Å	Quartz 4.26 Å	Quartz and Illite 3.33 Å	Orthoclase 3.24 Å	Plagioclase 3.20 Å	Calsite 3.04 Å	Dolomite 2.89 Å
1	—	w	w	m	vs	vw	vw	s	m
2	vw	vw	—	vw	vs	—	w	w	vw
3	—	vw	vw	w	vs	—	w	s	m
4	vw	w	w	m	vs	—	m	w	m
5	—	w	w	m	s	—	w	m	s
6	w	m	m	m	vs	vw	m	s	s
7	w	s	m	w	s	vw	v	v	—
7C	—	w	vw	m	vs	vw	m	vw	vw
8	vw	w	w	w	s	vw	w	m	vw
9	vw	w	m	m	vs	m	w	m	—
10	—	m	w	m	s	vw	w	s	s
11	vw	vw	w	w	m	—	vw	m	w
12	vw	w	w	m	vs	vw	w	s	m
13	—	w	w	vs	—	m	m	m	m
14	—	vw	vw	s	vs	vw	w	vw	vw
15	—	vw	vw	m	s	vw	vw	—	—

vs = very strong, s = strong, m = medium, w = weak, vw = very weak.

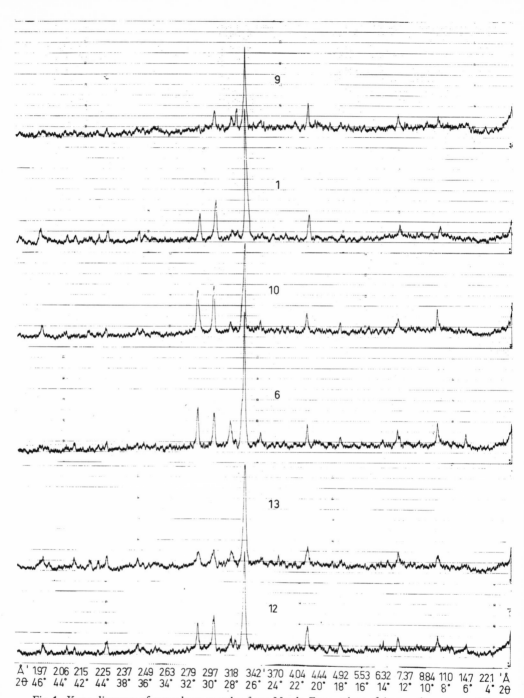

Fig. 1. *X*-ray diagrams of some loess samples from Mende. For numbers of the samples, see page 111.

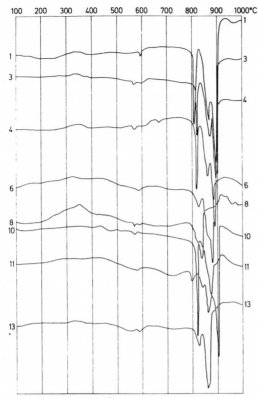

Fig. 2. Differential thermal curves of the samples of fresh loesses.

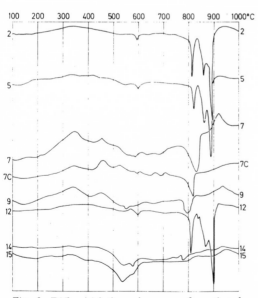

Fig. 3. Differential thermal curves of weathered loess layers.

INTERPRETATION OF DIFFERENTIAL THERMAL CURVES

In DTA, the measuring units were a Leeds & Northrup X-Y-recorder and temperature-regulating system, a specimen holder constructed by Geological Survey of Finland and furnaces made by Oy E. Sarlin Ab.

The samples analysed in DTA were 0.8 grams of ground raw material. The increase of temperature was 10°C/min in air atmosphere.

Differential thermal methods are very suitable for analysing carbonates. Small amounts such as 0.3 per cent dolomite (Rowland & Beck 1952) and 1 per cent calsite (Kulp, Kent & Kerr 1951) can be detected.

Endothermic decomposition is the most characteristic feature of carbonates (Webb & Heystek 1957 p. 330). Wide particle-size range causes broad endothermic peaks (Kulp, Kent & Kerr 1951). Peak temperatures vary depending on the calcium carbonate content. The temperatures of undiluted calcite vary from 860°C to 1010°C,

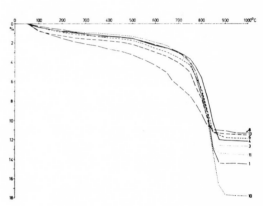

Fig. 4. Thermogravimetric curves of fresh loesses.

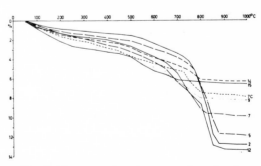

Fig. 5. Thermogravimetric curves of weathered loesses.

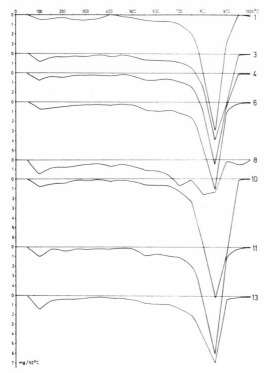

Fig. 6. Differential thermogravimetric curves of fresh
loesses.

and those of 20—25 per cent calcite from 830°C to 940°C
(Webb & Heystek 1957 p. 331).

Dolomite has two characteristic peaks whose temperatures are between 790°C and 940°C (Webb & Heystek 1957 p. 338).

Soluble salts are found to be a major cause of the variations in the position and shape of the lower-temperature endothermic deflection in the DTA curves of a member of low-iron sedimentary dolomites (Graf 1952).

The triple peaks which fell between 800°C and 900°C in the analyses show that the samples contain both calcite and dolomite (*e.g.* samples 1—4, Figs. 2 and 3). The second peak in sample 10 appears only as a slightly arched bend in the curve.

The first peak falls between 780°C and 840°C in the analyses. Samples whose DTA curves have only the first peak contain mostly only calcite (samples 7—9). The midmost peak falls between 840°C and 870°C and the third peak between 865°C and 905°C.

The last peak in samples 1, 6 and 10 is bigger than the others, which seems to mean that they contain plenty of dolomite.

The exothermic peak between 300 and 400°C is probably due to the burning of humus (Soveri 1951 p. 31).

The endothermic peak between 450 and 650°C is caused mostly by illite and to some extent by chlorite (Soveri 1951 p. 35—43). This analysis also proves that soil horizons have considerably more clay minerals than fresh loesses.

INTERPRETATION OF THERMOGRAVIMETRICAL ANALYSES

A Stanton Thermobalance HT-D was used in the TG analyses. The increase of temperature in the thermobalance was 6.6°C/min and the analysed amounts of material 200 mg.

The TG-curves show that the weight of fresh loesses decreases more on the average than that of the samples from soil horizons when they were heated to 1 000°C. This is directly proportional to the amount of carbonates.

The weight of weathered loesses decreases more evenly than that of fresh ones (samples 9, 14 and 15). Samples 2, 5 and 12 are much less weathered than the others, which is also shown clearly by the DTG diagrams (Fig. 7). Sample 8 has clear signs of weathering (Fig. 6).

CONCLUSIONS

The analyses showed that the mineral composition of loesses depends a great deal on their stage of weathering. As the soiling process continued the amount of carbonates decreased and that of clay minerals generally increased,

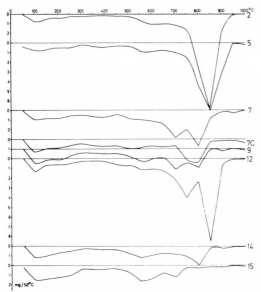

Fig. 7. Differential thermogravimetric curves of weathered
loesses.

even though clay and primary minerals occurred in all the layers.

The results of the X-ray analysis must be considered to be mainly qualitative. The amounts of calsite in particular cannot be detected clearly enough.

When comparing the total amounts of calcium and magnesium obtained by means of chemical analyses (Fig. 2 p. xxx) with the weight losses observed in the TG analyses, a high correlation is noticed between them ($r = 0.97$).

REFERENCES

GRAF, DONALD F. (1952) Preliminary report on the variations in differential thermal curves of low-iron dolomites. The American Mineralogist 37, 1—27.

KULP, J. LAURENCE, KENT, PURFIELD & KERR, PAUL F. (1951) Thermal study of the Ca-Mg-Fe carbonate minerals. The American Mineralogist 36, 643—670.

ROWLAND, RICHARD A. & BECK, CARL W. (1952) Determination of small quantities of dolomite by differential thermal analysis. The American Mineralogist 37, 76—82.

SOVERI, U. (1951) Differential thermal analyses of some Quaternary clays of Fennoscandia. Annales Academiae Scientiarum Fennicae Ser. A III: 23, 1—103.

WEBB, T. L. & HEYSTEK, H. (1957) The carbonate minerals. The differential thermal investigation o, clays (ed. Robert C. Mackenzie), London, 329—363·

37

Reprinted from *Rocznik Pol. Tow. Geol.*, **37**(2), 184–188 (1967)

Wind directions during the accumulation of the younger loess in East-Central Europe

HENRYK MARUSZCZAK

The author compiled a map of distribution of loess in East-Central Europe (Fig. 1), and elaborated an interpretation based upon the following assumptions:

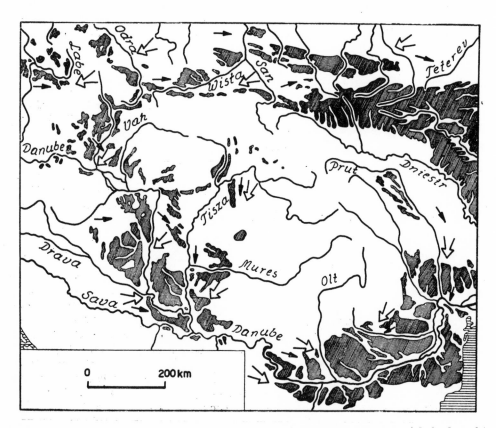

Fig. 1. Distribution of loess in East-Central Europe. Elaborated by the author on the basis of maps published by E. R ü h l e and M. S o k o ł o w s k a (1955), D. J a r a n o f f (1956), P. K. Z a m o r i j and G. M. M o l a w k o (1956), G. R a i l e a n u et al. (1959). J. S e k y r a (1960) and M. P é c s i (1962). Prevailing wind directions reconstructed for the main phase of loess accumulation during the last glaciation are marked schematically by a double arrow. Recen't wind directions marked by a single arrow

a) the loess was accumulated mainly by wind action, or was subject to aeolian transport immediately before deposition by other agents (e. g. slope processes);

b) the loess cover of the discussed area was accumulated mainly during the last glaciation ("younger loess");

c) after the end of accumulation of the loess the distribution and shape of the loess cover was not markedly changed by erosion.

Basing on these assumptions the following conclusions were drawn:

1) the loess covers are distributed in the discussed area not uniformly in the hypsometric range 0—500 m a.m.s.l.

2) frequently in the same geological and morphological conditions thick loess covers are neighbouring areas without any loess accumulation.

It follows from the above that the major part of the loess was transported on small distances in the lower part of the atmosphere. Otherwise the distribution of loess would be more uniform.

As the near-ground atmospheric currents are depending largely on the morphology, a relation should exist between the distribution of loess and the morphology. It can be stated, that in the northern part of the discussed area (Poland, Czechoslovakia, Ukraina) the loess covers are clearly related with dissected highlands and hilly areas while their are lacking on flat lands. Instead, in the southern part of the discussed area in the lower Danube basin, the loess covers are distributed mainly on flat lands, although they are found also on highlands (Northern Bulgaria). Between these two areas, in the middle Danube basin, the distribution of loess covers is intermediate (Fig. 1).

The northern part of the discussed area was lying in the periglacial zone during the accumulation of the younger loess and was characterized — similarly as in Recent times — by a greater wind velocity. It can be inferred therefore, that accumulation of dust occurred on a larger scale at places where orographic obstacles caused a decrease of wind velocity. Such conditions existed chiefly in highland areas, and especially in the zones of windward slopes separating the highlands from the low lands (Fig. 2). Instead, the southern part was lying at that time in a temperate zone where the average wind velocity was smaller. The accumulation of loess occurred there in areas characterized by local greater wind velocity, i. e. on flat lands.

Basing upon the above reconstruction of dynamic conditions of loess accumulation it is concluded that easterly winds prevailed in Poland during the formation of the younger loess cover. This is indicated especially by the distribution of loess on the highlands around the Sandomierz Basin. A rather continuous loess cover is present there on the highlands slopes on the NW and S, while it is nearly entirely lacking on the NE. Also an analysis of sandy intercalations in the loess covers indicated the predominance of easterly winds (H. M a r u s z c z a k, 1963, see also E. S c h ö n h a l s, 1953 and H. B r ü n n i n g, 1959). The change of the glacial air circulation (predominating easterly winds) into the post-glacial air circulation (predominating westerly winds) occurred in the area of Poland during the final phases of accumulation of loess, and before the formation of dunes (H. M a r u s z c z a k, 1963).

In the northern part of the middle Danube basin loess occurs chiefly on highlands and hills bordering the Alföld from W and NW, while it is almost entirely lacking on the NE and E side (Fig. 1). It is possible therefore that also there the loess was deposited mainly by north-easterly and easterly winds (see also P. K r i v a n, 1953 and I. M i h a l t z, 1953).

The conditions of deposition of loess in the southern part of the middle Danube basin are less clear. The distribution of loess covers seems to indicate both westerly and easterly winds. The results of investigations of eolian sediments in the southern part of the Danube — Tisza watershed (B. M o l n a r, 1961) suggest that westerly winds played a major rôle in the transport of loess dust. B. Z. M i l o j e v i ć (1950) assumed easterly and westerly winds for the Jugoslavian part of the middle Danube basin. The Pleistocene wind pattern in the southern and eastern part of the middle Danube basin probably did not differ from the Recent one. Only in the north-western part of the discussed basin a distinct change of wind pattern occurred and the north-easterly winds prevailing during the glaciation were replaced by north-westerly winds (Fig. 1).

In the lower Danube basin the loess forms an extensive continuous cover, while in the north-west of the discussed part of Europe the loess covers are much smaller and disseminated. It seems probable that these differences are significant and that they can be related with the character of the source of the loess dust. In the north-west part (Poland, Czechoslovakia, Northern Hungary) the dust was formed generally under periglacial conditions by frost weathering. The areas of alimentation and sedimentation were disseminated and closely intertonguing (A. J a h n, 1956). Instead, in the lower Danube basin in temperate climatic conditions the formation of dust in place by frost weathering was not important. Probably the major part of the dust was carried by the Danube and its tributaries from the periglacial stage of the mountains neighbouring the flat lands along the Danube valley. An analysis of the thickness of the loess cover confirms the conclusion on the varying rôle of river valleys as alimentation areas. In the Wisła basin in Poland the thickness of the loess ranges from a few m to 20—30 m. and does not show a distinct dependence on the position with regard to the river valleys. Instead, on the lower Danube the thickness of loess is clearly greatest along both shores of that river, amounting to 90 m. on the Bulgarian shore. The thickness of the loess cover decreases rapidly with increasing distance from the river, and does not exceed a few m. the peripheries of the loess cover. The grain size decreases in the same direction (D. J a r a n o f f, 1956; M. M i n k o v, 1960; Tr. N a u m and H. G r u m a z e s c u, 1954).

From the above it follows that the winds which blew out the dust from the lower Danube valley were parallel to the river. If the winds were perpendicular to the river, as assumed D. J a r a n o f f (1956), the loess cover would be asymmetric with relation to the river, extending chiefly on one side of it. In the western part of the Lower Danube basin the winds in question were probably westerly. This is indicated among others by the increase of the thickness of the loess cover east of the right Bulgarian tributaries of the Danube (M. M i n k o v, 1960). Instead, in the eastern part of the discussed basin the winds were probably northerly and north-easterly. This is indicated by the increase of the thickness of the loess cover south of the east-west stretches of the left tributaries of the Danube (H. M a r u s z c z a k, 1964). Therefore the reconstructed Pleistocene wind system in the lower Danube basin does not differ from the Recent one (Fig. 1). The wind system was also similar during the

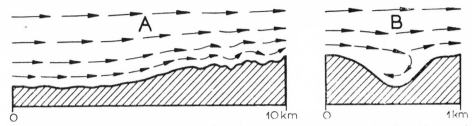

Fig. 2. Schematic cross-sections illustrating changes of velocity and patterns of near-ground winds perpendicular to: A-morphologic scarp separating a flat lowland from a dissected highland; B — axis of a small valley. The lenght of the arrows showing the wind direction is proportional to the wind velocity. The drawing A based upon data on wind velocity changes depending on altitude above ground and morphology; the drawing B is based upon schematic diagrams of wind patterns on glades elaborated by R. G e i g e r (1957)

period of formation of dunes superimposed on the loess cover. (Tr. N a u m and G. G r u m a z e s c u, 1954; H. M a r u s z c z a k and J. T r e m b a-c z o w s k i, 1960).

The areas in which the loess accumulation was related with easterly winds (Poland, Czechoslovakia, Northern Hungary) were situated within the periglacial zone. It can be assumed therefore that in this part of Europe the atmospheric circulation was influenced during the last gla-ciation by anticyclons related with the North European ice-cap. The influence of these anticyclons reached westward probably up to the area of East Germany. In West Germany and in Northern Switzerland the loess was accumulated under the influence of prevailing westerly winds (R. F. F l i n t, 1957; G. H. G o u d a, 1962). The south-western part of the described area characterized by the predominance of westerly winds was lying in a temperate climatic zone, where the atmospheric circulation was probably related with cyclones of the Mediterranean area. The results of the analysis are therefore conforming the reconstruction of the general atmospheric circulation during the maximum of last glaciation elaborated by H. C. W i l l e t (1950).

Department of Physical Geography
Maria Curie-Skłodowska University
Lublin

REFERENCES

B r ü n n i n g H. (1959), Periglazial-Erscheinungen und Landschaftsgenese im Be-reich des mittleren Elbetales bei Magdeburg, *Göttinger Geogr. Abh.*, 23.
B ü d e l J. (1960), Die Gliederung der Würmkaltzeit, *Würzb. Geogr. Arb.*, 8. Clima Republicii Populare Romine (1961—1962), Bucuresti.
F e d o r o v i c h B. A. Федорович Б. А. (1960), Вопросы присхождения лесса в связи с условиями его роспространения в Евразии (Problems of the origin of loess on the background of the conditions of its distribution in Eurasia) материали по геогрф. и палеогеогр. СССР 24, p. 96—117, Москва.
F l i n t R. F. (1957), Glacial and Pleistocene Geology, New York—London.
G e i g e r R. (1957), The Climate near the Ground, Cambridge, Massach.

G o u d a G. H. (1962), Untersuchungen an Lössen der Nordschweiz, *Geogr. Helvetica*, 17, p. 137—221.

G r a h m a n n R. (1932), Der Löss in Europa, *Mitt. Ges. Erdk.* zu Leipzig. 1930—1931, p. 5—24.,

J a h n A. (1956), Wyżyna Lubelska (Geomorphology and Quaternary history of Lublin Plateau) *Pr. geogr. Inst. Geogr. PAN*, 7, Warszawa.

J a r a n o w D. (1956), Losat i losowidnite sedimenti w Bałgarija (Le loess et les sédiments loessoides en Bulgarie), *Izw. na poczw. institut* 3, p. 37—78.

K r i v á n P. (1953), Die erdgeschichtlichen Rythmen des Pleistozänzeitalters, *Acta geol. Acad. Sc. Hung.*, 2/1—2, p. 79—90.

Magyarország éghajlati atlasza (1960), (Klimaatlas von Ungarn), Budapest.

M a l i c k i A. (1950), Geneza i rozmieszczenie lessów w środkowej i wschodniej Polsce (The origin and distribution of loess in Central and Eastern Poland), *Ann. Univ. MCS*, s. B, 4, p. 195—228, Lublin.

M a r u s z c z a k H. (1963), Wind direction during the sedimentation period of the upper Loess in the Vistula Basin, *Bull. Acad. Pol. Sc. serie sc. géol. géogr.*, 11/1, p. 23—28.

M a r u s z c z a k H. (1964), Conditions d'accumulation du loess dans la partie orientale de l'Europe Centrale, *Geogr. Polonica*, 2, p. 39—47.

M a r u s z c z a k H. (1965), Development conditions of the relief of loess areas in East-Middle Europe, *Geogr. Polonica*, 6, p. 93—104.

M a r u s z c z a k H., T r e m b a c z o w s k i J. (1960), Próba porównania wydm śródlądowych okolic Widina (Bułgaria) i Wyżyny Lubelskiej (Polska). (Attempt of comparing continental dunes of the Vidin region- Bulgaria-with dunes on the Lublin Plateau-Poland), *Czas. geogr.*, 31, p. 163—178.

M i h á l t z I. (1953), La division des sédiments quaternaires de l'Alföld, *Acta geol. Acad. Sc. Hung.*, 2/1—2, p. 109—120.

M i l o j e v i ć B. Ż. (1950), Les plateaux de loess et les régions de sable en Yougoslavie, *Mém. Soc. Géogr.*, 6, Beograd.

M i n k o w M. (1960), Losat i losowidnite sedimenti meżdu rekite Skomla i Ogosta (Der Löss und die Lössartigen Sedimente zwischen den Flüssen Skomlija und Ogosta), *Tr. geol. Bulgarie, ser. strat. tekt.*, 1, p. 249—294.

M o l n a r B. (1961), A Duna-Tisza közi eolikus rétegek felszini és felszin alatti kiterjedése (Die Verbreitung der äolischen Bildungen an der Oberfläche und untertags im Zwischenstromland von Donau und Theiss), *Földt. Közlöny*, 91, p. 300—315.

Monografia geografica a Republicii Populare Romine (1960), 1, Bucuresti.

N a u m T., G r u m a z e s c u H. (1954), Problema loessului, *Probleme geogr.*, 1, p. 154—192.

P o s e r H. (1951), Die nördliche Lössgrenze in Mitteleuropa und das spätglaziale Klima, *Eiszeit. Gegenw.*, 1, p. 27—55.

S c h ö n h a l s E. (1953), Gesetzmässigkeiten im Feinaufbau von Talrandlössen mit Bemerkungen über die Entstehung des Lösses, *Eiszeit. Gegenw.*, 3, p. 19—36.

W i l l e t H. C. (1950), The general circulation at the last (Würm) Glacial maximum, *Geografiska Annaler*, 32, p. 179—187.

38

Reprinted from *Stiinta Solului,* 6(2-3), 177-184 (1968)

Some Features of Loess-Soil Parent Material in Central and South Dobrodgea

ANA CONEA *

The soils of Dobrogea (region situated in the south-eastern part of Romania, between the Danube river and the Black Sea) are developed, in their greatest majority, on loess. An insignificant part of them developed on other parent materials : limestones, green schists, cristalline schists a.s.o. In this paper we shall not deal with the loose material, resulted from the weathering of hard rocks — material usually forming a surface layer no thicker than several tens of cm — but only with loess, — deposit reaching, sometimes, considerable thickness. We shall limit ourselves but to certain aspects concerning the loess of the last glaciation -— namely, of the Würmian —, much better represented than the older loess.

MATERIALS

The soil surveys and the analysis on which we, partially, based this paper, were made by researchers belonging to the State Committee for Geology. We, also, undertook separate researches concerning the loess and the buried soils of the region.

The soil map of the region was printed (10) at the scale 1 : 200.000. For the particle size distribution, we used the results of the soil and parent material sample analysis, collected down to 200—250 cm and from several sections of greater thickness.

DISCUSSION

The regions covered by loess are characterized, apparently at least, by uniformity — as regards the features and origin of surface deposits and the forms of relief. In fact, the loess cover is far from being as uniform as may be suggested by the general features of the rock. Firstly, in Dobrodgea, its continuity is interrupted by the outcrop of hard rocks or fossil soils, which were exposed by erosion (or, perhaps, were not covered by an other loess layer) on slopes or on ridges.

On the second hand, the loess deposits of Dobrodgea display no uniform features as regards the particle size distribution, both in horizontal and vertical directions, lime content, thickness of layers and their vertical succession. Neither its origin and age are the same on the whole area, or on the layer thickness.

---

* Principal Researcher, Geological Institute

Table 1: Mechanical composition of Loess

| Texture of the loess layer 200–500 cm thick | | Nr. of analysed samples | Values | Particle size distribution % | | | | | | | | | |
in the upper part	in the lower part			2–0.2 mm	0.2–0.1 mm	0.1–0.05 mm	0.05–0.02 mm	0.02–0.002 mm	<0.002 mm	<0.001 mm	2–0.05 mm	0.05–0.01 mm	<0.01 mm
heavy loam	heavy loam	44	average	0.07	0.6	3.0	30.4	30.0	34.4	30.7	4.0	47.4	49.0
			extreme	0.0–0.3	0.2–1	1–6	27–35	29–34	30–37	27–34	2–9	42–52	45–54
heavy loam	medium loam	55	average	0.08	0.6	5.0	33.1	29.0	32.1	28.8	5.5	48.1	46.2
			extreme	0.0–0.2	0.2-1	3–7	30–36	26–33	28–35	26–33	4–8	45–51	44–48
		49	average	0.09	0.8	6.7	35.1	28.9	27.9	24.9	7.7	51.5	41.2
			extreme	0.0–0.3	0.2–1	5–9	31–41	24–34	24–32	20–30	5–11	47–57	37–45
medium loam	medium loam	127	average	0.06	0.8	7.9	37.7	28.0	27.5	24.4	9.1	52.0	40,1
			extreme	0.0–0.3	0.2–3	5–12	32–40	25–31	21–32	18–29	6–14	47–57	32–45
medium loam	coarse loam	108	average	0.80	0.8	11.3	38.0	23.3	24.1	21.6	12.8	52.2	35.2
			extreme	0.0–0.2	0.2–2	8–16	33–42	21–32	20–30	18–27	9–17	47–56	30–41
		35	average	0.05	1.0	15.8	41.5	23.2	17.3	15.6	17.0	55.2	26.4
			extreme	0.0–0.1	0.4–2	12–20	39–45	19–26	14–20	13–18	13–21	51–59	22–30
coarse loam	loamy sand	23	average	0.02	1.0	11.5	48.0	22.2	19.0	16.6	12.1	59.7	27.5
			extreme	0.0–0.1	0.5–2	6–18	41–57	20–24	14–22	12–20	6–19	53–67	21–31
		8	average	0.0	1.4	13.2	51.6	19.2	13.2	11.4	15.2	55.6	18.7
			extreme	0.0–0.1	0.5–2	8–18	43–60	16–22	11–18	10–16	8–19	62–72	17–21

Particle size distribution

From the analytical data — synthetized in table 1 — it results that, according to Katchinski's [1]) classification, the last loess layer, covering the central and southern part of Dobrogea, may be classified into four textural classes, namely : loamy sand, coarse loam, medium loam and heavy loam, having, correspondently, an average of 18,7%, 26,4%, 40—41% and 46—47% [2]) particles < 0,01 mm, in the surface layer, thick of 200—250 cm [3]). Materials belonging to any mentioned textural class have very low contents of 2—0.2 mm ϕ particles (sometimes totally absent), not exceeding the average of 0.09%. The fraction with the ϕ of 0.2—0.1 mm reaches an average of 0.6% for the heavy loam loess and 1.4% for those with loamy sand texture. More important differences are to be found in the particle contents with the ϕ of 0.1 mm — and thiner. Thus, the quantity of the fraction with 0.1—0.05 mm ϕ is four times greater in the coarser textured loess, as compared to the finest textured one ; the clay fraction content (< 0.002 mm) increases from 13%, in the loamy sand, up to 34% for the heavy loam material ; as for the 0.5—0.01 mm ϕ fraction — characteristic for the loess — it decreases from 66% to 47%.

Therefore we distinguish a coarse loess — loamy sand —, an other, medium textured one, — coarse to medium loam — and an other one with fine texture — heavy loam. This last one, presenting only partially the specific features of the loess, may be considered rather a loesslike deposit (2). The typical loess has a loamy texture. Besides, it has the greatest extention not only in Dobrogea, but also in the Danube Plain from south Romania. On such materials were formed the actual soils of Dobrodgea, and namely :

1. Chestnut Steppe soils (mainly on coarse loess) ;
2. Chernozems (on typical loess) ;
3. Leached Chernozems (especially on fine loess) ;
4. Chestnut Forest and Grey Forest soils (on typical loess).

The loess of Dobrogea — as well as of other regions of Romania — has not the same particle size distribution on the whole thickness of the layer. One remarks, generally, a gradual transition from a finer texture to a coarser one, within the limits of the same textural class, or from a textural class to the immediately following one. There are, thus, loess which, along the whole thickness we are speaking about, has a heavy loam texture ; some other times only the solum or only the soil upper horizons have a fine texture, the underlying loess layer having a medium texture. On certain areas, the loess has, on the whole

---

[1]) Classification used in the Soil Science studies from Romania
[2]) In the table = 49,0% value resulting yet, by taking into consideration also the analytical data of the solum material.
[3]) According to the textural classes from U.S. Soil Survey Manual (9), it could be possible to distinguish a loess with a silty clay loam texture (with an average of 27.5—34.4% clay and 60—65% silt) and an other loess with silt loam texture (with an anverage of 13.2—24.1% clay and 60—70% silt).

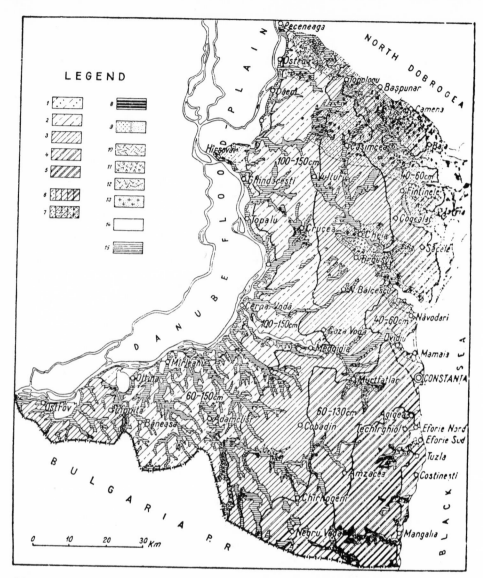

Fig. 1 — Soil Parent Materials in Central and South Dobrodgea, Loess and Loesslike deposits.

1. coarse or sandy loess (coarse loamy * over loamy sand); 2. typical loess (medium loam over coarse loam); 3. typical loess (medium loam); 4. typical loess (heavy loam over medium loam); 5. fine loess (heavy loam); 6. loess with coarse sand less than 0.2%; 7. loesslike deposits (with coarse sand ranging from 0.2 to 15% and more); clayey deposits : 8. red or reddish clays (fossil soils uncovered with loess); sands: 9 a. quartz sands, 9 b. calcareous sands; hard rocks: 10. green schists; 11. limestones ; 12. cristalline schists; 13. porphyries; alluvia ; 14. coarse to fine textured aluvia ; slop deposits: 15. loess, loesslike deposits a.s.o.

_____

* The thickness of the finer textured materials, forming the upper part of the loess layer, is indicated in cm on the map.

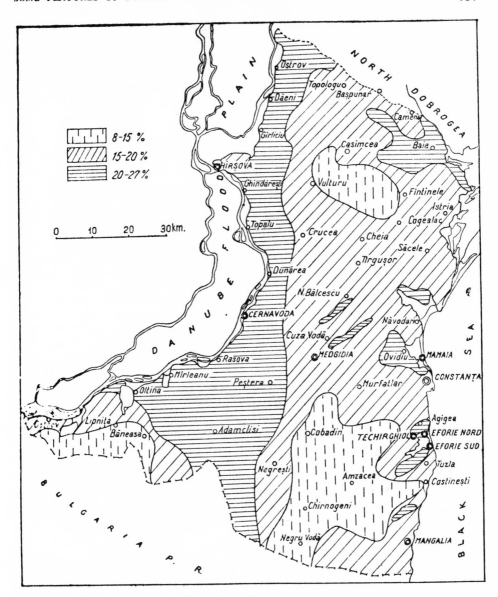

Fig. 2 — Lime Content of the Last Loess Layer.

thickness of the layer, a medium loam texture, with insignificant variations of content in different fractions, or changes from medium to coarse loam or, finally, from coarse loam in the upper part of the layer, to loamy sand in the lower part of it. In almost all the above mentioned cases, a textural change is to be noticed at the level of the Cca horizon — a feature met, also, at the soils developed on loess in the Danube Plain (2).

Besides the coarse loess, there are also, especially in central Dobrogea, deposits similar to it, but with a higher content (sometimes > 20%) of coarse particles (ϕ 0.2—2 mm and larger). The coarse fragments result either from green schists, frequently outcropping on the region, or from aeolian sands. They are classified in what we, also, call loesslike deposits.

Studying on the map (fig. 1) the distribution of the above discussed soil parent materials, we conclude the followings : a. on the territory of central and southern Dobrodgea, loess texture becomes more fine from NW towards SE ; b. the loamy loess has the greatest extention, as compared to the coarse and fine textured loess, which occupy small areas ; c. the loesslike deposits (loess with coarse fragments) cover large areas in central Dobrodgea — that is to say there, where relief is more dissected and hard rocks often outcrop and where the loess is settled in thick layers inside depressions.

The reasons determining the textural changes we spoke about, are connected both to the sedimentation conditions and to the transport agent and origin of the material. In the vertical change of the particle size distribution, within the limits of the solum, an important role was, surely, played by the soil forming process, by the translocation of the fine material along the soil profile or by argillisation in situ. As the origin is concerned (even if, subsequently or concomitently, the material has been removed), the loess of Dobrogea is predominantly aeolian. Considering the horizontal distribution of differently textured loess, it results that the north-western winds have blown — at least during the last part of Pleistocen — with an intensity and frequency perhaps greater than the north-eastern ones prevailing now-a-days.

Lime content

The last Würmian loess, forming the parent material of central and south Dobrodgea soils, contents, always, a high percentage of lime, mainly of Ca. From the analytical data — refering only to unweathered loess and not to solum or Cca horizon — it results that this material has between 8 and 27% lime content. Conventionally, we distinguished : a low calcareous loess : 8—15% ; a moderate calcareous loess : 15—20% ; and a rich one : 20—27%. Following the geographical distribution (fig. 2) of these three loess categories, a double relation is noticed : on one hand, between the loess texture and its lime content and, on the other hand, between the outcropping of limestones and the lime content of the neighbouring loess. The coarse textured loess (loamy sand to coarse loam) is richer in lime ; generally, the finer becomes the texture, the lower is the lime content exhibited by loess. In other words this content decreases from NW towards SE, and along this same direction, the texture of the deposit changes, too. The insular areas with higly calcareuos loess, occur, yet, wherever limestones outcrop ; so as an area with a low calcareous loess occurs in central Dobrodgea, due to the material coming from the weathered green schists, — uncalcareous rocks.

In central and southern Dobrogea there is, also, limeless loess, medium textured, sometimes mottled, but older : from low and medium Würmian or older. Due to the areas it occupies, it does not form a characteristic of the region. From this point of view, Central and South Dobrogea belong to the regions of loess deposition in the conditions of a dry climate (3, 4), in contrast with other parts of our — or other- countries, where uncalcareous loess is dominant and which had — and have now-a-days, also, — a more humid climate than Dobrogea.

Thickness of loess cover

The thickness of the Würmian loess cover is not uniform in Dobrogea. There are areas where the loess deposits are very thin and mixed to other materials, but there are, also, areas where their thickness exceeds 15 m. Sometimes, loess occurs like an uniform deposit, all the thickness deep, while sometimes strips of buried soils separate several layers (1, 5, 6, 7, 8).

When the Würmian loess was not removed or submitted to other processes which have taken away or, at least, weathered part of the material, it, generally, exhibits a succession of three very distinct loess layers, corresponding to the three Würmian stages. The loess horizons are separated by two strips of interstadial soils, each formed of two soils, much more slightly developed than the underlying interglacial one.

The mean thickness of the last layer of loess (including the present soil), in the area where the deposit is better preserved, is of 2,5—3 m, and the total thickness of the Würmian loess is 8—10 m.

This area coincides with the less dissected relief of Dobrogea, situated in the very proximity of the Black Sea. On the Danubian border, the Würmian loess frequently occurs as a continuous deposit, exceeding, here and there, 15 m in hickness, overlaying older soils — usually interglacial, limestones, or, more rarely, fluviatile gravel. Interstadial soils, which seldom occur, are slightly developed and badly preserved. In the center of the region, the loess cover varies both in thickness and number of buried soils. These last ones are discontinuous and are lacking on large areas.

CONCLUSIONS

The young loess, forming the soil parent material in Central and South Dobrogea, may be divided, according to the particle size distribution and lime content, into : coarse (or sandy) loess — characterized by a high percentage of fine sand and lime ; typical loess (for Romania), medium textured and relatively moderate calcareous, and a fine (heavy loam) loess, with lower lime content.

A loess free from carbonates, usualy mottled, has to be noticed, so as there are also loesslike deposits (loess with coarse fragments), generally calcareous.

REFERENCES

1. CONEA ANA. Problema solurilor fosile îngropate cu privire specială asupra Dobrogei sudice (The Problem of Buried Fossil Solis (with special view on South Dobrodgea), *D. de S. ale Sed. Inst. Geol., 1967, LII, part. III*

2. CONEA ANA ; GHIȚULESCU NADIA; VASILESCU P. Considerații asupra depozitelor de suprafață din Cîmpia Română de Est (Considerations on Surface Deposits from East Romanian Plain). *St. tehn. și ec., Seria C, Pedologie, 1963, 11.*

3. FINK J. Zur Korrelation der Terassen und Lösse in Osterreich. *Eisz. u. Gegenw., 1956, 7.*

4. FINK J. Die Gliederung der Würmeiszeit in Österreich *Rep. of the VI-th. Int. Congr. of Quaternary, 1961, INQUA, 1964, IV, Lodz.*

5. HAASE G. Stand und Probleme der Lössforschung in Europa. *Geogr. Berichte. Mitt. der Geogr. Gesellschaft der DDR, 1963, 27.*

6. HAASE G.; RICHTER H. Fossile Böden in Löss an der Schwarzmeerküste bei Constanța. *Pett. Mitteil. 1957, 3.*

7. MORARIU T.; POPOVĂȚ M.; CONEA ANA. New Contributions to the Knowledge of Periglacial Forms of the Black Sea Cliff South of Constantza. *Rev. Roum. de Géol., Géogr. et Gèoph., Série de Gèographie, 1965, 1.*

8. POPOVĂȚ M.; CONEA ANA; MUNTEANU I.; VASILESCU P. Loessuri și soluri fosile în podișul Dobrogei sudice (Loesses and Fossile Soils in the Tableland of South Dobrogea). *St. tehn. și econ., Seria C, Pedologie, 1964, 12.*

9. *** Soil Survey Manual. *U.S.Dep. of Agric. Soil Curvey Staff, Agr. Handbook, no. 18, 1952.*

10. *** Soil Map of Romania. sc., 1.200.000. Călărași, Mangalia and Constantza leaves *Inst. Geol. Bucharest, 1963—1965.*

Editor's Comments
on Papers 39, 40, and 41

Papers 39–41 concern the problem of the production of silt-sized quartz particles for the formation of loess deposits. It can be argued that without a silt-forming mechanism there can be no loess, and this simple test can be applied to contending theories of loess formation. There is no known way for silt-sized particles to form *in situ*, and this casts considerable doubt on the more extreme *in situ* theories. Glacial action is an obvious provider of silt-sized quartz, and the broken particles revealed by electron microscope examination of loess material are compatible with this mode of formation. The question here is, "Can desert action provide abundant silt-sized quartz particles?"

The desert loess problem is an elusive one; in the end it probably reduces to one of definition, as may the *in situ* problem. What gives point to the discussion is the existence of the large loess deposits in North China. Their association with the Ordos and Gobi deserts was so intimate and obvious that the dust source seemed plain; yet no known exclusivly desert process could provide the vast amount of silt required. With the revelation that there was widespread Quaternary glaciation in China, an alternative source for the silt was provided and the existence of a specific desert loess became doubtful. However, it cannot be denied that some desert areas do appear to be associated with loess or loess-like deposits (see Papers 30 and 31 and the associated discussion).

39

Reprinted from *Jour. Sed. Petrol.*, 38(3), 766–774 (1968)

THE FORMATION OF FINE PARTICLES IN SANDY DESERTS AND THE NATURE OF 'DESERT' LOESS[1]

I. J. SMALLEY AND C. VITA-FINZI
University College, London

ABSTRACT

Loess consists chiefly of quartz particles with diameters of about 20–50μ. It is commonly thought to have formed in periglacial areas and in deserts. Smalley (1966b) has suggested that direct glacial grinding could produce suitable particles. The existence of dust storms suggests that fine particles also form in hot, sandy deserts; these could be the result of interparticle contacts where a certain critical kinetic energy is involved.

There are no loess deposits within sandy desert areas, but they may occur at the desert margins. The major loess provinces to which a desert origin has been ascribed either are not true loess or are explicable in terms of glacial origin. Of the minor deposits, only that of the Negev seems to have a well established desert origin. The quartz particles in the Negev deposit probably formed by simple impact. This mechanism does not seem adequate for the Chinese loess, and it is suggested that the widespread Pleistocene glaciation of China was responsible for the bulk of the fine particles. Thus although loess particles can form mechanically in deserts, they have not done so on a scale sufficient to produce major loess deposits. Desert loess deposits should be distinguishable from glacial loess deposits by their size and their mineralogical nature.

Stresses produced by temperature changes can break rocks provided that the rock particles are not below a certain critical size. Most quartz is introduced into the sedimentary system as sand-size particles (about 500μ); if the critical thermal break size is above 500μ, then the thermal breakage process will not produce loess-size particles.

INTRODUCTION

Two types of classification are commonly applied to sedimentary rocks: descriptive and genetic. A third approach, proposed by Grabau (1904, 1913; see Krumbein and Sloss, 1963, p. 150), demands that each rock name should express "the texture and chemical composition, as well as the agent mainly responsible for the deposition of the sediment."

Many of the published definitions of loess fall into this third category, in that they lay equal stress on its eolian origin and on its physical characteristics. Those definitions which are purely descriptive boil down to "unbedded, well sorted silt" and so undermine the case for the continued use of the term loess with its well established eolian connotations. Even worse, they sometimes give rise to such grotesque hybrids as "fluvial loess" or "colluvial loess"; this may be achieved by implication, as in a recent study of the so-called loess of southeastern England (Tilley, 1961) where it is attributed to solifluction. Finally, a purely genetic definition without lithologic qualifications amounts to saying that the deposit is windblown; if we need a synonym for this we already have one in 'eolian.'

If the term loess is to have any value, it should be applied only to the original wind-laid deposits (which clearly throws the burden of demonstrating an eolian origin on to the field investigator). This is the basis of the definition adopted in this paper. Texture presents little difficulty, since it is generally agreed that loess is composed primarily of silt-sized particles. As regards composition, quartz usually predominates and is accordingly the core of our discussion of mechanisms of particle formation; but the possibility that other minerals can provide suitable raw material cannot be overlooked.

The literature of loess recognizes two major sources of loess material: glacial or periglacial areas, and hot deserts. It has already been shown (Smalley, 1966b) that glacial grinding could produce suitable particles; the question remains whether suitable particles could be formed directly and in sufficiently large quantities in hot deserts to give rise to major loess provinces.

DEFINITIONS

The definition of loess by Pettijohn (1957, p. 377–8) is descriptive *par excellence:*

Loess is an unconsolidated silt commonly buff in colour . . . characterised by its lack of stratification and remarkable ability to stand in a vertical slope. It is generally highly calcareous.

The statements by Charlesworth (1957, p.

[1] Manuscript received July 10, 1967; revised February 21, 1968.

512–5) and Flint (1947, p. 175–6) also stress composition, texture, and structure. Flint writes:

Loess is a buff-colored nonindurated sedimentary deposit consisting predominantly of particles of silt size. Commonly it is nonstratified, homogeneous, calcareous, and porous, and it may possess a weak vertical structure resembling jointing. . . .

Mineralogically it is made up principally of quartz, with smaller amounts of clay minerals, feldspar, micas, hornblende, and pyroxene. Carbonate minerals are variable, ranging as high as 40%. This composition is so elastic that it tells little about the rocks in which the minerals originated.

Carbonate content depends largely on post-depositional weathering and alteration (Lozek, 1965). The two crucial physical characteristics of loess would thus appear to be a high degree of sorting and a mode diameter of about 20–50μ for its constituent particles. They suffice to explain the survival of vertical loess bluffs (Smalley, 1966b), although calcareous cementation may help, and thus make this feature of loess landscapes redundant in any definition. The predominance of quartz is also characteristic.

Zeuner's definition complies with our requirements in that it specifies an eolian origin; loess is described as "wind blown dust which is finer than sand but coarser than clay" (Zeuner, 1959, p. 24). Admittedly the recognition of an eolian deposit is not easy. It cannot be based on faunal or lithological grounds alone. All but the most fragile organic remains can be redeposited by running water. A high degree of sorting may be inherited by alluvial formations or even produced during their accumulation; and, as Lyell (1872, I, p. 430) pointed out long ago, fluvial deposits may on occasion show no bedding (see, for example, Higgs and Vita-Finzi, 1966, p. 3), while it is well known that eolian ones often do. It is thus important to take into account the relationship of the deposit to the underlying topography, and to exploit to the full any negative evidence such as the absence of any trace of fluvial features. Circumstantial arguments should be supplemented by detailed mechanical analysis, in order to discover whether changes in grain size within the deposit could be explained by the effect of the existing topography on the former wind patterns (Hjulström, 1955). In the broad sense, however, one would expect any topographic effect on deposition to be more marked in the case of fluvial agencies. The point is made by Butler in his summary of the features generally regarded as typical of loess (Butler, 1956, p. 146):

(Loess) occurs as extensive, nearly uniform sheets, which blanket the landscape, covering both higher and lower elements. The character of the substrate, within broad limits, has no influence on the nature of the overlying sheet.

Our proposed definition is in effect a more emphatic version of Flint's later statement (Flint, 1957, p. 181):

Loess is a clastic deposit which consists predominantly of quartz particles 20–50μ in diameter and which occurs as wind-laid sheets.

In other words, once such material is redeposited by water, mass movement, or any other agency, it forfeits the right to be called loess, for it has reverted to being purely silt. If its origin as loess is to be stressed, it should be termed 'redeposited loess material'.

LOESS PARTICLES: FORMATION AND COMPOSITION

The conventional view is that loess particles are derived either from periglacial or from hot desert areas. The variations played on this two-fold theme may be seen in the following table:

Author	Loess Types	
Bryan (1945)	glacial	desert
Butler (1956, p. 147)	cold	hot
Butzer (1965, p. 194)	periglacial	desert or continental
Charlesworth (1957, p. 511)	periglacial	continental
Flint (1957, p. 183)	glacial	desert
Grahmann (1932)	glacial	continental
Holmes (1965, p. 766)	glacial	desert
King (1966, p. 172)	cold	hot
Obruchev (1945, p. 259)	cold	warm
Scheidegger (1961, p. 34)	glacial	desert
Scheidig (1934, p. 5)	glacial	continental
Thornbury (1954, p. 313)	glacial outwash	desert

Since the days of von Richtofen the secure status of 'hot' loess has been usurped by the cold variety, although few authors go beyond professions of uneasy doubt in the existence of desert loess. Butler (1956, p. 147) stated that "considering the vast areas of deserts in the world and our relative ignorance of 'hot' loess, the latter may be more hypothetical than real."

Rarely is the production of suitable particles discussed in detail; it may earn a paragraph heading and little more. Thus Scheidig (1934, p. 42) listed 20 theories put forward to explain the origin of loess, but the mechanisms whereby cosmic, volcanic, weathering, or any of the other agencies might form loess particles were not explored.

Smalley (1966b) has shown that, with quartz, glacial grinding could produce suitable material. This appears to be a more convincing mechanism than the nebulous "frost action" favoured by various authors, and such a process could account for the largest loess deposits shown on figure 1. The European (no. 1 on fig. 1), the North American (no. 5), the Chinese (no. 4),

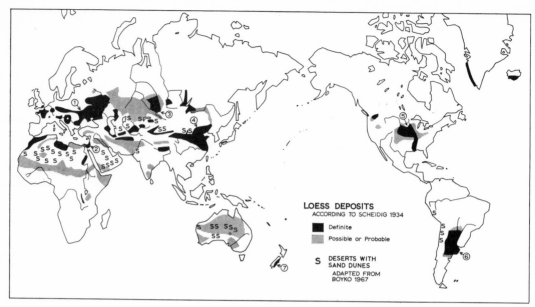

FIG. 1.—World distribution of loess deposits according to Scheidig (1934) and location of deserts with sand dunes (S) from Boyko (1967).

and the New Zealand (no. 7) loesses are composed of particles which were formed by glacial grinding. The Pleistocene glaciations of Europe and North America are well attested and have been fully investigated (Flint, 1957; Charlesworth, 1957), and it is generally accepted that they bear a causal relationship to the associated loess deposits. The Chinese glaciers are not so well established; the glacier-loess relationship in China is discussed in a later section. These are the three major loess deposits with an apparent glacial origin.

The New Zealand deposit is on a much smaller scale. The eastern part of the South Island has extensive loess deposits which are associated with the glaciers in the mountains along the west coast. Raeside (1964) has indicated that the periods of loess deposition corresponded to glacial stages. His observations show that the deposit is larger than it may at first appear, since considerable deposition has occurred on the continental shelf, now submerged. The presence of sponge spicules and other faunal debris in the land deposits suggests that some material has been blown back on to the land. There is a small amount of loess on the west coast of South Island; this can be linked to the local glaciation (Young, 1967). The question that still has to be answered is whether any of the remaining loess provinces shown on the map are of hot desert origin.

The stages in the formation of fine particles in deserts can be described in terms of three

basic events (see Strakhov, 1967, p. 1): formation of detrital particles (P), transportation (T), and deposition (D). This method has been applied to the formation of glacial loess deposits (Smalley, 1966b) and can be applied to desert material for comparison. The suggested stages are:

P1. Quartz sand formed from granitic (Smalley, 1966a) and other rocks, is

T1. transported, and then

D1. deposited elsewhere in the desert area.

T2. Winds move the sand grains by saltation.

P2. When impacts involving more than a certain critical kinetic energy occur, small-scale fracturing results giving rise to small quartz chips

T3. which are carried away in suspension by the wind that formed them.

D2. The winds carry the fine detritus away from the desert areas and may give rise to deposits beyond their margins.

This system comprises one controversial event (P2) and three of dubious character (P1, D1, T1). We are concerned with the first (P2).

Do grain-grain contacts produce loess-sized particles? Kuenen (1960) found that in the course of experimental eolian abrasion, no quartz particles of the size range $(20-50\mu)$ that dominates in loess were formed. In his opinion the quartz silt of loess is produced by "breakage of cracked igneous and metamorphic grains, ice scour, and primary production from phyllites and similar fine grained rocks", none of which are specifically desert processes. Yet, even if existing particles are not worn down to loess size, we should consider the chips and splinters produced during the rounding of the original grains. In a large sandy desert enough of these could form to produce a sizable dust cloud. Knight (1924) showed that sand grains of 1–2 mm diameter were considerably reduced in diameter during transportation by a 40 mph wind; after 50 miles of travel 34.5 percent by weight of the grains were reduced to less than 0.5 mm in diameter. In deserts the material abraded away would go to form dust clouds and might conceivably form loess deposits in due course.

Figure 2 shows some quartz particles from weathered granites; their outlines are based on those given by Moss (1966). If these were transported by the wind, the protuberances would tend to be broken off. Since the kinetic energy ($KE = \frac{1}{2}mv^2$) at impact is critical, wind velocity will be the controlling factor in chip formation. Provided adequate wind storms occur, fragments of up to about 50μ could be produced; larger fragments are unlikely because of the geometry and size of the original sand grains. Sand commonly exhibits a mode size of about 500μ owing to the manner in which it is formed (Moss, 1966; Smalley, 1966a; Blatt, 1967). Figure 2 shows how grains of this size will tend to produce fragments of up to about 50μ in diameter.

Particles formation by chipping is analogous to the secondary grinding process described by Smalley, Heaver, and McGrath (1967); the original particles retain their identity, and the final product is distinctly bimodal. Figure 3 shows the estimated size distributions of particles produced by the two mechanisms, glacial

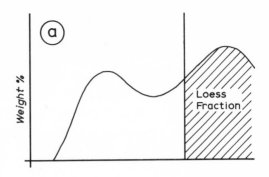

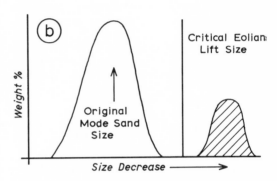

Fig. 3.—Size distributions for sands deformed to produce loess particles (a) by glacial grinding and (b) by eolian abrasion.

grinding (Smalley, 1966b) and desert particle interaction. The small particle product of the desert P-actions lies within the size range most easily transported in suspension by the wind (Bagnold, 1941, p. 4), so that P2 and T2 take place simultaneously. As Yaalon and Ginzbourg (1966) point out, the carrying capacity of strong winds is enormously enhanced in comparison with weaker winds. The small particle product of glacial grinding has to be separated by the wind from a deposit with a continuous range of particle sizes, even though it is distinctly bimodal. Hence the material sorted from glacial deposits has a larger mean size than the fine material formed in deserts. (See fig. 3.) It is worth noting that Grahmann (1932), who believed it should be possible to distinguish between desert and periglacial loess on the basis of particle size distribution, held that the latter should show a smaller dominant grain size, owing to double selection first by meltwater streams or frost weathering and then by wind deflation, whereas desert or continental loess, being only wind sorted, contains more clay and medium sand and has a less marked silt or fine sand fraction. Quite apart from the mechanism of desert particle formation described above,

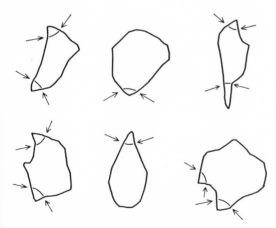

Fig. 2.—Quartz sand particles from weathered granites (after Moss, 1966). Likely impact fragments are indicated.

one would be more justified in expecting peri-glacial loess to be coarser, if only because the energy available for quartz particle breakage is greater than when sand collides during wind transport and will produce a wide range of sizes by primary breakage.

Primary breakage occurs when the particle which is broken is reduced to entirely new particles and the original particle is no longer identifiable. In secondary breakage the original particle retains its identity; small particles are chipped off, but major fracturing does not occur. Thus the slightly paradoxical situation arises in which the low energy process produces finer particles than a high energy process. The difference may be obscured by sorting during eolian transport of glacially-produced particles.

The kinetic energy at impact is higher, the greater the mass of a sand particle; thus the larger the sand particles, the more chipping and rounding they will suffer, as MacCarthy (1935) observed when he found a high correlation between size and roundness. In his classic paper on the rounding of sand grains, Mackie (1897) had also noted that the amount of rounding is proportional to the capacity for work of the grains (that is, $\frac{1}{2}mv^2$). In the Culbin Sands, which have a mode particle diameter of about 250μ, he found that grains of up to 1 mm in diameter were less rounded than those close to the mode, showing that winds of sufficient velocity to move them are exceptional at the present day. In other words one would not expect a loess deposit to be accumulating downwind of the Culbin Sands.

Loess particles can be formed in appreciable amounts by a mechanism such as that outlined above only if the available sand grains are relatively angular. (See fig. 2.) The problem of explaining the origin of the parent sand grains thus still remains. But one can at least propose a modification of Butler's view that "a newly developing desert . . . would provide a copious supply . . . of dust from the erosion of its soil mantle" (Butler, 1956, p. 147). In our view the existence of a soil cover is not a prerequisite of dust formation; in fact weathering would in all likelihood reduce the potential yield of dust particles produced by impact. Chemical weathering of rocks does not appear to produce silt-sized quartz particles in appreciable quantities, though sand and clay may occur in abundance. (See McKeague and Cline, 1963, p. 350.)

Thus we have two possible mechanical processes for the production of loess particles: glacial grinding and desert sand grain impact. Glacial grinding involves a large energy transfer and converts available quartz particles into fine fragments by primary breakage. Less energy is involved in grain-grain contacts in hot deserts, and a relatively small amount of fine material is produced by secondary breakage. Although it is difficult to visualize comparable situations, since glacier movement, the nature of the country rock, wind conditions, the nature of the parent sand, and many other variables are involved, one would expect the yield per unit of time to be far greater for glacial grinding than for sand grain impact. The resulting loess would show some correlation with glacial phases. Deserts would produce a fairly continuous, slow output which in the major deserts would not show marked climatic breaks, since they persisted, only slightly reduced in area, during the Pleistocene glaciations (Hare, 1961, p. 30); indeed it might pay to bear this possible criterion of desert origin in mind when developing stratigraphies in loess areas. Continuous production and the large areas of sand available might counteract the inefficiency of the mechanism. At the same time they would tend to discourage local deposition; witness the fact that much dust material from the Sahara is now blown out to sea (Wittschell, 1931). Deposition within desert areas would be discouraged by the instability of sand-silt mixtures: the fine material is disturbed by movement of the sand grains and is exposed to wind transport, by which it is conveyed to the desert fringes. The possibility that this process has contributed to the Be'er Sheva loess deposit described by Yaalon (1965) is examined later.

It has been suggested that the size reduction of quartz particles from the sand mode size of around 500μ to the loess mode size of around 50μ could be accomplished by thermal means. For this to happen, the stresses produced by thermal expansion and contraction would have to provoke significant breakage of sand grains. Quartz particles when initially supplied into the sedimentary system are sand-sized; this is determined by the dimensions of quartz particles in igneous rocks (Moss, 1966; Blatt, 1967). Bagnold (1941, p. 8) states that there is little or no evidence for temperature splitting of rock fragments smaller than 1 cm in diameter, which is 20 times the sand mode diameter. Yet Zeuner (1959, p. 26) has suggested that insolation weathering (as well as frost weathering) in arid and semi-arid climates can produce particles with sizes of the order of $10–100\mu$; below this limit, the stresses produced by insolation or frost are compensated by the elasticity of the material. Such a process would liberate vast quantities of fine material from sandy deserts. But the bimodal nature of desert particulates suggests that thermal fracturing is not significant, since it would yield a continuous range of particle sizes.

There remains the special case of volcanic material. Eolian deposits of volcanic origin are already endowed with names of their own, but within the terms of our definition there is nothing to exclude a well sorted, silt-sized volcanic deposit from the category of loess. This, in effect, is the claim made for the Pampean Formation of Argentina (no. 6 in fig. 1), a sediment consisting of silts or sandy silts (the sand being very fine) predominantly of volcanic origin (Teruggi, 1957). In the present state of knowledge Teruggi would seem justified in terming it loessoid: reworked or secondary loess is perhaps more abundant than primary 'loess'; and there are present both weathered volcanic material (montmorillonized shards) and unaltered volcanic ashes. Since the likely source area is Patagonia and the Cordillera, glacial grinding may have liberated at least some of the material for wind transport. Charlesworth (1957, p. 521) has in fact stated that periglacial loess underlies the South American pampas. The apparent contradiction that its deposition continued throughout the Pleistocene may be explained by the persistence of volcanic action since the Tertiary; it has still to be determined whether 'air fall tuffs' (in the terminology of Fisher, 1966) predominate in the interglacial parts of the Pampean formation at the expense of "volcanic loess."

LOESS DEPOSITS OF POSSIBLE DESERT ORIGIN

Figure 1 shows the distribution of loess deposits as indicated by Scheidig (1934). There has been much discussion and investigation of loess deposits since Scheidig's map was first published, but it has not been superseded as an indication of world distribution of loess, and was reproduced virtually unaltered by Woldstedt (1954, p. 172). Scheidig appears to have included everything that could possibly be called loess, so that 'probable or possible' loess (Löss wahrscheinlich oder möglich) covers vast areas; many of these deposits would be much better described as 'doubtful or unlikely.'

Loess deposits which are considered to be well established are numbered in figure 1; they correspond mostly to the deposits described by Scheidig as 'definite' (nachgewiesen), with certain exceptions, for example Tripolitania. The location of sandy deserts is also shown in figure 1. Scheidig located some loess deposits near them, whereas Penck (1909) remarked on the absence of loess around the Sahara and the deserts of America. In Tripolitania, Stella (1914, p. 108) had observed deposits of wind-blown

sand, 100μ. in diameter, and he was apparently the first to compare the deposit with the loess of China and North America; but it was Rathjens (1928) who enshrined this misconception in the literature; it appears as 'definite' on the Scheidig-Woldstedt map. Penck (1931) had already pointed out that the material was not true loess. Strictly speaking it is fine sand, albeit well sorted, with 90 percent of the particles over 100μ. in diameter and thus too large to be carried by the wind in suspension. The bulk of the material is only weakly cohesive, but the presence of a small proportion of silt and clay, combined with cementation by calcium carbonate, conspire to give the deposit the structural properties normally associated with classic loess deposits.

In China, the proximity of the loess deposits to the Gobi and Ordos deserts, coupled with the belief that there had been no widespread Pleistocene glaciation, encouraged the survival of Richthofen's thesis that this province (no. 4 in fig. 1) was of hot desert origin. It is the one major loess deposit placed by Charlesworth (1957, p. 542) in his desert or continental category. Particle size and thickness trends are said to confirm that the source lay in the deserts to the north (Liu and Chang, 1961). Now that the glaciation of much of southern China is well attested (Sun and Yang, 1961; Ching, 1965), the possibility that glacial grinding supplied the necessary particles has to be considered. Loess fringes the glacial moraines of the Tienshan, and the Lishih and Wucheng deposits suggest a southern source in that they thin to the north. Furthermore the Gobi is a predominantly rocky desert.

According to Obruchev (1945), loess covers the steppe region north and east of the Caspian Sea in the Kazakh, Uzbek, and Turkmen Republics, extending as far east as Lake Baikhash and south over the foothills of the Pamir-Alai and Tien Shan mountains. Obruchev states that most Russian workers believe the loess in European Russia and Siberia to be a direct consequence of Pleistocene glaciation. Chernyakhovskiy (1966) is of this opinion as regards the loess of Central Asia (no. 3 in fig. 1), in that he ascribes it to the 'cold pluvial' conditions of glacial times.

As regards possible loess deposits in Australia, it is sufficient to quote Butler (1956): "No material having all the characteristics of loess has been reported for Australia." The so-called loess dunes have a high clay content and do not occur as extensive sheets.

The most likely loess deposit of desert origin appears to be that of the Negev (no. 2 in fig.

1). The deposits in the Be'er Sheva basin cover an area of about 1600 km²; according to Ginzbourg and Yaalon (1963), mineralogical and faunal evidence indicate that this material is derived mainly from the weathering residues of calcareous rocks in the Sinai desert. The distance covered during transport was small, and deposition took place at the desert margins where the particles were trapped by a steppe-like vegetation (Yaalon, 1955). In his study of the soils of the northern Negev, Ravikovitch (1953) showed that their texture becomes finer towards the north and northeast and away from the boundary between the Negev and the Sinai desert, due to large-scale sorting of the various textural fractions, this giving rise, in turn, to sandy soils, sandy loams (loess-like soils), and clay loams (loess soils). He also suggested that the dust-bearing south-westerlies could be disturbed by westerly winds carrying fine particles from the Sahara into Sinai, whence they could be blown up into the Negev. It should be noted that some of the loess deposits show traces of fluviatile erosion and redeposition (Ginzbourg and Yaalon, 1953). This could account, in part at least, for the fact that, of the samples analyzed by Ravikovitch (1953, table III, p. 412), none exceeded a silt content of 29.7 percent (as compared with 37.9 for a sample collected during a dust storm), which he attributed to post-depositional weathering into clay.

The Be'er Sheva deposits contain some quartz (Ginzbourg and Yaalon, 1953). It is suggested that this may be of mechanical origin according to the following scheme:

P1.   Fine quartz particles (about 20–50μ) are formed in the sandy regions of the Sahara by impact interaction,

T1.   are carried away in suspension by the wind which formed them, and

D1.   are then deposited within the desert boundary.

T2.   Since the sand-loess material mixture is unstable, the loess material is lifted and carried again and

D2.   deposited in the desert again;

T3.   raised and carried again . . .

Dn.   deposited in Sinai desert, and

Tn+1. blown, along with assorted calcareous detritus, up to the Negev area,

Dn+1. where it settles as a segregated, fine particle deposit which is fairly stable.

If the criterion suggested by Penck (1931) is to be applied and loess is to be thought of as rock powder ("der Löss ist ein Gesteinsmehl")

rather than a weathering product (Verwitterungsrückstandes), then it appears that the Negev loess, fitting into the second category, is not a true loess. If a predominance of silt-sized quartz particles, produced by some specific pulverizing action, is a requirement for a material to qualify as loess, then the Negev deposit again falls short. The fact that quartz loess particles do not dominate the loess of the Negev suggests that, even where geographical factors are most favorable to the accumulation of hot desert loess, the process of formation of fine particles in sandy deserts is so inefficient that loess deposits should not be expected at the desert fringe unless some other source of fine material is available. This means that 'desert' loess deposits will inevitably be mineralogically distinct from glacial loess deposits; this, allied to the areal difference, may be sufficient grounds for a separate and distinct name, as suggested by Butler's use of 'parna' for eolian deposits in Australia (Butler, 1956).

TENTATIVE CONCLUSIONS

1. In areal and mineralogical terms we might expect two types of 'loess' deposit: extensive glacial deposits with high quartz content and small deposits of hot desert origin with a lower quartz content.

2. Quartz loess material is formed by the reduction of particles from about 500μ to a 'rock powder' of about 50μ, and it would appear to be accomplished in nature solely by mechanical means and not by temperature splitting or chemical weathering.

3. In areas where there were large continental Pleistocene glaciers there are now extensive loess deposits; they include North America, Europe, and China. The deposits have been formed in each case near the glacial limits. A similar situation may exist in Argentina, but the picture is complicated by the presence of volcanic material.

4. Of the minor deposits, that of New Zealand appears to be glacial, while that of the Negev appears to be of desert origin. Some desert fringe deposits, for example Tripolitania, consist not of loess but of fine sand; in other words, they are composed of small primary quartz particles rather than fine debris produced by secondary breakage.

ACKNOWLEDGEMENTS

We are grateful to Dr. R. U. Cooke and to Dr. D. H. Yaalon for their critical comments on the first draft of this paper.

REFERENCES

BAGNOLD, R. A., 1941, The physics of blown sand and desert dunes. Methuen, London, 265 p.

BLATT, HARVEY, 1967, Original characteristics of clastic quartz grains: Jour. Sedimentary Petrology, v. 37, p. 401–424.

BOYKO, HUGO, 1967, Salt water agriculture: Sci. American, no. 3, v. 216, p. 89–96.

BRYAN, KIRK, 1945, Glacial versus desert origin of loess. Am. Jour. Sci., v. 243, p. 245–248.

BUTLER, B. E., 1956, Parna-an aeolian clay: Australian Jour. Sci., v. 18 p. 145–151.

BUTZER, K. W., 1965, Environment and archeology. Methuen, London, 524 p.

CHARLESWORTH, J. K., 1957, The Quaternary Era. 2 vols., Arnold, London, 1700 p.

CHERNYAKHOVSKIY, A. G., 1966, Origin of Central Asian loess: Biul. Komiss. Izuch. chetvert. Perioda S.S.S.R., no. 31, p. 21–31. (in Russian).

CHING TSAI-JUI, 1965, The relics of the Quaternary glaciation and the classification of glaciation epochs of the plateau area of western Hupei: Acta Geol. Sinica, v. 45, p. 237–242. (in Chinese)

FISHER, R. V., 1966, Textural comparison of John Day volcanic siltstone with loess and volcanic ash: Jour. Sedimentary Petrology, v. 36, p. 706–718.

FLINT, R. F., 1947, Glacial geology and the Pleistocene epoch. John Wiley & Sons, Inc., New York, 589 p.

———— 1957, Glacial and Pleistocene geology. John Wiley & Sons, Inc., New York, 553 p.

GINZBOURG, DOV., AND YAALON, D. H., 1963. Petrography and origin of the loess in the Be'er Sheva basin: Israel J. Earth Sci., v. 12, p. 68–70.

GRABAU, A. W., 1904, On the classification of sedimentary rocks: Am. Geologist, v. 33, p. 228–247.

———— 1913, Principles of stratigraphy. A. B. Seiler & Co., New York, 1185 p.

GRAHMANN, RUDOLF, 1932, Der Löss in Europa: Mitt. Ges. Erdk. zu Leipzig, v. 51 (for 1930–31), p. 5–24.

HARE, F. K., 1961, The causation of the arid zone, in STAMP, L. D., ed., A history of land use in arid regions: UNESCO, Paris, p. 25–30.

HIGGS, E. S., AND VITA-FINZI, C., 1966, The climate, environment and industries of Stone Age Greece: Part 2: Prehistoric Soc. Proc., v. 32, p. 1–29.

HJULSTRÖM, FILIP, 1955, The problem of the geographic location of windblown silt—an attempt of explanation: Geog. Annaler, v. 37, p. 86–93.

HOLMES, ARTHUR, 1965, Principles of physical geology. 2nd. ed., Nelson, London, 1288 p.

KING, L. C., 1966, Morphology of the earth. 2nd. ed., Oliver and Boyd, Edinburgh, 726 p.

KNIGHT, S. H., 1924, Eolian abrasion of quartz grains [abs.]: Geol. Soc. America Bull., v. 35, p. 107–108.

KRUMBEIN, W. C., AND SLOSS, L. L., 1963, Stratigraphy and sedimentation. 2nd ed., W. H. Freeman & Co., San Francisco, 660 p.

KUENEN, PH. H., 1960, Experimental abrasion 4: eolian action: Jour. Geology, v. 68, p. 427–449.

LIU TUNG-SHENG, AND CHANG TSUNG-HU, 1961, The "huangtu" (loess) of China: Rept. 6th. INQUA Congress, v. 4, p. 503–524.

LOZEK, VOJEN, 1965, Das Problem der Lössbildung und die Lössmollusken: Eiszeit. u. Gegenwart, v. 16, p. 61–75.

LYELL, CHARLES, 1872, Principles of Geology. 2 vols., John Murray, London, 1322 p.

MACKIE, WILLIAM, 1897, On the laws that govern the rounding of particles of sand: Geol. Soc. Edinburgh Trans., v. 7, p. 298–311.

McCARTHY, G. R., 1935, Eolian sands: a comparison: Am. Jour. Sci., v. 230, p. 81–95.

McKEAGUE, J. A., AND CLINE, M. G., 1963, Silica in soils: Adv. Agronomy, v. 15, p. 339–396.

MOSS, A. J., 1966, Origin, shaping and significance of quartz sand grains: Geol. Soc. Australia Jour., v. 13, p. 97–136.

OBRUCHEV, V. A., 1945, Loess types and their origin: Am. Jour.. Sci., v. 243–262.

PENCK, ALBRECHT, 1909, Die Morphologie der Wüsten: Geog. Zeit., v. 15, p. 545–558.

———— 1931, Zentral-Asien: Zeit. Ges. Erdk. zu Berlin, (for 1931), p. 1–13.

PETTIJOHN, F. J., 1957, Sedimentary rocks. 2nd. ed., Harper & Bros., New York, 718 p.

RAESIDE, J. D., 1964, Loess deposits of the South Island, New Zealand, and soils formed on them: N. Z. J. Geol. Geophys., v. 7, p. 811–838.

RAVIKOVITCH, SHLOMO, 1953, The Aeolian soils of the Northern Negev, in Desert Research: Research Council of Israel, Spec. Pub. 2, p. 404–433.

RATHJENS, CARL, 1928, Löss in Tripolitanien: Zeit. Ges. Erdk. zu Berlin, (for 1928), p. 211–228.

SCHEIDEGGER, A. E., 1961, Theoretical geomorphology. Springer-Verlag, Berlin, 333 p.

SCHEIDIG, ALFRED, 1934, Der Löss und seine geotechnischen Eigenschaften. T. Steinkopf, Dresden, 233 p.

SMALLEY, I. J., 1966a, Formation of quartz sand: Nature, v. 211, p. 476–479.

———— 1966b, The properties of glacial loess and the formation of loess deposits: Jour. Sedimentary Petrology, v. 36, p. 669–676.

SMALLEY, I. J., HEAVER, A. A., AND MCGRATH, LEONARD, 1967, Variation of cohesion with fineness of mineral particles and the development of cohesion during grinding: Inst. Min. Metall. Trans. Section C, v. 76, p. C183–C188.

STELLA, AUGUSTO, 1914, Geologia, in La Missione Franchetti in Tripolitania (Il Gebel). Soc. Ital. stud. Libia, Firenze-Milano, p. 89–126.

STRAKHOV, N. M., 1967, Principles of lithogenesis. Oliver and Boyd, Edinburgh, v. 1, 245 p.

SUN TIEN-CHING, AND YANG HUAN-JEN, 1961, The great ice age glaciation in China: Acta Geol. Sinica, v. 41, p. 234–244. (in Chinese)

TERUGGI, M. E., 1957, The nature and origin of Argentine loess: Jour. Sedimentary Petrology, v. 27, p. 322–332.

THORNBURY, W. D., 1954, Principles of geomorphology. John Wiley & Sons Inc., New York, 618 p.

TILLEY, P. D., 1961, The significance of loess in Southeast England: Rept. 6th INQUA Congress, v. 4, p. 591–596.

WITTSCHELL, LEO, 1931, Über Sand- und Staubstürme und ihre Bedeutung für die Morphologie der Erdoberfläche: Zeit. Geomorphologie, v. 6, p. 1–18.

WOLDSTEDT, PAUL, 1954, Das Eiszeitalter. v. 1, F. Enke, Stuttgart, 374 p.

YAALON, D. H., 1965, Source and sedimentary history of the loess in the Beer Sheva Basin, Israel: INQUA 7th Congress abstracts, p. 514.

YAALON, D. H., AND GINZBOURG, D., 1966, Sedimentary characteristics and climatic analysis of easterly dust storms in the Negev, Israel: Sedimentology, v. 6, p. 315–332.

YOUNG, D. J., 1967, Loess deposits of the west coast of the South Island, New Zealand: N.Z. Geol. Geophys., v. 10, p. 647–658.

ZEUNER, F. E., 1959, The Pleistocene period. 2nd ed., Hutchinson, London, 447 p.

Reprinted from *Jour. Sed. Petrol.,* 39, 1631–1633 (Dec. 1969)

ORIGIN OF QUARTZ SILT[1]

PH. H. KUENEN

Geologisch Instituut, Universiteit Groningen, Melkweg 1, Groningen

ABSTRACT

Eolian action does not produce quartz particles between 20 and 50 microns in mass, either by abrasion of larger grains or by spalling of chips in that size range. Aqueous spalling is insignificant compared to disintegration of phyllite or similar material and of coarser crystalline rocks. Glacial action is particularly important.

EOLIAN ACTION

In a recent paper Smalley and Vita-Finzi (this Journal, 1968), claim that protuberances of irregular quartz grains must tend to be broken off during wind transport, thus producing some loess material. I had come to the opposite conclusion in an experimental study (Kuenen, 1960).

The above-mentioned authors postulate that fragments of up to about 50 microns could be formed, provided adequate wind storms occur, but without suggesting what strength of wind appears to be required. However, they do not produce evidence or arguments to support that spalling does actually take place. (Spalling is here used in the sense of chipping by loss of relatively large flakes, say more than a few percent diameter of the mother particle.) They quote my experimental results that no quartz particles were produced of the size range that dominates in loess (20–50 microns). The word "produce", used by me in a wide sense, they interpret as referring only to the smallest size to which a larger grain can be reduced by abrasion. They then go on to call attention to the smaller size of the particles that may be chipped off larger grains, ignoring the negative experimental results I reported on this opposite aspect of "production".

[1] Manuscript received March 12, 1969.

In my paper it was pointed out (p. 445 and 446) that spalling produced some chips smaller than 40 microns but that nearly all of the abrasion dust was smaller than 2 microns. If medium silt sizes had been abundantly formed the particle size range of 5 to 10 microns should have been much more plentiful than that of 20 to 30 microns, because the latter requires an impact that is about 10 times more effective. The scarcity of the smaller sizes in the dust shows convincingly that loess-sized fragments were very rarely produced. Maximum wind force used was the equivalent of Beaufort scale 10 (= 25 m/ sec, whole gale), covering the strongest desert winds of sufficiently frequent occurrence to play a part of any significance. The experimental grains consisted of crushed quartz up to 1 millimeter with shapes that must tend to spall much more effectively than natural grains. Needle-sharp points, knife edges, and thin chips abound in crushed quartz, but are not encountered in weathered crystalline rocks. The scarcity of loess-sized spallings in the experimental dust is strong evidence that they play a negligible part in natural eolian abrasion.

This conclusion is strengthened by the absence in desert sands of fresh shiny corners to frosted irregular grains as far as my observations go. Although such freshly chipped edges would be etched in their turn, there should be distinct marks due 'o corner-breakage on a number of grains, if spalling were quantitatively important.

The question can be raised how quartz manages to escape the loss of all but minute chips in eolian abrasion. The chief reason is evidently the great toughness of this mineral. Feldspar is more fragile owing to its cleavage and produces loess-sized particles in plenty during experimental eolian abrasion. This, by the way, shows that the absence in quartz experiments must be real and cannot be attributed to loss of some kind. The spectacular resistance to experimental abrasion of rounded quartz grains with a marked polish (Kuenen, 1960, p. 438) is obviously likewise attributable to resilience.

A second reason for resistance to abrasion of

they do, on spalling of quartz grains to account for its presence. For it can hardly be doubted that in non-glacial deserts these sizes will occur ready-made in some of the disintegrating rocks.

Smalley and Vita-Finzi claim that disintegration of phyllite is not "specifically a desert process." This is hardly a cogent argument against a modest amount of production in deserts or elsewhere, if suitable rocks are present.

There is, moreover, another primary source of quartz silt, for even weathered coarse-grained crystalline rocks produce enough silt-sized quartz to be of quantitative importance in the resulting sedimentary material. In the accompanying table some analytical results on weathered rock from South Africa are brought together.

Percentages of some silt fractions in weathered rocks from South Africa. One metamorphic schist and seven granites and granite gneisses

									Average	Quartz
60–50 micron	0.5	0.5	0.6	0.2	0.4	0.1	0.4	0.3	0.4	—
50–42 micron	1.3	3.3	1.6	0.9	3.3	1.0	2.1	0.7	1.8	1.1
42–35 micron	1.8	2.9	0.4						1.7	—
35–25 micron	6.4	1.0	0.3						2.6	—
25–16 micron	0.8	1.5	0.1						0.8	—
16–0 micron	0.9	17.3	2.7						7.0	—
Silt 60–10	11	10	4						8	

quartz in eolian transport is that the force of impact is mitigated by two factors. In the first place the particles tend to spin with the upper side moving down-wind. The relative velocity on the striking surface when a grain lands is hereby reduced. In the second place the impactee is normally free to move and thus to reduce the force of the blow for both grains. As spalling requires not a head-on collision but a grazing blow, the tendency to convert part of the energy into spin must help save the grains from losing chips.

Collision of particles off the ground must involve contrary spinning of the impacting surfaces, but relative velocities will usually be moderate and the ability to give way will be at its maximum. Hence, impacting should be less severe than on the ground.

The evidence against the unsupported opinion presented by Smalley and Vita-Finzi is sufficient to maintain my former conclusion that quartz particles of 20 to 50 microns in eolian deposits are not newly formed by spalling under eolian action.

The comparatively small amount of loess-sized material in non-glacial areas is an interesting fact to which Smalley and Vita-Finzi draw attention. There appears to be no need to call, as

The amount of quartz is an estimate of the average after staining the feldspar grains of fraction 50–42 microns. These data warrant the conclusion that of the quartz particles formed by weathering of coarse acid rocks several percent are within the range dominating in loess.

AQUEOUS ACTION

Recent attempts have been made to account for the silt-sized quartz particles in sediments by spalling of pebbles (Rogers et al. 1963; Schubert, 1964). But the amounts produced can hardly be of more than local importance, because the total volume of material abraded from pebbles, and specially when limited to those formed of quartziferous rock material, is very small compared to the volume of all quartz sands. Besides, most of this abrasion dust coming off rolling pebbles is below silt size. This was demonstrated experimentally by Daubrée a century ago. Admittedly there is a catch in Daubrée's and Schubert's experimental results. They used ball-mill type abraders, and I argued (1956) that the type of movement is a poor reproduction of natural action. Moreover, as there is only a small volume of water involved and because of the grinding nature of the experimental movement in a heap of pebbles, the chances of a spalled-off chip to be ground down to clay size in these experiments is much higher than in nature. In a stream, fast enough to roll pebbles, a newly formed small particle will be wafted away safely, and the pebbles roll as separate individuals, not like grinders pressed together in a ball-mill. Unhappily I omitted to analyze the abrasion dust of my experiments with current-rolled pebbles. But much of it was obviously very fine flour.

On pebble beaches some silt is probably formed because the rigour of the pounding is more severe than in a river. However, the quantity must again be very small.

SUMMARY

My former conclusions can be upheld and amplified, firstly that disintegration of fine grained schists with primary silt-sized quartz particles (phyllites) and of coarser crystalline rocks is a much more abundant source of quartz silt than aqueous or eolian spalling, and secondly that in certain periods of earth history glacial crushing has fashioned large volumes of silt-sized quartz particles. This latter contention has been given strong support by the findings of Smalley and Vita-Finzi. These two sources of quartz silt were overlooked by Rogers *et al.*, the former by Smalley and Vita Finzi. The deduction of Rogers *et al.* that the two size populations of quartz particles must be due to two processes can be replaced by the tentative conclusion that they stem from two sources. Abrasion and solution do shape the particles but have no significant influence on their size distribution.

REFERENCES

DAUBRÉE, A., 1880, Synthetische Studien zur Experimental-Geologie. Vieweg, Braunschweig, 192 p.
KUENEN, PH. H., 1956, Experimental abrasion of pebbles. 2. Rolling by current: Jour. Geology, v. 64, p. 336–368.
———, 1960, Experimental abrasion. 4. Eolian action. Jour. Geology, v. 68, p. 427–449.
ROGERS, J. J. W., KRUEGER, W. C., AND KROG, MARILYN, 1963, Sizes of naturally abrared materials: Jour. Sedimentary Petrology, v. 33, p. 628–632.
SCHUBERT, CARLOS, 1964, Size-frequency distributions of sand sized grains in an abrasion mill: Sedimentology, v. 3, p. 288–295.
SMALLEY, I. J. AND VITA-FINZI, C., 1968. The formation of fine particles in sandy deserts and the nature of "desert" loess: Jour. Sedimentary Petrology, v. 38, p. 766–774.

41

Copyright © 1970 by the Society of Economic Paleontologists and Mineralogists

Reprinted from *Jour Sed. Petrol.*, **40,** 1367–1368 (Dec. 1970)

ORIGIN OF QUARTZ SILT: COMMENTS ON A NOTE BY PH. H. KUENEN[1]

C. VITA-FINZI
Department of Geography, University College London
AND I. J. SMALLEY
Department of Civil Engineering, University of Leeds

Kuenen (1969) has recently challenged our suggestion (Smalley and Vita-Finzi, 1968) that silt-sized quartz particles could form in deserts by spalling during wind transport. Our main concern had been to show that the widely accepted dual classification of loess into cold and hot (or glacial and desert) categories could not be upheld. In an effort to envisage a possible mechanism for the production of loess particles in deserts we had suggested that some silt-sized spalls might be broken off larger quartz particles by impact. Kuenen (1969, p. 1631) states that in the dust he obtained by laboratory simulation of wind abrasion there was a 'scarcity' (though not a total absence) of silt-sized spallings. Thus, although Kuenen's arguments against the workings of our proposed mechanism are persuasive, his experimental findings do not wholly invalidate it.

At the same time Kuenen's findings reinforce our case for the rejection of the term 'desert loess' and its synonyms, since the one uniquely desert mechanism that can be envisaged for its production turns out to be even less effective than we had thought. The concept it embodies hangs on in the literature with great tenacity: in a recent textbook it is categorically stated that "much loess has formed in relatively warm climates around the margins of some present day deserts" (Embleton and King, 1968, p. 572). The question of mechanisms apart, this contradicts the early observation of Penck (1909) that loess is absent from the margins of sandy deserts. Penck (1931) also pointed out that the so-called loess of Tripolitania is really a fine sand; the fact that it is still referred to as loess (Smith, 1968) shows how difficult it may be to dislodge an assertion once it gets into print.

Kuenen (1969, p. 1632) also takes us to task for claiming that the disintegration of phyllite to produce silt particles is not "specifically a desert process," and states that this "is hardly a

cogent argument against a modest amount of production in deserts or elsewhere, if suitable rocks are present." At the risk of becoming semantic bores we reiterate our contention that this is insufficient basis for recognizing a special desert brand of loess. If the particles themselves are at issue, they are better called silt; if their phyllitic origin is followed by eolian deposition, then 'phyllite loess' would be more appropriate since it is their lithological rather than their climatic character that distinguishes them from other kinds of wind-laid silt deposits. We would seem to agree that glacial grinding (Smalley, 1966b, Smalley and Vita-Finzi, 1968) is probably responsible for producing the bulk of fine quartz particles (that is, if we take the 'ice scour' and 'glacial crushing' of Kuenen (1960, p. 446, 1969, p. 1633) to amount to the same process). It seems unlikely that much loess of glacial origin will occur in the hot sandy deserts under consideration. Hence the only significant source of non-glacial silt-sized material is the weathering of fine- and coarse-grained quartzose rocks as postulated by Kuenen, with a possible minor contribution from impact spalling. Only the latter is climatically specific.

Kuenen (1969, p. 1632) also comments on the suggestion by Rogers *et al.* (1963) that the two size populations of quartz particles (sand and silt) are due to two different processes of formation. The inadequacy of the abrasion mechanism for the formation of sand and silt proposed by them and by Schubert (1964) has already been demonstrated (Smalley, 1966a). Kuenen's own tentative conclusion is that sand and silt stem from two sources. This could be the case where the silt is produced from phyllitic rocks, but as Kuenen notes that "coarse-grained crystalline rocks produce enough silt-sized material to be of quantitative importance in the resulting sedimentary material," it is clear that a single source could yield a bimodal population. This is conceivable in the case of rock disintegration, and demonstrable where

[1] Manuscript received April 28, 1970.

glacial grinding is involved (Smalley, 1966b). And since, as Kuenen acknowledges, the bulk of quartz silt is produced by glacial grinding, silt production is arguably *process*-dependent to a much greater degree than it is *source*-dependent.

It has been observed (cf. Pettijohn, 1957, p. 47, Tanner, 1958) that certain particle sizes appear not to be represented in the sedimentary system. This can be ascribed to the nature of the processes of particle production. Three separate classes of particle may be expected from the breakdown of rocks. (We are concerned with crystalline units; in the case of clastic sedimentary rocks, breakdown will simply lead to the liberation of particles whose size distribution is inherited from an earlier phase of primary particle production). Pebbles tend to be multigrain or multimineral fragments, so that their size range is difficult to define in mechanistic terms; but at least it is clear that once a certain size is attained, their release into the sedimentary system does not lead to further breakage unless exceptional forces come into operation. The sand size limits are determined by source, i.e. the dimensions of the quartz grains in granitic rocks

(Smalley, 1966a). As we have seen, most silt particles are process-dependent. The gap between sand and silt size ranges is due to the peculiarities of glacial grinding (Smalley and Perry, 1969, p. 35), during which the sands that do not escape the crushing zone are pulverized.

We reach the following conclusions:

1. The bulk of the quartz silt in the sedimentary system, much of which occurs as loess deposits, is produced by glacial grinding.

2. There are no specifically desert processes which produce quartz silt in appreciable quantities. Some may be liberated by the weathering of crystalline rocks, as suggested by Kuenen (1960, 1969). The mechanism suggested by Smalley and Vita-Finzi (1968), whereby silt-sized fragments are produced by interparticle impact in hot deserts, appears to be unimportant.

3. The two size populations of quartz (sand and silt) are essentially due to one particular source (quartz crystals in granitic rocks) and one particular process (glacial grinding). The two-process theory of Rogers *et al.* (1963) and the two-source theory of Kuenen (1969) are not satisfactory.

REFERENCES

EMBLETON, CLIFFORD AND KING, C. A. M., 1968, Glacial and Periglacial Morphology. Arnold, London, 608 p.
KUENEN, PH.H., 1960, Experimental Abrasion, 4: Eolian Action: Jour. Geology, v. 68, p. 427–449.
——, 1969, Origin of Quartz Silt: Jour. Sed. Petrology, v. 39, p. 1631–1633.
PENCK, ALBRECHT, 1909, Die Morphologie der Wüsten: Geog. Zeit., v. 15, p. 545–558.
——, 1931, Zentral-Asien: Zeit. Ges. Erdk zu Berlin, (for 1931), p. 1–13.
PETTIJOHN, F. J., 1957, Sedimentary Rocks, 2nd ed. Harper, New York, 718 p.
ROGERS, J. J. W., KREUGER, W. C., AND KROG, MARILYN, 1963, Sizes of naturally abraded materials: Jour. Sed. Petrology, v. 33, p. 628–632.
SCHUBERT, CARLOS, 1964, Size-frequency distributions of sand sized grains in an abrasion mill: Sedimentology, v. 3, p. 288–295.
SMALLEY, I. J., 1966a, Origin of Quartz Sand: Nature, v. 211, p. 476–479.
——, 1966b, The properties of glacial loess and the formation of loess deposits : Jour. Sed. Petrology, v. 36, p. 669–676.
——, AND PERRY, N. H., 1969, The Keilhack approach to the problem of loess formation: Leeds Phil. and Lit. Soc. Proc. (Sci. Section), v. 10, p. 31–43.
——, AND VITA-FINZI, C., 1968, The formation of fine particles in sandy deserts and the nature of 'desert' loess: Jour. Sed. Petrology, v. 38, p. 766–774.
SMITH, H. T. U., 1968, Nebraska dunes compared with those of North Africa and other regions, in Schultz, C. B., and Frye, J. C., eds., Loess and Related Eolian Deposits of the World, Inqua 7th. Congress Proc., v. 12: Univ. Nebraska Press, Lincoln, p. 29–47.
TANNER, W. F., 1958, The zig-zag nature of type I and type IV curves: Jour. Sed. Petrology, v. 28, p. 372–375.

Editor's Comments
on Papers 42 and 43

42   LOŽEK
     The Loess Environment in Central Europe

43   LUGN
     Excerpts from *The Origin of Loesses and Their Relation to the Great Plains in North America*

The 1965 INQUA Congress was held in Boulder, Colorado, and volume 12 of the *Proceedings* was devoted to loess and related eolian deposits. Most of this material was concerned with stratigraphy, but there were some important contributions on loess lithology and deposition. The paper by Ložek is of interest because the author manages to survey the loess environment of a considerable proportion of Europe. One of Ložek's major contributions to the study of loess (perhaps his most important) has been his work on the molluscan fauna. This has an important part in Paper 42, but is emphasized in his paper in *Eiszeitalter und Gegenwart* (1965), which is the most comprehensive statement of his views on snails and loess formation.

The *Proceedings* contain an interesting paper by Kes, who discusses loesses and related mantle deposits in Russia and China. He provides a slightly unusual Russian view of loess in that eolian origins are recognized and considered to be important. Lugn recognizes eolian origins but proposes a "desert" model for the formation of the Great Plains loess deposits. He has been concerned with the loess of Nebraska in most of his investigations and has produced a series of papers presenting his theory of origin (Lugn, 1962, gives more details; Lugn, 1969, gives an up-to-date version with some discussion of recent papers). His views on the origin of the Nebraska loess are in distinct contrast with those of some other workers (e.g., Frye and Leonard, 1951; Swineford and Frye, 1951; Leonard and Frye, 1954) who argue for floodplains and against a sandhills source. The Lugn paper has been shortened but the essential elements of the argument are retained.

The paper by Obermaier cited by Ložek in Paper 42 is available in an English version (Obermaier, 1935). The reference Ložek, 1965, has the wrong page numbers; these should be 61–75.

REFERENCES

Frye, J. C., and Leonard, A. B. 1951. Stratigraphy of the late Pleistocene loesses of Kansas. J. Geol. 59, 287–305.

Leonard, A. B., and Frye, J. C. 1954. Ecological conditions accompanying loess deposition in the Great Plains region of the United States. J. Geol. 62, 399–404.

Ložek, V. 1965. Das Problem der Lössbildung und die Lössmollusken. Eiszeitalter Gegenwart 16, 61–75.

Lugn, A. L. 1962. The origin and sources of loess. University of Nebraska Studies no. 26, Lincoln, Neb. 105p.

——. 1969. The geomorphology of loess in North America; its sources and distribution. *In* Etudes sur le Quaternaire dans le Monde, v. 1, 8th INQUA Congress, Paris, 1969, p. 77–84.

Obermaier, H. 1935. The formations of "loess" in Europe and their importance in determining the chronology of fossil man. Res. Prog. (Berlin) 1, 111–117.

Swineford, Ada, and Frye, J. C. 1951. Petrography of the Peorian loess in Kansas. J. Geol. 59, 306–322.

42

Reprinted from *Loess and Related Eolian Deposits of the World,*
Proc. 7th INQUA Congr., Boulder, 1965, C. Schultz and J. C. Frye, eds.,
University of Nebraska Press, Vol. 12, 1968, pp. 67–80

The loess environment in Central Europe

VOJEN LOZEK, *Geological Institute of CSAV*
Quaternary, Prague, Czechoslovakia

INTRODUCTION

Loess and related deposits have always attracted the attention of geologists owing to their properties and a comparatively uniform development over vast areas. In the voluminous literature on their origin, there is a wide divergence of opinions. Most authors, however, are inclined to the view that not only deposition of eolian dust but also a special, concomitantly acting process called loessification, are responsible for the formation of loess. A comprehensive survey of these problems was presented by V. S. Obruchev, in 1948, in one of his last loess papers (Gerasimov, 1964). Loessification demands the existence of particular environmental conditions at the site of loess deposition, which I shall try to explain.

The character of the environment where loess was formed has so far been inferred mainly from its petrographic properties and the mode of occurrence. Paleontological criteria, if used at all, served particularly mammals, although mollusca were long known to be the most abundant and most typical loess fossils. That little attention was paid to the molluscan fauna can be accounted for by the inconclusive results of the earlier, not sufficiently precise, analyses of loess malacocoenoses (Geyer, 1927). The discussions bearing on cold and warm loesses, for instance, which have reappeared in various modifications in the literature, generally resulted from the use of differing, one-sided criteria (Ambroz, 1947; Obermaier, 1935; Werner, 1949; etc.).

The systematic study of the mollusca of loess complexes which I have carried out during the post-war years has shown an interdependence between the sedimentation and soil-forming process, on the one hand, and the mollusc assemblies on the other (Lozek, 1964; 1965). The critical analyses proved a specific character of the loess molluscan fauna, which, in correlation with other criteria, can substantially contribute to the knowledge of the natural environment at the time of loess formation. Recently, analogous results have been achieved by means of pollen analysis (Frenzel, 1964; 1965), which helps complete the general picture. It is the objective of this paper to review briefly the results obtained thus far in central European countries.

MOLLUSCAN FAUNA OF LOESS

As mentioned above, molluscs are the most abundant loess fossils. In many regions the shells are so numerous that their distribution is traceable both in vertical and lateral directions. It should be stressed that they are found in the loess itself, while bones are often joined to the horizons of another origin, inserted in the loess complex (Lugn, 1962, p. 23–25). Numerous "loess finds" made by earlier collectors do not originate from pure loess, but from other layers of the loess series. For this reason we must base our investigation only on new finds, recovered from safely determined sediments. The mass collections from the richly differentiated loess complexes, when studied in detail, make it possible to correlate loess assemblies with those which inhabited the same environment during the intervals on non-deposition of loess (Lozek, 1964).

On the basis of quantitative analyses of the molluscan faunas from different parts of central Europe, the following characteristics of the malacocoenoses can be stated:

1. A small number of species is represented by prolific populations.

2. The presence of special species and races is characteristic of loess environment.

3. A special composition of loess assemblages distinguishes them from other recent and fossil malacocoenoses.

4. Species of loess environment are either eurythermic or cold-loving.

5. They live, or are able to live, in open habitats, and the woodland species are missing.

6. The assemblages consist of xerophilous and mesic elements.

7. The loess thanatocoenoses represent monotonous assemblages which preserve a uniform composition over vast areas, extending from France to the Volga River.

This concise outline of the fundamental features of the loess molluscan fauna should be completed by additional data. The loess assemblages, when viewed from the standpoint of the modern ecology of associated species, prove to be a peculiar mixture of elements whose ecological requirements are nowadays occasionally so contradictory as to prevent them from living in the same habitats. Thus, for instance, the prevalently hygrophilous *Succinea oblonga Drap.*, the species of lowland steppe *Helicopsis striata* (Müll.) and rocky steppe species of *Pupilla—P. sterri* (Vth) and *P. triplicata* (Stud.)— forming closed assemblies in the loess, are found at present in different, frequently far apart, habitats. The penetration of petrophilous elements to the loess habitats is not a sporadic phenomenon. Apart from the above-mentioned Pupillas, this is valid also for *Clausilia dubia Drap., Cl. parvula Fér.*, and *Orcula dolium* (Drap.), which we would seek in vain in the present-day loess areas. We must, of course, keep in mind that all the above petrophilous elements live today in an intimate contact with a fresh, usually calcareous, substratum. This suggests that their presence in loess was due to the fact that pure loess likewise provided them with an unweathered carbonate environment. Their absence from the fossil soils of loess complexes can be interpreted in the same terms.

The mixture of ecologically very different species gives evidence that the loess habitats were of such a special character that they complied with apparently contrary requirements of associated molluscs. The methods of detailed quantitative research rule out the possibility that the faunas concerned could have been secondarily mixed up (Lozek, 1964).

The broadly monotonous complex of loess faunas bears signs of local distributional effects, resulting in the local occurrence of some species, e.g., *Orcula dolium* (Drap.) in the Carpathian Basin, and *Clausilia parvula Fér.* in the Rhône valley (in southwest Germany), and in the Jizera Valley in Bohemia. On the basis of representation of the index species, we can distinguish pronouncedly cold Columella faunas, the medium-tolerant Pupilla faunas, and the Striata fauna, which represents the warmest facies of the loess molluscan fauna. This differentiation indicated temperature differences which, between the Columella and Striata faunas could have attained 3–4°C. in the annual mean. The Striata assemblages, even if designated "warm loess fauna," correspond to climatic conditions with a mean annual temperature about 0°C. but relatively warm summers. Within the whole of the climatic cycle, the above conditions are indeed glacial, but, within the scope of loess phases, they represent the warmest interval. Because of this, undoubtedly, the Striata faunas are commonly restricted to thin, often-contaminated loess interlayers inside the soil complexes which are matched with the minor cold fluctuations—as, for instance the Early Würm cold interval that separated the Last Interglacial from the Early Würm interstadials.

The so-called marshy loesses were distinguished by faunas of periodic marshes. Although their occurrences within the complexes with terrestrial faunas were conditioned mainly by topogenic relations, as is shown by their deposition in subhorizontal loess layers in the proximity of the flood plains of major streams, they give clear evidence that the loess "steppe" was not a scorched desert but an area with at least temporarily wet grounds. The retaining of water was probably due to the frozen ground conditions.

Before we proceed to the correlation of the results of loess paleontological investigation with other data, it is necessary to mention briefly the zonation of loess formations, which cannot be omitted in solving the problem discussed.

ZONATION OF LOESSES

In central Europe the true calcareous loesses are confined to the low-lying areas. The upper limit of their occurrence is drawn at an altitude of about 300–400 m., which in the driest areas is scarcely surpassed. In more humid regions, on the windward sides of the mountain ranges or in the intramontane depressions, the limit drops to 200 m. or less (the Ostrava area, the Tisa Plain). The loess zone can be divided into subzones which at a first sight differ in the development of fossil soils (Fink, 1956). In general, the loess zone is divisible into the dry subzone, with fossil Chernozems and richly differentiated soil complexes that are intercalated with loess beds that contain the Striata fauna, and into the humid subzone, where the sectors corresponding to Chernozems and loess interlayers usually consist of wash and solifluction complexes. or weakly developed Brown soils. Between them, a transition subzone intervenes, which under favorable conditions can be fairly large in extent.

It is noteworthy that this difference also is reflected in the composition of fauna of the loess itself. Whereas in the dry subzone the species *Helicopsis striata* (Müll.) is a common component not only of the Striata faunas but also of the drier facies of the Pupilla faunas, in the humid subzone proliferous *Trichia hispida* (L.) and *Arianta arbustorum* (L.) are characteristic, and *H. striata* (Müll.)—as well as the Striata fauna—are lacking. A common feature of the humid subzone is poor assemblages, composed of one to three Pupilla species and *Succinea oblonga* (Drap.).

The problem of the loess equivalents in the more elevated or moist areas is most relevant. The loess zone passes into the zone of dust loams (*Staublehm*); i.e., loams of eolian

origin which show an analogous mode of deposition and
granular composition as true loesses. They lack, however, calcium carbonate—to a large extent or completely—so that their
texture is altogether unaffected by it. The calcium carbonate
is replaced by the iron compounds. The pseudogley process
occurs currently, and is often observable throughout the
thickness of these deposits as at least moderately developed.

The character of the dust loams reveals that they represent a loess facies from the more humid environment. The
boundary between the loess zone and the dust loam zone was
not constant. Its slight oscillations can be inferred from the
presence of both calcareous loess and dust loam layers in
some sections. The maximum extent of loesses falls into the
phase of maximum loess deposition within the loess zone.
Therefore, the sector corresponding to the loess maximum
is not unfrequently developed as calcareous loess, whereas
other parts of the complex are already formed of lime-
deficient dust loam. Such loess "islands" within the dust
loam areas have been established, for instance, southeast of
Prague (between Říčany, Davle, and Zbraslav), in the Sázava
Valley, and along the middle course of the Vltava (Cholín)
River.

Because of the lime-deficient environment, the area of
dust loams yields only scarce paleontological finds. In the
above loess "islands," only poor Pupilla faunas are preserved,
which, moreover, are not decisive for the reconstruction of
the environmental conditions. Because they derive from the
intervals and sites of loess formation, they reflect the state
of the loess zone.

The zone of dust loams passed upward into a zone which
is distinguished by the lack of extensive sheets of sorted,
wind-blown loams. These are replaced by slope deposits containing a variable admixture of loamy matrix whose nature
corresponds to that of dust loam. The amount of the admixture is not uniform; it is largest on the leeward hillsides,
which, within the zones of loess and dust loam, would be

most suitable for the deposition of eolian material. Under favorable conditions, the eolian component can predominate, giving rise to patches of dust loam which grade continuously into deposits with a prevalence of slope material. This zone, which can most conveniently be termed as the zone of mixed slope sediments, is followed upward by the zone of montane slope deposits—with coarse blocks of solid rocks—where the eolian component recedes to the background.

For completeness' sake it should be added that, in the dissected rocky areas, the mixed sediments consisting of coarse slope material and the matrix of loess nature can also be found within the loess zone. This suggests that the zonation (in textural [stoniness], not chemical composition!) is not only controlled by climatic conditions but is also affected by the relief configuration and the parent material, whose effects are marked only at particularly suitable places. They are, however, most important in connection with the paleontological considerations.

The paleontological evidence for the zones of dust loam and mixed and montane slope deposits is everywhere hindered by the lack of calcareous sediments that are derived safely from the loess phase. The distribution of loesses and dust loams is independent of the parent material, as can be easily proved by the presence of loess in areas built up of carbonate-free rocks and, on the other hand, by the absence of loess from the areas which are composed of pure carbonate rocks but lie outside the loess zone (e.g., the Nitrica Valley). The paleontological evidence thus is afforded only by pollen analysis, which so far has been carried out only in the foreland of the Alps (Frenzel, 1964; 1965). Although the existing data are far from complete, they indicate that the zone of dust loams was substantially more humid than the loess zone, allowing a sparse growth of tolerant shrubs and trees (relictual forests). Faunal remains can be found only in places where, in the loess phase, mixed sediments with a large component of local carbonate material were laid down, i.e., chiefly under the limestone scarps or in the cave entrances. It is, of course, very difficult to attest that the respective sediments actually correspond in time to the true loess or dust loams. This question is most relevant because the interrelationship between the loesses and slope deposits shows that— at the time of loess optimum—the slope deposition was very limited, even where coarse scree rapidly accumulated during other phases (e.g., in the warm Postglacial [Lozek, 1963]).

It follows from the above that only fauna, provided its composition corresponds to the fauna of the loess zone, remains as a main evidence. Slope deposits with a loess admixture and the corresponding fauna were found, for example, in the Upper Hron Valley (Farkašovo) or in the south Slovak Karst (Hrhov, Jasov). A few differences that have been ascertained, such as the occurrence of *Vertigo alpestris Ald.*, which is absent from the true loess, are in fact meaningless. Otherwise, those finds proved that the fauna of the zones of dust loam and mixed slope deposits was very similar in composition to that of the loess zone, not exceeding the framework of loess habitats.

As concerns the reconstruction of general environmental conditions, it can be objected, of course, that—as the fauna was preserved only on the calcareous material—it must have been affected by this factor. It provides evidence solely of the general character of climate, but not of soil conditions on the deposits of dust loams and slope sediments, which can be reasonably inferred as originally lime-deficient or lime-free. From this it may be deduced that, while the process of loessification occurred in the loess zone, in the dust-loam and higher zones processes of decalcification and weak pseudogley formation took place. On the other hand, the interpretation of the lack of carbonate as a secondary phenomenon; i.e., that the dust-loams were decalcified, should also be taken into consideration.

The relics of calcareous loesses support this latter alternative with marked faunas which are preserved in travertine fissures within the zone of dust loam or mixed slope deposits (Bojnice, Bešenová), and do not differ either in composition or paleontological content from the common loess. The carbonate parent material in the surroundings enabled their preservation but did not cause their origin.

The process of dust loam formation requires further detailed study, based on additional criteria. Generally it can be said that the environmental conditions in the zones of dust loam and mixed slope deposits were similar to those of the zone of calcareous loess. These monotonous relations, existing over vast areas during the loess phase, will be examined below.

LOESS ENVIRONMENT

In restoring the loess habitats, we shall proceed from the correlation of the paleontological and sedimentological data.

From the analysis of molluscan fauna completed by the mammalian finds and from the results of pollen analysis, the following characteristic features were deduced.

1. Open country;

2. Low temperature;

3. Relatively rich biocoenoses, composed, however, of a small number of tolerant species; and

4. Striking diversity of loess biocoenoses from the contemporaneous fauna and flora.

These points should be considered in relation to the sedimentological statements mentioned below:

A. Loess forms sheets and drifts that are composed mainly of sorted eolian material that is independent of the underlying rocks.

B. It also occurs in those places where, in other periods, deposition of quite a different type prevailed (e.g., coarse scree in warm periods).

C. Some features of the loess—the texture and form of calcium and iron compounds in particular—cannot be explained as due to sedimentation factors but to a special process called loessification.

The correlation of these individual items leads to the following considerations.

An open country (1 [above]) provides very favorable conditions for the accumulation of dust (A) that is blown in from places where the deflation is not hindered by a dense vegetation cover—not even a steppe with a continuous sod mat. The loessification process (C) is conceivable only in open steppe conditions, where long periods of drought and freezing prevent strong weathering and favor precipitation of salts in the surficial layers. A low temperature (2) is attested by the presence of cold-loving molluscs and mammals, as well as by pollen analyses. The absence of thermophiles serves as indirect evidence of this. This conclusion is consistent with the assumption that the eolian activity (A) could have developed when large areas were deprived of a continuous vegetation cover or buried under fresh aggradation that was susceptible to deflation. In European conditions, such a state is plausible mainly as a result of glacial climatic effects—at the time of a strong deterioration of climate. The formation of deserts under a warm climate is—in Europe—hardly conceivable.

The range of temperature at which loesses originated, as evidenced by the differences in the molluscan assemblies, plays only a subordinate role. The Striata fauna, which pronouncedly lacks cold-loving species and consists of elements that—up to the present—live on the xerothermic habitats of central Europe, differs from similar contemporaneous malacocoenoses in the absence of true thermophiles (Smolíková and Lozek, 1964). It suggests a mean temperature of about 0°C., thanks to comparatively warm summers.

It can therefore be concluded that central-European loesses universally originated under glacial conditions, with long, severe winters and relatively warm, dry summers. This climatic pattern is also suggested by the mixture of cold-loving and steppe elements in the faunal assemblages.

A rich malacofauna (3) points to a densely inhabited

surface of the generating loess, which in turn implies a well-developed vegetation cover (Frenzel, 1964; 1965). The terrain could not have been a barren, arid steppe but, on the contrary, must have been overgrown with vegetation which enabled the existence of assemblages with a strong component of mesic, relatively moisture-loving species. This assumption is in keeping with the presumed character of areas where the wind-blown dust (A) accumulated, and with the repression of other types of sedimentation (B) at the time of loess deposition. The relatively slow accretion of eolian dust did not impede the evolution of the respective biocoenoses; on the contrary, it supported those plants and molluscs which preferred raw soils that were well supplied with salts (Frenzel, 1964; 1965).

The conditions enabling the development of relatively rich vegetation and mollusc populations (3) are somewhat at variance with the character of loess as a soil type (C). It should be stressed again that the loess must be considered as soil, and the loessification as a soil-forming process, because the organisms living on its accruing surface affected its properties and composition. Thus, for instance, the roots of steppe vegetation undoubtedly took part in the development of the loess structure and concretions; and shells, when present in a sufficient amount, controlled the $CaCO_3$ percentage. The loess also lie near the present-day desert soils (Gerasimov, 1964) whose habitats differ widely from those suggested by the paleontological content of loesses, both in climatic and vegetation conditions. The deserts comparable to the Pleistocene loess areas exist nowadays in central Asia, whereas in Europe no modern analogies can be found. From this discrepancy it can be inferred that, at least within the European loess zone, loessification was a rather particular process, conditioned by a special environment in one phase of the glacial stages, whose designation as the loess phase is fully warranted (Lozek, 1965).

Although the annual climatic cycle cannot yet be reconstructed precisely, a few fundamental data have been recognized. The climate of the loess phase was continental and cold, with the annual temperature mean between −3 and 0°C—the loess optimum being nearer to −3°C. Winters were long and severe, and the summer season short but relatively warm. In addition, an intervening humid and fairly warm phase is to be presumed, which would enable a rich plant life on the loess steppe, fading out gradually during a dry period in the late summer. There were no abrupt changes of the weather, i.e., no abrupt recurrent changes in temperature or alternation of short, dry periods with sudden downpours. The cold period also was dry, so that—in spite of favorable temperature conditions—cryoturbation was of small extent. In our opinion, this tranquil course of climatic cycle is the only reason which can account for the strong limitation of denudation and deposition processes (*B*)—except for the wind activity, so distinctive for the loess phase.

The special character of the loess fauna and flora (4), which is difficult to compare with the present-day assemblages, is closely connected with the above-stated relationships.

CONCLUSION

From our investigation we can conclude that loesses in central Europe originated under continental, cold climatic conditions, at a mean annual temperature of about −3°C. to −2°C.—rising in the warmest phases toward 0°C. The weather had a tranquil annual course. A long, severe, but relatively dry winter was succeeded by a humid and fairly warm period that became summer (July averaged more than 10°C., probably about 15°C.!); then a dry period persisted until winter.

At this time the loess steppe was covered by a rich steppe vegetation in which species that loved fresh soils, with a sufficient amount of mineral salts, occurred in abundance. The rich molluscan fauna consisted of large populations of highly tolerant species. The soil conditions were very favorable for the molluscs, but only some species were capable of adapting to the loess climate.

At present, analogies of the loess environment are difficult to find. In the main, there were steppes that passed—in the humid tracts—into tundra formations, which, however, were not identical with the modern tundras of northern Europe or with the Alpine barrens of the high-mountain ranges of central Europe. The special character of the environment is evidenced not only by the composition of the plant and faunal assemblages that were composed of species that differed in ecologic requirements, but also by the species and races of molluscs that adapted to the loess environment. Therefore we are justified in using the term "loessic" in a wider ecologic sense, and in speaking of the loess environment.

The conditions we have described existed in a wide belt, spreading from the Atlantic Ocean to the Volga River, which in the north reached to northern Germany, the southern half of Poland and the U.S.S.R., and in the south to Bulgaria and through the Rhône Valley to the Mediterranean. In the more humid and higher-lying areas, it passed into the zone of non-calcareous dust loams and mixed slope deposits, which, irrespective of a higher humidity, showed very similar environmental conditions.

The task for future research is to follow the fading out of the loesses at the southern and northern boundaries of the loess zone, to identify their facies equivalents in other climatic belts, and to carry out a correlation of European loesses with loesses and related deposits on other continents. Other problems, especially whether there is a warm loess or a modern analogy of the European loesses, also await solution. A closer approach to the solution requires much study, particularly paleontological study, but the results will be invaluable for the recognition of the paleoclimatology and paleogeography of the Quaternary.

REFERENCES

Ambrož, V., 1947, Spraše pahorkatin (The loess of the hill countries): Sborník Státního geologického ústavu CSR, Prague, v. 14, p. 225–280.

Fink, J., 1956, Zur Korrelation der Terrassen und Lösse in Osterreich: Eiszeitalter und Gegenwart, Ohringen, v. 7, p. 49–77.

Frenzel, B., 1964, Zur Pollenanalyse von Lössen: *ibid.*, v. 15, p. 5–39, 1 beil.

———, 1965, Uber die offene Vegetation der letzten Eiszeit am Ostrande der Alpen: Verhandlungen der Zoologisch-Botanischen Gesellschaft in Wien, Vienna, v. 103/104, p. 110–143.

Gerasimov, I. P., 1964, Loess genesis and soil formation: Report of the VIth Internatl. Cong. on Quaternary, Warsaw, 1961, v. 4, p. 463–468, 9 pl.

Geyer, D., 1927, Unsere Land- und Süsswassermollusken: Stuttgart, 3d ed., 224 p., pl. I–XXXIII.

Ložek, V., 1963, K otázce tvorby svahových sutí v Ceském krasu (On the formation of the slope [deluvial] material in the Bohemian karst) Ceskoslovenský Kras, Prague, v. 14, p. 7–16.

———, 1964, Quartärmollusken der Tschechoslowakei: Rozpravy Ustředního ústavu geologického, Prague, v. 31, 374 p., 32 pl., 4 tab.

———, 1965, Das Problem der Lössbildung und die Lössmollusken: Eiszeitalter und Gegenwart, Ohringen, no. 16, p. 1–15.

Lugn, A. L., 1962, The origin and sources of loess: Univ. Nebraska Studies, new ser., v. 26, 105 p., 6 pl.

Obermaier, H., 1935, Lösse und Lössmenschen in Europa: Forschungen und Fortschritte, Berlin, v. 11, no. 6, p. 71–74.

Obručev, V. A., 1948, Lëss kak osobyj vid počvy, ego genezis i zadači ego naučenija (Loess as a special type of soil, its genesis and tasks of its study): Bjulleten, Komissii po izučeniju četvertičnogo perioda, Moscow, v. 12, p. 5–17.

Smolíková, L., and Ložek, V., 1964, The holocene soil complex of Litoměřice: Sbornik geologických věd, A–Antropozoikum, Prague, v. 2, p. 41–56, 3 pl.

Wernert, P., 1949, Le problème des loess anciens à faune forestière et subtropicale: Sédimentation et Quaternaire, Bordeaux, p. 285–292.

43

Reprinted from *Loess and Related Eolian Deposits of the World*,
Proc. 7th INQUA Congr., Boulder, 1965, C. Schultz and J. C. Frye, eds.,
University of Nebraska Press, Vol. 12, 1968, pp. 139–152, 180–182

The origin of loesses and their relation to the Great Plains in North America

A. L. LUGN, *Department of Geology (Emeritus),
University of Nebraska, Lincoln, Nebraska, U.S.A.*

INTRODUCTION

GENERAL AND RIVER FLOODPLAINS

The writer in several earlier publications proposed a "regional source area" or "desert source" explanation for the source of most of the loess deposits in Nebraska and extensive adjacent areas. He has stated (1939b, p. 873–874; 1935; 1960; and 1962) that "Most of the Peorian and later loess of Nebraska has come from the Sand Hills Region located in the central and central-western parts of the state [Nebraska]. It is an area of about 20,000 square miles, and the dune sand has been derived by wind action mostly from older Tertiary (Ogallala group) sandy formations. . . . The dune sand is the material left behind after the fine silt and clay had been sifted out and carried eastward and southeastward to become the yellowish and yellow-gray loess, which was spread over a fan-shaped area of tens of thousands of square miles in southern and eastern Nebraska and areas farther eastward." To this area of distribution should be added much of northern and eastern Kansas. Further, the eolian reworking of exposed areas of Cretaceous and earlier shales, including "Red Beds," may well have supplied loessic materials for the Loveland loess as well as for later loess deposits (Lugn, 1935, p. 128–167).

River floodplains are sources of loess dust, and in certain regions most if not almost all of the existing loessic material has been blown from river floodplains. Undoubtedly this is the best explanation for the source of most of the loess deposits in the United States east of the Mississippi River. The long extension of loess occurrence from south of the Ohio River through western Kentucky, Tennessee, and Mississippi, almost wholly east of the Mississippi River, which are in continuity with the loesses of the upper Mississippi Valley region, is a complex of several distinct deposits, which may and probably do correspond to the loess complexes of Loveland and Peorian loess in other areas. Also, there appears to be a thin light-yellowish to very light gray loess unit in Iowa and in large areas east of the Mississippi River in Illinois, Tennessee, and southward into Mississippi which may be more or less equivalent to the Bignell loess of the central Great Plains. The problem of the precise zonation of the loesses will not be discussed in this paper; it has been dealt with by the writer elsewhere (1960, p. 224; 1962, p. 4–8).

The writer believes that the lower Mississippi Valley loesses have been deposited by wind action, which has winnowed the loessic materials out of the floodplain alluvial sediments on the extensive Mississippi River floodplain and low benches. Other nearby floodplains and low benches in southern Arkansas and Louisiana may have made some contributions.

Much of the loess of western Germany may have been derived from the Rhine River floodplain, and loess in Poland has apparently been derived from floodplains, spillway valleys, and abandoned glaciated areas. Much of the loess of the Soviet Union (in Europe) appears to have been similarly derived from river floodplains and, more regionally, from extensive till plains. Most of the loess of the Argentine Pampa Plain may well have been blown from the Gran Chaco Plain at the northwest.

The relatively thin loess deposits in Illinois have been quite convincingly demonstrated to have largely if not almost entirely originated from river floodplains (Smith, 1942, and later writers). The intricate relations of outwash floodplains, spillways, and abandoned till plains as sources of local loesses in Illinois have been worked out very ably in Illinois, and in general these explanations are acceptable to the present writer.

However, competent Pleistocene geologists should take a new, honest, and unprejudiced look at the loess west of the Mississippi River. The loess in Iowa had some new sources—floodplains, of course, and the early Wisconsin (Iowan) till surface; and, farther west in Iowa, more increments of the loess came not only from the Missouri River floodplain but from areas still farther west, from such places as the Nebraska Sand Hills and from other regions of the Great Plains. This is even more true of the loesses in Nebraska and Kansas. The stripping of the alluvial Tertiary cover from large areas of the High Plains from late Illinoian and Sangamon time must have resulted in much wind-blown loessic material being carried out of that region and spread widely over great distances in easterly directions. It seems most probable that large quantities of loess deposits, older and of more ancient dating than any which presently exist, may have been deposited and, subsequently, completely removed by erosion from extensive areas during the earlier Pleistocene ages. Eolation in the Great Plains also was greatly aided and facilitated by profound fluvial erosion from time to time, possibly even during dust-blowing periods, when wind action was most vigorous.

The river floodplains of Nebraska and Kansas have made contributions to loess sedimentation, especially in producing narrow "loess lips" adjacent to such valleys on the leeward sides. However, apart from this very interesting feature of loess sedimentation, river-valley floodplains in this region have been wholly inadequate to have supplied the quantity and to have been responsible for the distribution of the loess as it occurs in Nebraska, Kansas, and in other Great Plains areas.

LOESS LIPS AND MISSOURI VALLEY LOESS

A "loess lip" occurs close to the southeast of the Platte River Valley. Wright (1947, p. 14) commented: "The area enclosed by the 30-foot (thickness) contour mainly across parts of Polk and Butler counties (Nebraska) seems to be conclusive proof that the floodplain (Platte-Loup River) contributed to this locally thicker loess deposit. This 'loess lip' occurs at the edge of the tableland to the southeast of the (valley) floodplain which is 12 to 18 miles wide, opposite the junction of the Loup and Platte rivers.

"Textural analyses of the loess from this area indicate that the sand fraction is very fine sand and in an amount that normally is present in the loess some miles to the northwest, as in Boone, Greeley, and Wheeler counties. Coarse and medium sands were not carried up and out of the valley."

The floodplain influence affecting both the thickness and the coarseness of the loess in the "loess lip" persists in the direction of the prevailing winds (to the southeast) for only a short distance. The higher and coarser sand content in the loess is present only relatively close to the southeast margin of the floodplain; the wind lost competency very quickly beyond the eddying areas, and even medium sand was not carried very far up the southeast valley wall.

This area of loess that is thicker than 30 feet (actually, up to 50 feet) contains about 235 square miles, and the thickness of loess that would normally have been expected in this area is about 20 feet or less, the same as it is on the north side of the river valley. If the excess thickness of the loess in the loess lip area over the normal thickness of 20 feet were to be spread out evenly over the 2,390-square-mile area enclosed locally by Wright's 20-foot thickness contour close southeast of the Platte Valley, it would increase the thickness of loess in this area by only about 7 feet or up to 27 feet. This is the amount of loess which can be considered as having been originated from the wide Loup-Platte River floodplain (see Lugn, 1962, p. 31–32).

The relatively great width of the Loup-Platte River floodplain to the northwest of the loess lip area discussed above very definitely indicates that a floodplain must be wide relative to the depth of the valley, and also accessibly situated

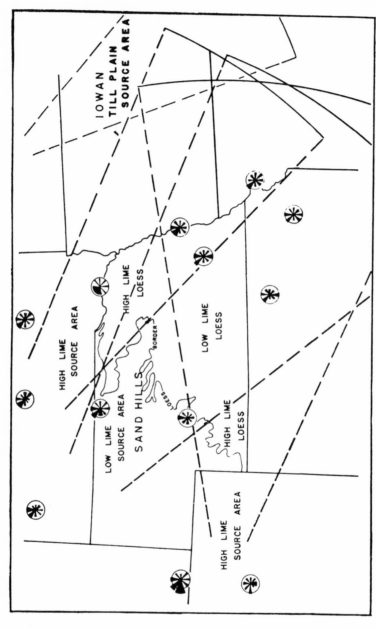

Figure 9–1. Map of an area (Nebraska, Iowa, and adjoining areas) in which loess is widespread. It illustrates Udden's general theory of wind action, of source areas for the loess, and its fanwise distribution in the direction of the prevailing winds during the winter and spring seasons. It is modified from "The Calcium Carbonate Content of the Peorian Loess in Nebraska" by J. Maher (Master's thesis, University of Nebraska, 1937), and C. G. Bates, "Climatic Characteristics of the Plains Region," *Possibilities of Shelterbelt Planting in the Plains Region* (U.S. Forest Service, 1935), p. 85.

relative to winds in order to be an effective source area for loessic materials; and, of course, loessic materials must also be available.

Other loess lips have been described and quantitatively evaluated by the writer in other publications (1960; 1962, p. 31–36), including some discussion and evaluation of the thick loess on both sides of the Missouri River in Iowa and Nebraska. The thicker loess on the Iowa side of the Missouri Valley must in part be attributed to increments of loessic sediment derived from the Missouri River floodplain. It is a "loess lip" development. However, there is also an apparent "loess lip" of thick loess also on the Nebraska side of the Missouri River Valley. Attention is here directed to a wind rose (see fig. 9–1), which is located athwart the Missouri River Valley in northeast Nebraska in the north-central part of the map. It indicates a strong preponderance of winds from the northwest, trending in the same direction as the northwest to southeast direction of the wide floodplain of the Missouri and Sioux River valleys in this area during the main dust-blowing season of the year. This strongly suggests a valid basis for the origin of the thick loess increments to the southeast on both the Iowa and Nebraska sides of the Missouri River Valley as having been derived from the floodplain of the wide valley floor here, as well as from more remote source areas.

It now seems increasingly apparent that the location of floodplains, their trends, the depth of a valley floodplain below adjacent upland levels, as well as the composition of the floodplain sediment, and especially the prevailing wind direction or directions during dust-blowing seasons, all are important factors in determining if loessic materials in any quantity will be contributed, or had been contributed in the past, to the eolian load from any river valley floodplain.

Loess deposits become thinner to the east and in south-central Iowa (less than 8 feet), but in east-central Iowa they thicken to more than 32 feet in proximity to the border of the Iowan substage till of early Wisconsin age, a different kind of source area and material. It is difficult to evaluate the quantitative importance of the Missouri River floodplain as a source of the thicker loess deposits on both sides of the valley and to the eastward in Iowa. It is the considered but tentative opinion of the writer that from 25 to 50 percent of the loess in western Iowa, perhaps up to 150 miles east of the Missouri River, and in the extreme northeastern part of Nebraska, may have been blown from the floodplain of the Missouri River.

On the other hand, the writer's observations on the great dust storms of the mid-1930's (1935, p. 165) led him to express the thought that "Depressions and large, wide river valleys like the Missouri River Valley served as sedimentation traps for the loess dust, as it was swept across the country by strong westerly winds." Many of the great dust storms, the "black blizzards," extended up to heights of 10,000 to 15,000 feet above the surface of the ground.

PROPOSED MULTIPLE SOURCES

Condra, Reed, and Gordon (1947; 1950, p. 42–45) state that Nebraska loesses originated from fresh till sheets, White River formations, badlands, the High Plains, the Sand Hills Region, old till sheets, and from alluvial deposits (river floodplains). This appears to indicate a variety of sources and source materials. All of the badlands of South Dakota and Nebraska occur on White River formations, mainly developed out of the Chadron beds. Very little dust, as such, is now blown directly from any known badlands areas, and such as there is originates almost entirely from floodplain materials that have been loosened by weathering and washed by rain from the surface of the somewhat indurated Chadron and Brule formations. Most of this reworked material is now disposed of by fluvial action.

Loess that may have originated from fresh till sheets in Nebraska (mostly the Nebraskan and Kansan till sheets when they were fresh) have not been recognized, and any such loesses do not appear among the Pleistocene formations of Nebraska (Condra, Reed, and Gordon, 1947; 1950, p. 46), nor are any loesses known in Nebraska that have been assigned to any known "old till sheet" source (*ibid.*). Mention has been made of the earlier "Iowan" or older Peorian loess of eastern and southeastern Iowa having been blown from the surface of the Iowan till sheet (early Wisconsin) when it was freshly exposed.

Almost all areas of sand hills or dunes which occur in the High Plains (Great Plains Region) from Nebraska south into Texas and New Mexico, either of recent or ancient (Pleistocene) development, have originated from underlying Ogallala (Pliocene) sediments. Very little Ogallala remains anywhere in the Great Plains north of Nebraska.

The Sand Hills Region of Nebraska (and also many other sand dune areas) is underlain by Ogallala sediments or formations, so that, prior to fluvial and eolian reworking, this region was an intact part of the Great Plains; in fact, it still is a part of the High Plains. Areas of sand hills or dunes seldom occur on High Plains areas that are underlain by other Tertiary formations, such as Sheep Creek, Marsland, Harrison, Monroe Creek, Gering, or the formations of the White River group. Therefore it appears that almost all of the eolian silt (loess) that derives from or has originated in any way from the High Plains or the Sand Hills Region of Nebraska or other sand dune areas has originated from the Ogallala fluvial sediments, so that High Plains and Sand Hills Region are essentially synonymous. It seems that the apparent multiple sources for the loess indicated by Condra, Reed, and Gordon (1947; 1950) actually narrow down to only two kinds of sources advocated by the writer. These are the Sand Hills Region and other dune areas (High Plains and the Pliocene Ogallala group, which is actually alluvial in character), and river floodplain alluvium.

EOLIAN ORIGIN OF LOESS

The eolian origin of loess is here accepted without reservation and it is believed to be fully applicable to all loess deposits from and including the Loveland loess, the complex of "Peorian" loesses, and the younger Bignell loess in Nebraska and Kansas and in broad areas to the eastward. Massive eolian silt deposits are extensive and thick in the central Great Plains, especially in Nebraska and Kansas, and such deposits occur in every state in the Great Plains in some quantity. Thinner eolian silts (loesses) are widespread in Iowa, Missouri, Illinois, and eastward in Indiana and in Ohio (Smith, 1942; Leighton and Willman, 1950; Lugn, 1962).

The fundamental facts supporting the eolian origin of loess have been well stated by Swineford and Frye (1951, p. 316–317), and they have been reviewed by the writer (1962, p. 9–12). They will not be restated here.

UDDEN'S GENERAL THEORY

J. A. Udden's conclusions (1898) regarding atmospheric or eolian erosion, transportation, and deposition are particularly interesting and appropriate at this point (see Lugn, 1935, p. 162–163; 1962, p. 13–17). Udden stated: "The work of the atmosphere begins with erosion. This erosion is confined to much smaller areas than atmospheric sedimentation. One such area of erosion may be regarded as one of the corners of an isoceles triangle, pointing against the wind. Between the two equal sides of this triangle transportation and sedimentation are taking place. The quantity of work performed is greatest near the area of erosion." Sorting takes place in the prevailing wind direction.

Udden continued: "It is evident that the place of greatest deposition is never far from the place of greatest erosion, when the eroded terrain consists of coarse as well as fine materials. It is generally marked by the accumulation of dune sand. From this point deposition decreases, owing to the transversely horizontal and the vertical dispersion of the load by spreading winds and owing to the previous settling of the coarser particles. A limit is sooner or later reached, where aqueous erosion is more rapid than the accumulation of atmospheric sediments. Beyond this limit the latter will of course not appear."

The Sand Hills Region of Nebraska can be considered as a source area at "one of the corners of an isoceles triangle pointing against the wind," in this case westward. Transportation and deposition have taken place between the two sides of the triangle eastward from the source area, and the loess has thus been spread over a very large area in eastern, southeastern, and southern Nebraska, and into Iowa, Missouri, and northern and eastern Kansas, after having been sifted from the Sand Hills Region and from other dune areas in southwestern Nebraska and northeastern Colorado.

Udden's general theory of source areas for the loess and its distribution stated above is illustrated in figure 9–1. The proportionate time and velocity of the wind from each direction for the winter period (six months, November to April, in the main the dust-blowing part of the year) are shown by the width and length, respectively, of the wedgelike sectors of the wind roses. The radii of the circles denote a velocity of 10 miles an hour (Maher, 1937; Bates, 1935). Other wind roses outside the map area located farther to the northwest show winds during this period to be predominantly from the northwest.

Udden also stated (1898, p. 67): "It seems probable that the Western Plains and the Mississippi Valley maintain the windward-leeward relation to each other. Dust which is stirred up over the plains must be carried east by the prevailing winds, and a part of it no doubt settles over the great central valley. The loess and surface silts, which are spread over most of the territory of this valley, resemble atmospheric sediments considerably in their mechanical composition. It is generally finer in the east and coarser in the west, and it decreases in thickness from west to east." Udden (1914, p. 720–726, 728) supported this general thesis by the publication of 85 analyses of samples of loess and wind-blown dust.

Udden's statement in the paragraph above, though written more than 65 years ago, quite correctly states the findings and conclusions of many other students of the loess down to the present time. The long list of such persons includes J. E. Todd (1906), G. E. Condra (1908), F. J. Alway (1916), J. C. Maher (1937), A. L. Lugn (1935; 1939b; 1962), R. E. Bolen (1945), J. J. Wright (1947), Ken C. Wehrman (1961), R. H. Castellano (1961), and many others.

DUST STORMS AND LOESS

Dust storms during 1934 and 1935, especially in March and April of 1935, in 1936, and more recently from 1950 to 1956, have attracted considerable attention and done no little damage in certain areas in South Dakota, Nebraska, Colorado, Kansas, Oklahoma, Texas, and New Mexico.

Abnormally low precipitation for prolonged periods caused extended droughts in certain areas in the states noted above in the middle 1930's and again in the 1950's. This caused topsoils to crumble and lose solidity and consistency; soil particles were pulled apart and soil structure in many areas was broken down. Even dehydration of the soil colloids took place. This prepared such material for wind erosion, and dust-blowing took place wherever and whenever the wind had sufficient force. Not only the soils but also Cretaceous and older shales, including the "Red Beds" and the younger Ogallala sediments, loosened and were blown out of the "dust bowls" in the recent dust storms. It is of special interest that dust became available for transportation directly from the parent-materials formation, without first being reworked by fluvial processes into local floodplain alluvium.

It has been shown that the dust from modern dust storms

is identical to ancient loess (Swineford and Frye, 1945, p. 245–255; Udden, 1914, p. 720–726, 728; Wright, 1947).

During any of the great dust-blowing periods of recent years, the quantity of dust which fell in Nebraska in any dust fall was generally less at the east and southeast, and the amount increased to the west, north, and northwest (Lugn, 1935; 1962, p. 18–23). This is similar to loess thicknesses in the same areas. The Peorian (and also the Loveland loess) thickness increases in eastern Nebraska from east to west and from southeast to northwest, and from south to north in south-central Nebraska. Thus, both loess thicknesses and the quantity of dust-fall increase from any point in the direction toward the source areas. This is believed in both cases to be because of the great differences in the rate of deposition and not due to any greater eolian activity in the places where loess is thicker or the dust falls greater in amount. Further, field and laboratory evidence seem to indicate clearly that deposition of all loesses in eastern and especially in south-eastern Nebraska, in general, was less in amount in any dust-fall period, took place under more mesic conditions, and was subject to more immediate weathering and soil development than in any areas farther west, where conditions always were more arid and accumulations more voluminous.

Ancient loess deposits contain few vertebrate fossils, and these are found almost exclusively in the lower few feet of any loess, which was deposited at the very beginning of the age of dust-blowing, or in old loess soils at the top of loesses, which were developed during intervals of non-deposition (Lugn and Schultz, 1934; Schultz and Tanner, 1957, p. 77). Apparently, the mammals endured the dust-blowing and the scarcity of water and forage as long as possible; then many of them died and left their bones entombed in the first (lower) few feet of the dust deposit. Those which survived long enough migrated to more hospitable localities, and ultimately the population may have reached quite distant areas. Later generations returned to repopulate the loess areas when conditions again became favorable, and there was forage and water and soil development.

Quantitative data have been secured by the writer (1935) and some of his students, and by others (Russell, 1936; Stone, 1934; Warn and Cox, 1951) which demonstrate the quantitative importance of eolian transportation, and that if dust-blowing and deposition in certain areas were to continue at the same rates observed in the middle 1930's for sufficient periods of time, new loess deposits comparable to those of the Pleistocene period, could again form in the same places in periods of time comparable to those estimated for the accumulation of the several ancient loesses.

In view of the evidence of the prolonged droughts, the dust storms experienced in certain areas of the Great Plains during recent years (especially since about 1934), and the quantitative data which have been assembled (Lugn, 1960; 1962, p. 18–25), no one can very well doubt the efficacy or the quantitative importance of the wind as a geologic agent for the transportation and deposition of earth materials, nor that the wind can and does transport dust in great quantities over very great distances.

Data seem to indicate that the amount of precipitation received in Nebraska during 1934 and the early part of 1935, and in certain other areas then and in more recent years, was near the generally recognized "absolute annual minimum." It seems that if precipitation were to be permanently reduced to these minima, with attendant high temperatures, eolian erosion in "Dust Bowls" and dust-blowing and deposition would become the permanent regimen in almost every area of the Great Plains.

Further, the dust storms of about 30 years ago, which usually originated in the "Dust Bowl" areas of the western, central, and southern Great Plains, except for their violence and intensity, were comparable to the annual, spring periods of dust-blowing and "dust showers" in Mongolia and northern China, where the atmosphere for many weeks "is yellowed with the suspended dust from the Gobi" (personal communication from Dr. David Y. P. Chou, Chairman, Department of Chemistry, Lenoir Rhyne College, who lived in Peking for a number of years). Dr. Chou has indicated that the Chinese for many generations have recognized these annual dust falls in the Peking area, and the heavy dust clouds which drift on into Manchuria, Korea, and out to sea, as originating in the Gobi Desert. The dust is winnowed from the desert sands by the dry winds which originate in the great, permanent high-pressure configuration (anticyclone) over the central Asian desert plateaus.

The Gobi region is dominated from fall to spring by the permanent winter "high." The winds are cold and dry, and they blow out of the Gobi to the southeast, carrying great quantities of desert dust, comparable to the winter and spring dust-blowing from the western Great Plains, especially in drought times. A further valid comparison may be made between the western Great Plains and the Gobi Desert. The Gobi, like other Asiatic high-desert plateaus, is ringed and sheltered by high mountainous regions at the west and southwest which intercept all moisture-bearing westerly winds, just as the Rocky Mountains often intercept moisture-bearing westerly winds from delivering precipitation to the central Great Plains. Meteorological conditions in Mongolia and northern China are comparable to the meteorological conditions in the western Great Plains during the dust-blowing seasons, especially in prolonged drought periods.

Also, there is every good reason to believe that the large and violent dust storms of the "Dust Bowl" period of the mid-1930's not only simulated the Dust Bowl and dust-blowing periods of the loess-forming ages in the Pleistocene but that they attained almost identical proportions and vigor. However, they did not continue for such long periods of time. Likewise, the "black blizzards," the violent dust storms of the mid-1930's, equaled in intensity, violence, and scope even the largest desert sand and dust storms in any modern desert.

[*Editor's Note:* Material has been omitted at this point.]

REFERENCES

Alway, F. J., 1916, The loess soils of the Nebraska portion of the Transition Region: Soil Science, v. 1, p. 210, 407–409.

Bates, C. G., 1935, Climatic characteristics of the Plains region: p. 83–110, section 11, *In* Possibilities of shelterbelt planting in the Plains region, U.S. Forest Service, p. 1–201.

Bollen, R. E., 1945, Characteristics and uses of loess in highway construction: Amer. Jour. Sci., v. 243, p. 283–293.

Buffington, J. W., 1961, Tertiary geology of the gangplank area, Colorado and Wyoming: Unpublished master's thesis, Univ. Nebraska.

Castellano, R. H., 1961, A study in clay mineralogy and the relationship of the clays to soils and texture in selected exposures of the Loveland and Peorian formations in eastern Nebraska and western Iowa: Unpublished doctoral dissertation, Univ. Nebraska.

Condra, G. E., 1908, The sand and gravel resources and industries of Nebraska: Nebraska State Geol. Survey, ser. 1, v. 3, pt. 1, p. 58.

——, Reed, E. C., and Gordon, E. D., 1947, Correlation of the Pleistocene deposits of Nebraska: Nebraska Geol. Survey Bull., no. 15, 73 p.

——, and Reed, E. C., 1950, Correlation of the Pleistocene deposits of Nebraska: Nebraska Geol. Survey Bull., no. 15A (a revision of Bull. 15), 74 p.

Cronin, J. G., and Newport, T. G., 1956, Ground water resources of the Ainsworth unit, Cherry and Brown counties, Nebraska: U.S. Geol. Survey, Water Sup. Paper, no. 1371, 120 p.

Fenneman, N. M., 1917, rep. 1921, Physiographic divisions of the United States: Annals Assn. Amer. Geogr., v. 6, p. 19–98, pl. I.

Frye, J. C., and Leonard, A. B., 1951, Stratigraphy of the late Pleistocene loesses of Kansas: Jour. Geol., v. 59, no. 4, p. 287–305.

——, 1952, Pleistocene geology of Kansas: State Geol. Surv. of Kansas Bull., no. 99, 230 p.

——, and Swineford, A., 1956, Stratigraphy of the Ogallala formation (Neogene) of northern Kansas: State Geol. Survey of Kansas Bull., no. 118, 92 p.

Johnson, W. D., 1900, 1902, The High Plains and their utilization: U.S. Geol. Survey, 21st Ann. Rept., pt. 4, p. 601–741; *ibid.*, 22nd Ann. Rept., pt. 4, p. 631–669.

Leighton, M. M., and Willman, H. B., 1950, Loess formations of the Mississippi valley: Jour. Geol., v. 58, no. 6, p. 599–623.

Lueninghoener, G. C., 1934, A lithologic study of some typical exposures of the Ogallala formation in western Nebraska: Unpublished master's thesis, Univ. Nebraska.

Lugn, A. L., 1935, The Pleistocene geology of Nebraska: Nebraska Geol. Survey Bull., no. 10, ser. 2, 223 p.

————, 1939a, Classification of the Tertiary system in Nebraska: Geol. Soc. Amer. Bull., v. 50, p. 1245–1276.

————, 1939b, Nebraska in relation to the problems of Pleistocene stratigraphy: Amer. Jour. Sci., v. 237, p. 851–884.

————, 1960, The origin and sources of loess in the Great Plains in North America: Internatl. Geol. Cong., Proc. 21st Sess., Norden, pt. 21, p. 223–235.

————, 1962, The origin and sources of loess: Univ. Nebraska Studies, new ser., no. 26, p. xi, 1–105.

————, **and Lugn, R. V.,** 1956, The general Tertiary geomorphology in Nebraska and the northern Great Plains: Compass, v. 33, no. 2, p. 98–114.

————, **and Schultz, C. B.,** 1934, The geology and mammalian fauna of the Pleistocene of Nebraska (part I by Lugn; part II by Schultz): Nebraska State Mus. (now Univ. Nebraska State Mus.), v. 1, Bull. 41, p. 319–393.

Maher, J. C., 1937, The calcium carbonate content of the Peorian loess of Nebraska: Unpublished master's thesis, Univ. Nebraska.

Robbins, H. W., 1941, The Pleistocene geology of Portales Valley, Roosevelt County, New Mexico, and certain adjacent areas: Unpublished master's thesis, Univ. Nebraska.

Russell, R. D., 1936, The mineral composition of atmospheric dust collected at Baton Rouge, Louisiana: Amer. Jour. Sci., v. 31, p. 50–66.

Schultz, C. B., and Tanner, L. G., 1957, Medial Pleistocene fossil vertebrate localities in Nebraska: Univ. Nebraska State Mus., v. 4, no. 4, p. 57–81.

Smith, G. D., 1942, Illinois loess—variations in its properties and distribution; a pedologic interpretation: Univ. Illinois, Agr. Exp. Stat. Bull., no. 490, p. 137–184.

Stone, R. G., 1934, Dust storm, April 10 and 11, at Sioux City, Iowa: Bull. Meteorological Soc., v. 15, p. 196–198.

Swineford, A., and Frye, J. C., 1945, A mechanical analysis of wind-blown dust compared with analyses of loess: Amer. Jour. Sci., v. 234, p. 249–255.

———, 1951, Petrography of the Peorian loess in Kansas: Jour. Geol., v. 59, no. 4, p. 306–322.

Thorpe, J., and Smith, H. T. U., 1952, Pleistocene eolian deposits of the United States, Alaska, and parts of Canada (ED Map): Geol. Soc. America.

Todd, J. E., 1899, The moraines of southeastern South Dakota and their attendant deposits (see Loess, p. 95–103): U.S. Geol. Survey Bull., no. 158, 171 p.

———, 1906, More light on the origin of the Missouri River loess: Iowa Acad. Sci. Proc., v. 3, p. 187–195.

Udden, J. A., 1898, The mechanical composition of wind deposits: Augustana College Lib. Pubs., Rock Island, Ill., no. 1, p. 1–69.

———, 1914, Mechanical composition of clastic sediments: Geol. Soc. Amer. Bull., v. 25, p. 655–744; see esp. p. 720–726, 728.

Warn, F. G., and Cox, W. H., 1951, A sediment study of dust storms in the vicinity of Lubbock, Texas: Amer. Jour. Sci., v. 249, p. 553–568.

Wehrman, K. C., 1961, A study of the transition zone between the Loess Hills and Sand Hills in central Nebraska: Unpublished master's thesis, Univ. Nebraska.

Wright, J. J., 1947, A textural and thickness study of the Peorian loess in Nebraska: Unpublished master's thesis, Univ. Nebraska, 30 p. and maps.

Editor's Comments
on Paper 44

44   KUKLA and KOČÍ
      End of the Last Interglacial in the Loess Record

In January 1972 conference on the present interglacial was held at Brown University, Rhode Island. The resulting papers were published in a special issue of the journal *Quaternary Research*. Several of these papers offered useful considerations of loess, in particular the paper by Kukla and Koči, reprinted here as Paper 44. Fairbridge (1972, p. 284) considered glacial cycles and observed that older river terrace deposits are found to be progressively more leached and otherwise modeified by pedogenic processes than the younger ones, and that such processes require repeated intervals of mild, humid climatic conditions. He further proposed that

> Studies of loess, notably in the American Midwest, but later also in [southeast] Europe, showed a similar alternation, a second demonstration of cyclicity. After much study and controversy, the loess itself became recognized as a glacigene eolian steepe deposit, a product of an extremely cold-dry climate; comparable dust storms today are experienced in some parts of the Gobi Desert. Loess is usually found to overlie alluvial gravel beds, such as we now identify with the heavy stream loading and which were associated with solifluction in the transition stages before and after a full glacial condition. Then the topmost layer in the loess is found always to have been converted into a red or red-brown paleosol, penetrated by worm tracks, rabbit burrows, and various "krotovina" (Russian name for mole holes). At Red Hill (Cerveny Kopec), near Brno, Czechoslovakia, there are ten complete cycles of loess, separated by interglacial horizons (Kukla, 1961).

The work of Kukla is mostly concerned with the stratigraphic aspects of loess, but various important material factors are involved and these are considered in Paper 44. Kukla and Koči are associated with the

Lamont-Doherty Geological Observatory and the Geological and Geophysical Institutes of the Czechoslovak Academy of Sciences.

REFERENCES

Fairbridge, R. W. 1972. Climatology of a glacial cycle. Quaternary Res. 2, 283–302.

Kukla, G. J. 1961. Quaternary sedimentation cycle. Survey of Czechoslovak Quaternary, Institut Geologiczny, Prace 34, 145–154.

44

Reprinted from *Quat. Res.,* 2(3), 374–383 (1972)

End of the Last Interglacial in the Loess Record [1]

GEORGE J. KUKLA [2] AND ALOIS KOČÍ [3]

Received July 15, 1972

In the loess series of Central Europe the last interglacial is recorded by the para-brownearth soil accompanied by land snails and pollen of thermophilous deciduous forests. The termination of forest environment is marked by restricted eolian sedimentation and development of chernozemic steppe soil, followed in turn by rapid deposition of calcareous air-borne silt in the so called "Marker" horizon. Hillwash loams and loess interrupted by the weak rendzina-type soil *B1f* were then formed. The sequence is overlain by the interstadial soil complex which is correlated with the Barbados II Terrace because of its stratigraphic position and warmth-loving snail fauna. Through most of the section bracketed by the interglacial and interstadial soils, the sediments display the reversed declination but positive inclination. The top of the reversed interval is here informally called the Brno magnetostratigraphic horizon. It correlates reasonably well with the upper boundary of the Blake event estimated to be 108,000 yr old.

The vertebrate and the snail faunas of chernozemic soil and of the loess together with the pedogenetic character of strata point to the harsh continental climate with the large temperature variation, dry seasons and partly, with torrential summer rains.

If the remarkably periodic deposition of loess series has to continue, following the pattern observed through the last 350,000 yr or so, then the shift to expressed continental climate in this part of Europe is to be expected soon.

INTRODUCTION

In the neighborhood of Prague and Brno, Czechoslovakia, the richly subdivided sequences of alternating windblown loess, hillwash loams and different soils cover the flanks of river valleys. The climatostratigraphic information is obtained from the changes in lithology and pedogenesis, from the land snail assemblages, pollen and verte-brate finds. Chronostratigraphic control is achieved by radiocarbon dates and paleomagnetic horizons (cf. Demek-Kukla, 1969; Kukla, 1970).

As this part of Central Europe was never glaciated, the semicontinuous depositional records reach deep into the past. The oldest soils at Červený Kopec in Brno are of Jaramillo age (about 1 million YBP) and the system could have been extended up to the top of Olduvai event (about 1.7 million YBP) in neighboring Krems, Austria.

The purpose of this article is to review shortly the basic facts about the last cycle in loesses and then to analyse in detail the interval which immediately followed the disappearance of interglacial deciduous forests from the area.

---

[1] Lamont Doherty Geological Observatory contribution No. 1837.

[2] Czechoslovak Academy of Sciences, Geological Institute, Trojanova 13, Praha 2, Czechoslovakia. At present: Lamont-Doherty Geological Observatory, Palisades, N.Y. 10964.

[3] Czechoslovak Academy of Sciences, Geophysical Institute, Bočni II, Praha–Spořilov, Czechoslovakia.

STRATIGRAPHY OF THE LAST GLACIAL CYCLE

In Fig. 3. the stratigraphic sequence of the last glacial cycle is plotted against the time scale. Similar sequence of strata, with few locally missing minor soils at some sites, was observed in more than 30 localities in the western half of Czechoslovakia. The principal locality where the radiometric determination and the depth plots in our diagram come from is Věstonice (location in Fig. 1, detailed description in Klíma *et al.,* 1962). The rich snail assemblages serve as sensitive indicators of vegetational and climatic changes throughout the shown interval. They are most completely studied in Kutná Hora (KH) and Věstonice (Ložek, 1964). The paleontologic independent check supports the paleoclimatic reconstructions obtained from soil micromorphology (Smolíková, 1967, 1968a) and lithology (Kukla, 1961a). Additional climatostratigraphic criterion is provided by palynology. Even that the probability of redeposition of pollen grains in this type of sediments is very high, the well-expressed abundance peak of ther-

mophilous broadleaf arboreal elements in the oldest soil of the glacial cycle demonstrates its synchroneity with the interglacial environments (Frenzel, 1964).

The sequence shown in Fig. 3 illustrates also the principles of the stratigraphic subdivision of the loess series.

The Cyclic Subdivision of the Loess Series

The striking feature of the loess series is their cyclic development. The first order "Glacial cycles" and second order "Stadial cycles" were recognized as early as 1960 and form the basis of the local lithostratigraphic subdivisions (Kukla, 1961). Only later it was found that eight glacial cycles were completed within the Brunhes paleomagnetic epoch (Bucha *et al.,* 1969), same as in the deep-sea sediments (Kukla, 1970).

The cyclic units form the frame of the lithostratigraphic subdivision of the loess series. The principal boundaries and units are recognized as follows:

Marklines: The principal lithostratigraphic boundaries, separating the individual glacial cycles are called "Marklines." They are defined as the boundaries between the

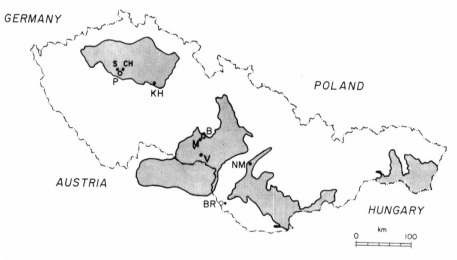

FIG. 1. Location of sites discussed in text. Full circle: localities Modřice (M), Sedlec (S), Chabry (CH), Kutná Hora (KH), Věstonice (V), Nové Mesto (NM). Open circle: towns Praha (P), Brno (B) and Bratislava (BR). Areas of calcareous loesses dotted.

thick layer of the "cold" type windblown loess and the hillwash loam occurring closely below the level of the deciduous forest soil (Kukla: in Demek and Kukla, 1969). This boundary could be recognized even when masked by the subsequent development of decalcified forest soil (cf. Fig. 2). Marklines are labeled ML and numbered consecutively backwards with roman numerals.

The last Markline (ML-I) in the described region is about 10,000 yr old. The dating was indirectly obtained by tracing the

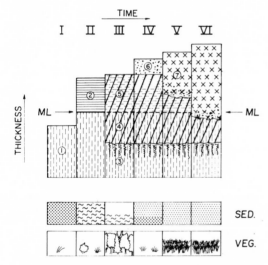

FIG. 2. Development of the soil complex in loess areas shown schematically in six subsequent stages. The major lithostratigraphic boundary, Markline, could be precisely fixed only in stages II–V. 1—Windblown loess, 2—hillwash, 3—Ca-horizon on the base of decalcified soil, 4—decalcified forest soil with preserved loess fabric, 5—ditto, with hillwash fabric. In upper part of that horizon the original structure could be destroyed by clay movement, 6—eolian dust with small stones (skellet of older soil), 7—biogenic soil high in humus. Stones moved to the base by the burrowing of worms and rodents. ML—Markline. In the upper horizontal band (SED.), the prevailing type of sedimentation is shown. By *full dots*: eolian loess; *fine dots*: eolian fine dust; *thick wavy line*: hillwash deposition; *fine wavy line*: slow hillwash deposition. In the lower horizontal band (VEG.), the inferred type of vegetation is schematically shown.

last marks of loess deposition in Late Paleolithic Magdalenian sites (cf. Valoch, 1968). There is no doubt that the Marklines are the expression of the same rapid environmental change at the end of a glacial, which produced the "Terminations" in the deepsea sediments (Broecker and Van Donk, 1970) and that the Marklines and Terminations in a broad sense timely correlate.

Sub-marklines: These are defined as boundaries between the windblown loess of any kind and the hillwash loam or soil of at least interstadial character (including brown earth and chernosem).

Glacial cycle: This is the unit delimited defined by *Marklines*. Starts with hillwash sediments and soils related to the interglacial climate (similar to present conditions) and ends with the potent loess member, which contains arctic-type frost–gley soils and cold-resistant snails of *Collumella collumella* and *Pupilla muscorum* group.

Glacial cycles are labeled from the present backwards with the capital letters A to R. Present glacial cycle (first of the Holocene age) is labeled A, the last one completed is labeled B. In the intercorrelations with the systems from outside the loess region, the prefix L- should be used to distinguish the loess units from similarly labeled other strata (e.g., L-B for the last glacial cycle).

Stadial cycle (also subcycle): It is the unit delimited by *Sub-marklines*. It starts with hillwash sediments and soils related to the interstadial climate and ends with the deposition of the windblown loess related to either cold or mild climate of continental type.

The stadial cycles are labeled by majuscules referring to the corresponding glacial cycle and by number. Originally (Kukla, 1961) only three stadial cycles, namely $B1, B2, B3$, were distinguished within the last glacial cycle B. However, it is now possible and useful to subdivide further the unit $B2$ into three units $B21, B22$ and $B23$.

The recognized strata: On the localities described in detail (e.g., Klima *et al.,* 1962; Kukla *et al.,* 1962) all recognized layers were designated by small letters.

More frequently only soils are separately recognized. This is done by a symbol consisting of the designation of the stadial cycle with the suffixed small latin letters.

As miscorrelations within the loess area are locally possible, the complete reference to any given soil should include a prefix which refers to the corresponding locality (cf. Kukla: in Demek and Kukla, 1969).

The Chronostratigraphic Correlations

The sedimentation in most of the selected sites seems to be essentially continuous, but the sedimentation rate varies greatly. Thus 1 cm of loess may have been deposited in about 20 yr and the pellet sands still faster, but in the forest soils accretion of 1 cm have taken 500 yr and more.

The chronostratigraphic interpretation is therefore based on the ^{14}C-dated section reaching to 30,000 YBP and the correlation of three strongly developed soil groups *B1b-c; B2a-b* and *B2e-g* with the Barbados Terraces I–III (cf. Kukla, 1970). It must be admitted, that the interpretation of the section between 30,000 and 80,000 YBP is at present highly tentative, if not hypothetical.

The general sawtoothlike pattern of ^{18}O cycles in deep-sea sediments (Broecker and Van Donk, 1970) which could be also clearly recognized in the faunal and soil development of the loess glacial cycle, as well as existing eight cyclic units within the Brunhes epoch, are strong supports of the interpretation present in Fig. 3.

Paleoclimatic Record of the Last Glacial Cycle

The paleoclimatic criteria are paleopedological, lithological and paleontological.

(1) The *B1b* soil is developed as a typical parabrownearth (lessivé). It has

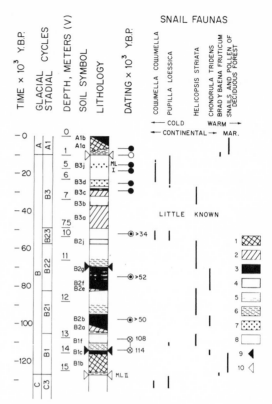

Fig. 3. General stratigraphy of the last glacial cycle in Czechoslovakia, plotted against the time scale (Kukla 1969, 1970). Diagram is based on the Věstonice (V) section (Klíma *et al.,* 1962). Snail faunas basically from Kutná Hora (KH) and after Ložek (1964). Few minor layers, not known in Věstonice, are shown in their corresponding position observed on other localities (principally Chabry (CH), Mšené (MS) and Nové Mesto (NM).

Designation of glacial cycles after Kukla (1961), subdivision of Stadial cycles after this paper. Soil symbols after Kukla (1970).

Chronological control: *Full circle*—deposits ^{14}C dated on charcoal and soil humus. *Open circle* —indirectly correlated ^{14}C date. *Open circle with dot*—infinite ^{14}C date. *Crossed circle*—paleomagnetic correlation with Blake event (Smith-Foster, 1969). Symbols in lithological column: 1—soil of deciduous forests with clay skins (parabrownearth), 2—decalcified brownearth, 3—intensive chernozem (steppe soil), 4—poor in humus biogenic steppe soil, 5—hillwashed loams, 6—pellet sands (cf. text), 7—frost-gley soils, 8—loess, 9—marker, 10—Marklines I and II.

a thick brown to reddish-brown *B*-horizon of columnar structure with frequent clay skins. The soil of this type is forming today in loess areas only below deciduous forests. The rich snail assemblage of *Helicigona banatica* characteristic for the deciduous forests and the deciduous trees pollen, accompany the *B1b* soil, supporting the conclusion on its interglacial age.

(2) The decalcified brownearths under conifers do not show clay motion. The coniferous forests may have been related to the formation of soil *B3a*.

(3) The chernozemic steppe soils are intensively mixed and reworked by worms. They form today under the continental climate, which is relatively warm but dry (cf. Southern Russia, etc.). Similar conditions probably existed during the formation of *B1c, B2b* and *B2g,* perhaps even shortly at the time of formation of *B2j.* Their characteristic gastropod is *Chondrula tridens.*

(4) The greenish-grey soils *B3d-B3j* were interpreted as frost–gley soils, indicative for the existence of deep frozen ground.

(5) Loess is widely recognized as the windblown sediment of mostly local or regional origin, which forms in climates with long dry seasons in areas very sparsely vegetated by patches of grass.

Gastropod assemblages are principally of two types. *Pupilla* and *Collumella* faunas point to cold rough continental conditions ("cold loess"), whereas *Helicella striata* assemblages indicate continental but not extremely cold conditions ("warm loess," cf. Ložek, 1964).

(6) *Pellet sands* are rapidly sedimented psammites built of redeposited fragments of soil material. As has been demonstrated by recent observations (Kukla-Ložek, 1961), the pellet sands today are laid down during torrential rains coming after a period of relatively hot dry weather. The redeposition is so fast, that the hardened particles cannot

soften during the transport. The large areas stripped of vegetation must be present. blown dust, mostly calcareous, separating the

(7) *Marker* is the thin bed of wind-decalcified humous soils from pellet sands Because of its fine granulation, sharp bound aries, calcareous content and stratigraphi position, it is interpreted as the far trans ported deposit of dust storms (Kukla 1961a).

THE STRATIGRAPHY OF SUBCYCLE *B1*

The typical sequence of soils and sediments in the subcycle *B1* is demonstrated in Fig. 4 showing the section in Modřice.

Location

The section is located in the abandoned front of the excavation above the second extraction floor in the NE segment of the active brickyard Modřice, some 200 m west of the main road Brno–Bratislava, 5 km south of town Brno. The scale shown in Fig. 4 starts at the second extraction floor (at about 200 m above the sea level).

The lithologic description of the strata is given in an appendix to Fig. 4. In Modřice is one of the well-developed sequences. Practically all important strata shown in Fig. 3 with exception of soil *B2e* and the upper Marker are here present (cf. with neighboring section Modřice II described in Kukla and Ložek, 1961).

Paleomagnetic Investigation

The section was sampled for paleomagnetic investigation in 5-cm intervals (samples 1–40). Closer sampling was performed on the interglacial soil *B1b* and in the underlying loess (samples 50–90). These latter samples were taken about 100 m toward the west for technical reasons. The continuity of soil *B1b* between both sites was visible in the front of the excavation.

The samples of loess were found to have

the NRM (natural remanent magnetization) close to about $I_r = 28.10^{-7}$ e.m.u. samples of soil *B1b* showed $I_r = 20.10^{-6}$ e.m.u.

The a/f demagnetization was performed by the rotation of samples around three perpendicular axes in the field of 200 Oe for 30 sec. About two-thirds of NRM was removed. The demagnetization curves for samples 13 and 17 are shown in Figs. 5 and 6, corresponding plots of *D* and *I* in Fig. 7.

Samples were measured on the spinner magnetometer with a sensitivity of 1.10^{-8} e.m.u.

The results plotted in Fig. 4 show the change of declination by about 160° completed between samples 13 and 17. Reversed declination continues downward with one interruption almost to the base of the measured interval. No substantial and consistent change in inclination was observed.

No special investigation of mineralogic composition of sampled sediments was performed.

The results are interpreted as demonstrating the local fossil record of the disturbance of Earth's magnetic field, described in deep-sea sediments as the *Blake event* (Smith and Foster, 1969). This correlation is based on the general stratigraphic position of the reversed interval in the relative close of Termination II alias Markline II, and the estimated duration of reversed interval, both in the deep-sea sediments and in the loess series on the order of about 5000 yr. We did plot in Fig. 4 Smith's and Foster's age estimates for the upper and lower boundary of the Blake event. The timing is only tentative because of two reasons: (1) the anomalous age claimed by Smith and Foster for the top of the *-x*-zone in their cores and (2) for the geographically restricted recognition of this reversed interval in the deep-sea floor.

We must, however, admit that the timing fits very well into the stratigraphic frame based on independent correlation of *B1b* soil with Barbados III Terrace and *B2a* soil with Barbados II Terrace.

Brno Magnetostratigraphic Horizon

For correlation purposes, the level of deposition of sample 15 (Fig. 4), halfway at the transition from negative to positive declination, is called here *Brno magnetohorizon*. It occurs within the period of dry, relatively cold continental climate between the end of the last interglacial and the start of the next major interstadial. Its correlation with the upper boundary of the Blake event is highly probable.

Paleoclimatic Implications of the Subcycle B1

The following conclusions could be drawn from the Modřice section:

(1) Major change in the local environments occurred at the level where *B1c* steppe soil started to develop over the *B1b* forest soil. Calcareous dust must have been deposited over the top of the forest soil at that time (compare horizon 6, Fig 4). Gleying and mechanical disturbance of the forest soil, previous to development of the *B1c* steppe soil was revealed by micromorphological studies of Smolíková (1968a, 1968b).

Retreat of the forest could have been time transgressive, however in most of the places where Marker is found, the development of chernozem *B1c* points to its similar duration on the order of 1–2 millenia. This estimate is based on its visual comparison with Holocene chernozemic soils, archeologically dated.

At least in Modřice, the change from forest to steppe occurred before the start of the paleomagnetic event correlated with Blake.

If the correlation and the estimated timing of the Blake event are correct, then the shift in Modřice occurred before 115,000

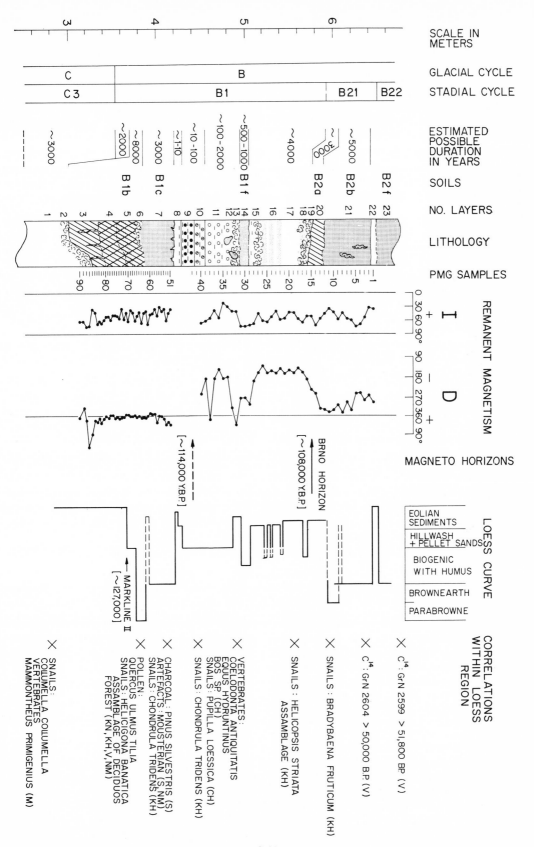

349

YBP, probably between 115,000 and 116,000 y. a. (cf. next point).

(2) The deposition of Marker and the start of badland development under the largely restricted vegetational cover, as indi-cated by pellet sands, occurred on the order of 1–2 millenia later than the disappearance of forest, but still at the time of normal polarity (i.e., before approx. 114,000 YBP). This conclusion is based on the observation

Fig. 4. Section in the Modřice brickyard showing the paleomagnetic stratigraphy of the soils and sediments of the last interglacial and of the next cold interval.

The estimated possible duration of individual layers is based mostly on the comparisons with Holocene soils archeologically dated, on sedimentation speeds of the different loess types as found in radiometrically dated Věstonice section, and on the recent observations of the hillwash deposition (Kukla-Ložek, 1961). Soil symbols same as in Fig. 3 (cf. text). Lithology of layers 1–22 described below. The tentative timing of the paleomagnetic event is based on the correlation with the Signs + and − stand for normal polarity. Blake event of Smith and Foster (1969).

In the loess curve the individual layers are ordered from left to right after their supposed occurrence in the transition from the continental to the maritime climates (more detail in Kukla, 1970). Lithologic description of the strata:

1—Yellowish-brown calcareous loess with dense impregnation of secondary carbonate along the root holes.

2—Whitish-yellow Ca-horizon of the overlying decalcified soil

3—Rusty brown decalcified base of parabrownearth with preserved fabric of autochthonous loess. Rare vertical fissures coated with braunlehmplasma (clay).

4—Reddish-brown B-horizon of well-lessivated parabrownearth with columnar structure and clay coatings. Original loess fabric preserved inside the columns.

5—Reddish-brown B-horizon of parabrownearth, aas in layer 4, but with the horizontal streaks of coarse sand. Probably weathered zone of hillwash sediments.

6—Reddish-brown loam, structureless, calcareous spots and small corroded concretions of secondary carbonate. Ca-horizon of overlying chernozem developed at the top of parabrownearth.

7—Dark brown-blackish, high in humus chernozem, secondarily decalcified.

8—Light grey-brown, in places primarily calcareous fine grained silt without the typical grass-root fabric of loess, locally with platty carbonate concretions. Marker. Sharp lower boundary, signs of redeposition by running water on slopes toward the upper boundary.

9—Dark brown pellet sands. Grain size around 1 mm. Interlayers of light brown eolian dust, possibly redeposited, similar to marker, at the base. Slightly calcareous.

10—Reddish-brown pellet sands, mostly of reworked B-horizon corresponding to layers 4 and 5. Calcareous.

11—Light brown pellet sands and subhorizontally banded loams, calcareous.

12—Discontinuous horizon of large carbonate concretions in the highly calcareous loam, mostly with the hillwash structure.

13—Light grey-yellowish loess, impregnated by secondary carbonate, numerous dark brown krotowinas penetrating downward.

14—Brown humous biogenic soil, intensively reworked by worms, moderate content of humus.

15—Light grey-yellowish loess with horizon of large carbonate concretions; some are platty. Streaks of light pellet sands.

16—Light grey-yellowish loess as above, but at least three interlayers of dark grey color, higher in humus. Possibly initial soils.

17—Light grey loess, few interlayers of pellet sands.

18—Dark grey, probably moderately high in humus, initial soil or hillwash interlayer. Intensively Ca-impregnated horizon at the base suggests dependence on soil formation.

19—Light whitish-grey Ca-horizon of the above lying soils.

20—Rusty brown, slightly recalcified B-horizon of brownearth. No signs of the clay redeposition. In the upper part spots of humus soil from above: dotty texture originated by burrowing worms.

21—Dark brown decalcified chernozem with subvertical corroded carbonate concretions.

22—Light grey, calcareous fine grained loam, probably of eolian origin. Transitional upper boundary.

382 KUKLA AND KOČÍ

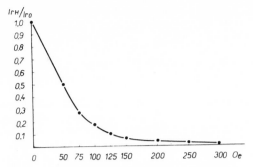

FIG. 5. Demagnetization curve of the sample 13.

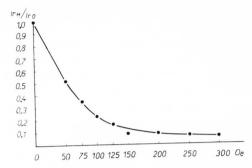

FIG. 6. Demagnetization curve of the sample 17.

that the biologically mixed soils such as chernozems tend to indicate the polarity of the time when their development stopped.

(3) The maximal deterioration of climate probably occurred before the forma-

tion of *B1f* soil, as indicated by the presence of *Pupilla loessica* and *Coelodonta antiquitatis* (wooly rhinoceros) at this level in Chabry (cf. Fig. 1 for location). The correlation of *B1f* soil with the Roedeback

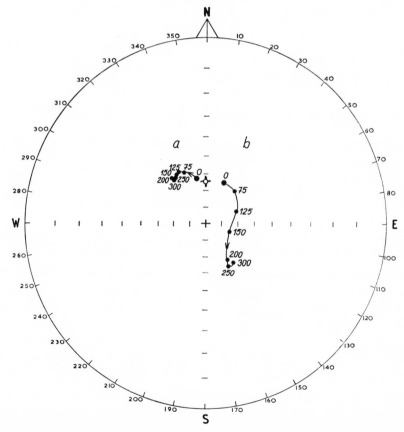

FIG. 7. Declination and inclination plots in the individual stages of demagnetization of samples 13 (=a) and 17 (=b) Recent *D* and *I* on the site marked by open circle with cross.

interstadial is highly probable (Demek and Kukla, 1969).

At present, on the sites undisturbed by man the mature Holocene parabrownearth still continues in development.

If the periodic pattern of the loess series has to continue, then the natural shift toward the greater continentality is the first event to come. If furthermore the Holocene warm interval will not last longer than the last interglacial, then the expected shift is due very soon.

ACKNOWLEDGMENT

This report was made possible by NSF Grant GX-28671-K. Careful reading of the article by N. D. Opdyke and C. D. Ninkovitch is appreciated.

REFERENCES

BROECKER, W. S. and VAN DONK, J. (1970). Insolation changes, ice volumes and the O^{18} record in deep-sea cores. *Reviews of Geophysics and Space Physics* **8**, 169–198.

BUCHA, V., HORÁČEK, J., KOČÍ, A., KUKLA, J. (1969). Paläomagnetische Messungen in Lössen: Periglazialzone, Löss und Paläolithikum der Tschechoslowakei." (J. Demek, and J. Kukla, Eds.), pp. 123–131. INQUA VIIIth Congress, Brno.

DEMEK, J. and KUKLA, J. (1969). Periglazialzone, Löss und Paläolithikum der Tschechoslowakei. INQUA VIIIth Congress, Brno.

FRENZEL, B. 1964. Zur Pollenanalyse von Lössen. *Eiszeitalter und Gegenwart* **15**, 5–39, Ohringen/Württ.

KLÍMA, B., KUKLA, J., LOŽEK, V. and DE VRIES, H. (1962). Stratigraphie des Pleistozäns und Alter des paläolithischen Rastplatzes in der Ziegelei von Dolní Věstonice. *Anthropozoikum*, **11**, 93–145, Praha.

KUKLA, J. (1961). Lithologische Leihorizonte der tschechoslowakischen Lössprofile. *Věstník Ustředního Ústavu Geologického*, **36**, 359–372, Praha.

KUKLA, J. (1961a). Quaternary sedimentation cycle. Survey of Czechoslovak Quaternary. Czwartorzed Europy Srodkowej i Wschodniej. *INQUA VIth Congress, Institut Geologiczny, Prace*, **34**, 145–154, Warszawa.

KUKLA, J. (1970). Correlations between loesses and deep-sea sediments. *Geologiska Föreningen i Stockholm Förhandlingar* **92**, 148–180.

KUKLA, J. and LOŽEK, V. (1961). Loess and related deposits. Survey of Czechoslovak Quaternary, Czwartorzed Europy Srodkowej i Wschodniej, *INQUA VIth Congress, Institut Geologiczny, Prace*, **34**, 11–28, Warszawa.

KUKLA, J., LOŽEK, V. and BÁRTA, J. (1962). Das Lössprofil von Nové Mesto im Waagtal. *Eiszeitalter und Gegenwart* **12**, 73–91.

LOŽEK, V. (1964). Quartärmollusken der Tschechoslowakei. *Rozpravy Ústředního Ústavu Geologického*, **31**, pp. 374.

SMITH, J. D. and FOSTER, J. H. (1969). Geomagnetic reversal in Brunhes normal polarity epoch. *Science* **163**, 565–567.

SMOLÍKOVÁ, L. (1967). Zur Mikromorphologie der jungpleistozänen Böden von Sedlec bei Praha. *Časopis pro Mineralogii a Geologii* **12**, 277–286.

SMOLÍKOVÁ, L. (1968a). Genese mladopleistocénních půd v Modřicích u Brna na základě půdní morfologie. *Časopis pro Mineralogii a Geologii* **13**, 199–209, Praha.

SMOLÍKOVÁ, L. (1968b). Mikromorphologie und Mikromorphometrie der pleistozänen Bodenkomplexe. *Rozpravy ČSAV* **78**, 2, 3–47, Praha.

VALOCH, K. (1968). Evolution of the Paleolithic in Central and Eastern Europe. *Current Anthropology* **9**, 351–368, Chicago.

Editor's Comments
on Papers 45 and 46

The basic idea expressed in these two papers is that it is possible to recognize a sequence of events involved in the formation of a deposit and that each of these events may be significant, worthy of study, and possibly may influence the nature of the eventual sediment. A secondary suggestion is that an unprejudiced look at the totally unreconcilable eolian and *in situ* theories of loess deposit formation may reveal possible compromises. It is possible to compromise in science; after all, the electron can be a particle or a wave, depending on one's point of view (or particular discipline), and, in some senses, it may be possible for both loess theories to be right. If Gerasimov's offering (Paper 33) is accepted and it is recognized that *in situ* theorists will allow some eolian deposition, we are left with a reasonable sequence of events: formation of primary mineral particles, eolian transportation, and then postdepositional activity that (in the philosophy of the *in situ* theorists) transforms the material into loess. If geologist A says that the fresh eolian deposit is loess and soil scientist B says that the same deposit is only loess after a few mineralogical changes brought about by percolating groundwater, their differences are not too great—back to the old problem of definitions.

Paper 46 takes the story on past the immediately postdepositional changes (loessification?) and considers the movement of loess material by fluvial agencies. It should be accepted that large deposits of primary loess must be in the drainage basin of a large river, so some range of secondary deposits should exist. An obvious example is the North China plain i.e., material from the inland loess deposit emplaced by the Yellow River. But what about the Danube deposits? Might Kolbl (1930) have been right in suggesting that these were fluvial deposits? Berg listed Kolbl as one of his supporters, but the material at Krems may be Alpine loess placed there by the Danube after a normal eolian deposition nearer to the source of the quartz particles. To judge by the geographical position of the Krems loess, it certainly seems likely to be composed of material produced by distant glaciers, with some fluvial transportation involved somewhere in the extended deposition cycle.

REFERENCES

Kolbl, L. 1930. Studien über den Löss (Umgebung von Krems). Mitt. Geol. Ges. Wien 23, 85–120.

45

Reprinted from *Earth-Sci. Rev.,* 7(2), 67–68, 81–85 (1971)

"IN-SITU" THEORIES OF LOESS FORMATION AND THE SIGNIFICANCE OF THE CALCIUM-CARBONATE CONTENT OF LOESS

I. J. SMALLEY

Department of Civil Engineering, University of Leeds, Leeds (Great Britain)

ABSTRACT

SMALLEY, I. J., 1971. "In-situ" theories of loess formation and the significance of the calcium-carbonate content of loess. *Earth-Sci. Rev.,* 7(2): 67–85.

It is generally accepted that loess deposits form by eolian action and yet the concept of "loessification" is still encountered, particularly in some European literature. Loessification often seems to be invoked when the deposit being investigated is rich in calcium carbonate because the use of the concept depends upon an eclectic definition of loess. Loessification is irrelevant if loess is defined in terms of the fine quartz material, but may be meaningful if the loess is defined in terms of calcium carbonate. The Russell theory of in-situ formation can possibly be reconciled with deposition by eolian action and production of quartz material by glacial grinding but the Berg theory cannot, and appears to be based on false premises. The "soil science" approach to the problem of loess formation may be responsible for the overvaluing of the significance of the carbonate content; the "geological" approach, with more emphasis on transportation problems, leads logically to the concept of eolian deposition but has, so far, failed to focus attention on the problem of the formation of the actual loess material. The fine quartz material is mostly formed by glacial grinding; there appear to be no other natural forces powerful enough to produce appreciable quantities of loess-sized quartz particles. Deposition of the secondary carbonate is usually the last of the significant stages in the formation of the deposit; scanning-electron-microscope studies suggest that it exists as discrete encrustations (gnarls) rather than as continuous coatings on the quartz grains.

INTRODUCTION

There have been many theories of loess-deposit formation and in the early days of loess investigation there was considerable disputation by the protagonists of rival theories. However, when the dust had finally settled, it appeared that the theory of eolian deposition held the field. Flint remarked in 1941 that three decades had elapsed since the last serious opposition had been voiced (FLINT, 1941, p.27). But one other theory has shown some signs of life; this is the "in-situ" formation theory, in which loessification takes the place of eolian deposition. A version of this was proposed by Berg in about 1915 in Russia and reiterated and developed by him in later years; but only recently, with the publication of an English translation of his major works (BERG, 1964), has it become more widely known. Another major supporter of in-situ formation was Russell, who developed his theory during years spent with the Mississippi loess and produced a very influential and much-cited paper (RUSSELL, 1944a) which was the focus of discussion during the late nineteen-forties. In 1949 Zeuner published his theory of loess formation which allowed in-situ development.

The translation of Berg's work was reviewed in *Geographical Journal* by THOMASSON (1965) who gave sympathetic treatment to the non-eolian theories advanced. This modest support was reflected by the treatment of loess by Ollier in his book on weathering (OLLIER, 1969, p.78). KELLER (1970), commenting on the Ollier section on loess, wrote of "Professor Ollier's eclectic review of the origin of loess in which he gives considerable attention to non-eolian processess". That a well-informed earth scientist could, in 1969, give equal weight to in-situ and eolian theories of loess-deposit formation suggests that the in-situ formation idea is not so dead as Flint thought.

The RUSSELL (1944a) paper forms the core of this study of the in-situ formation concept for several reasons; despite the late appearance of the English translation, the Berg theory was fully developed by 1932 so the Russell theory is, in fact, the latest full-scale in-situ loess formation theory. ZEUNER's (1949) paper is important, but contains more of a suggestion than a theory. It is interesting to note that Ollier presented the Berg hypothesis as a modern theory, there is no hint in his text that he is dealing with a fifty-year-old idea. There is a useful body of criticism of the Russell theory. Soon after it appeared, a critical note by HOLMES (1944) was published, to which RUSSELL replied (1944b). Subsequent papers by WASCHER et al. (1948), DOEGLAS (1949), LEIGHTON and WILLMAN (1950) and FISK (1951) also contain discussion, and more recently the Mississippi Valley loess has been considered in detail by KRINITZSKY and TURNBULL (1967) who also briefly discuss the Russell paper. And although the offerings of Berg are many many times more voluminous than that of Russell, they suffer from a certain lack of clarity and a very definite insularity—if that term can be applied to anything Russian! Berg is really writing about Russian loess, and while it is true that Russell is essentially writing about Mississippi loess his vision does encompass certain European deposits.

The widespread acceptance of the eolian theory of loess-deposit formation suggests that Russell was in error; Krinitzsky and Turnbull, considering the same deposit as Russell did, wrote that "the eolian hypothesis provides the only generally suitable explanation for the origin of loess". It certainly provides an explanation for the occurrence of large, essentially unstratified deposits of fine quartz particles, but it does not refer directly to the calcium-carbonate content of the loess and it is evident that supporters of in-situ theories tend to regard the carbonate as being of great significance. The nature and influence of the carbonate in loess will be considered in this paper, as well as the best-documented and most convincingly argued of the in-situ theories, that of RUSSELL (1944a and b).

[*Editor's Note:* Material has been omitted at this point.]

MORE DISCUSSION AND SOME CONCLUSIONS

The Berg hypothesis is untenable. It is untenable for one particular reason: it cannot account for the fact that loess consists largely of quartz particles with diameters of the order of 20–50 microns. One feels that the failure was largely due to the influence of the ideas of soil science which overruled the more applicable ideas of geology. A geologist looking at a sediment, particularly a clastic sediment, thinks in terms of provenance, and transportation, and deposition. The soil scientist thinks more in terms of "soil formation" in which the concept of transportation is less significant. Soil science was virtually invented in Russia by Dokuchaiev (see OLLIER, 1969, p.137) and the influence of soil science has been very great in much Russian loess investigation. It is obviously no coincidence that the first example of the revolutionary concept of horizonation offered by Dokuchaiev was the development of a chernozem soil formed on a loess deposit. The soil-science influence was (and is) very persistent (cf. GELLÈRT, 1962; GERASIMOV, 1964) and, in studies of eastern-European material and by eastern-European scientists, has led to the overvaluing of the carbonate fraction and an undervaluing of the critical quartz material. It may seem odd to a geologist that the major part of a deposit could be virtually ignored, but it can happen; there is in fact a directly applicable geological parallel in that for decades geologists discussed the transportation aspects of the loess problem and ignored (with a few very notable exceptions like Keilhack) the important problem of how the quartz material was formed. The in-situ theorists' approach simply led to another important stage being virtually ignored.

In the same way that some of Russell's observation can be applied to the very last stage in the formation process of a calcareous loess, so Berg's work may have some relevance in a somewhat similar context. But some of the relevance which has been found for Berg's work is illusory. To quote Thomasson's review:

"...Berg voices a number of shrewd criticisms of the popular aeolian theory. It is difficult to find a contemporary wind deposit which closely resembles loess in texture. The theory that loess is deflated from dry or unvegetated surfaces, and 'trapped' in moister or warmer areas by a vegetation cover may be difficult to sustain in continental regions where vegetation implies a soil cover with appreciable organic matter." (THOMASSON, 1965.)

These criticisms are not justified; HOBBS (1931) has described a contemporary wind deposit associated with the Greenland glacier which closely resembles loess in texture, in fact he called it loess; and CEGŁA (1969) has shown that the old idea of vegetation being needed to "trap" loess is erroneous, since ground moisture is the critical factor.

Thomasson concluded his review of Berg's work with these words:

"It follows that either some peculiarly effective sorting mechanism is at work or that all loess has some common source. We still await a global exposition of the origin of loess." (THOMASSON, 1965, p.415.)

This is what Keilhack said in 1920; was there really no advance in loess knowledge between 1920 and 1965? In fact, of course, many papers were published on most aspects of loess but with a distinctly local emphasis. The global view, as Thomasson remarked, was rather lacking, and yet enough data exists for a global view to be taken and some generally applicable principles to be suggested; the assumption being made that the quartz vs. carbonate controversy has been resolved and the loess definition used acknowledges the predominance of the quartz.

Most of the fine quartz particles appear to be produced by glacial grinding. The mechanism of glacial grinding is such that a distinctly bimodal system of quartz particles is produced (see SMALLEY and PERRY, 1969, p.35 for discussion). A mixed deposit is formed when the glacier melts and the wind sorts out the fine material and carries it away in suspension. The wind direction is largely influenced by the glacial anticyclone as described by HOBBS (1943a, b). The deposition of the particles depends largely on the ground moisture (CEGŁA, 1969). The initial deposit is stable because of the high cohesive forces between the small particles (SMALLEY, 1970b) but may be eroded by impacting sand grains or fluvial action. Where there is the interaction of a large river and a loess deposit, a lot of the loess material is carried away and may be redeposited, as in the case of the North-China plain or the floodplains of the lower Mississippi Valley. Carbonate may be lost, or gained, or lost and gained. Detrital carbonate may be dissolved and redeposited. The carbonate is definitely a transient material and the carbonate content of a loess deposit may vary widely over a relatively small distance. Because quite a high carbonate content can be generated (up to perhaps 40%), this tends to assume great significance in the areas where it forms.

The major deposits appear to have reasonably well-established glacial connections. Wide-ranging field work by Chinese geologists during the early nineteen-sixties (see SMALLEY, 1968, for review) has shown that there was widespread glacial cover and multiple glaciations in China during the Quaternary. The glaciers, tending in general to spread from the west across southern China, could provide the mechanism for the production of the Chinese loess material. The position of the inland primary deposit appears to bear the same relationship to the northern fringe of the continental ice sheets as the major North-American loess deposit does to the southern fringe of the North-American glaciers. A glacial connection is apparent for the European, North-American and New Zealand loess deposits, but the South-American material presents problems. It could be argued that Andean glaciation provided the requisite grinding mechanism for particle production but it appears (TERUGGI, 1957) that much of the material is volcanic, although CHARLESWORTH (1957, p.521) claimed that periglacial loess underlies the South-American pampas. CASTELLANOS (1962) also stated that much of the pampean material is typical loess and he obviously considered that it was closely related to loess material in other parts of the world. He considered the Russell hypothesis at some length but he appears to have done it via HOLMES' (1944) criticism rather than with reference to the original RUSSELL (1944a) paper.

The tiny deposit of English loess at Pegwell Bay in Kent was thought by PITCHER et al. (1954) to be composed of material produced locally by frost weathering but FOOKES and BEST (1969) have recently suggested that it is in fact a normal eolian loess composed of glacially produced material. Loess is found in north Italy (OROMBELLI, 1970) and this material could presumably be largely due to the Alpine glaciers. There are many minor deposits to which the name loess has been applied, some of which do not even fall within the quartz-based definition of loess, but once the name has been applied it tends to stick. For example, the material in Tripolitania described by RATHJENS (1928) as loess was correctly identified as fine sand by PENCK (1931) but it is still being referred to as loess (SMITH, 1968).

The basic outlines of the processes leading to the formation of the world's major loess deposits are thus discernable and some stages of the processes have been identified; what we have to do now is to establish a detailed history of each deposit from the formation of the quartz material to the last stage of particle deposition or carbonate formation. It is still also necessary finally to establish at what stages of the Quaternary the successive loess deposits were formed. It seems very unlikely that they were formed at glacial maxima, as BÜDEL (1953) and WOLDSTEDT (1967) have suggested. The glacial-grinding mechanism points to an identification of loess-deposit formation with the time of glacial retreat, as VISHER (1922) proposed. At the time of a glacial maximum all the rock debris, apart from some in the end moraine, will be covered by the ice sheet and trapped. When the ice sheet disappears the bulk of the rock detritus is released and the fine material is eventually lifted by the wind and carried away to become a loess deposit.

In simple terms: loess is glacial, loess is eolian, loess is quartz *and* calcium carbonate, loess is essentially "Gesteinsmehl" and not "Verwitterungsrückstand" (PENCK, 1931). "Das ist der Löss."

ACKNOWLEDGEMENTS

I thank Dr. A. Warren of University College, London, Dr. F. H. Wittmann of the Technische Hochschule, Munich, Dr. Jerzy Cegła of the University of Wroclaw and Mr. J. G. Cabrera of Leeds University for their generous assistance. I also thank Dr. J. Sikorski and Mr. T. Buckley of the Department of Textile

Industries, University of Leeds, for making the electron-microscope investigations possible. Financial support from the Arthur Haydock bequest of the British Association for the Advancement of Science and the overseas travel fund of the University of Leeds is gratefully acknowledged. And, last but definitely not least, for comments, criticisms, co-operation and various sorts of help I thank Dr. David Krinsley of Queens College, City University of New York and Dr. C. Vita-Finzi of University College, London.

REFERENCES

BERG, L. S., 1932. The origin of loess. *Beitr. Geophys.*, 35: 130–150.
BERG, L. S., 1964. *Loess as a Product of Weathering and Soil Formation.* I.P.S.T., Jerusalem, 205 pp.
BRYAN, K., 1945. Glacial versus desert origin of loess. *Am. J. Sci.*, 243: 245–248.
BÜDEL, J., 1953. Die "Periglazial"-morphologischen Wirkungen des Eiszeitklimas auf der ganzen Erde (Beiträge zur Geomorphologie der Klimazonen und Vorzeitklimate IX). *Erdkunde*, 7: 249–266. (English translation: *Intern. Geol. Rev.*, 1: 1–16.)
CASTELLANOS, A., 1962. El Holoceno en la Argentina. *Univ. Nac. Litoral, Inst. Fisiog. Geol., Publ.* 45, 78 pp.
CEGŁA, J., 1969. Influence of capillary ground moisture on eolian accumulation of loess. *Bull. Acad. Polon. Sci., Sér. Sci. Géol. Géograph.*, 17: 25–27.
CHARLESWORTH, J. K., 1957. *The Quaternary Era.* Arnold, London, 1700 pp.
DOEGLAS, D. J., 1949. Loess, an eolian product, *J. Sediment. Petrol.*, 19: 112–117.
FISK, H. N., 1951. Loess and Quaternary geology of the Lower Mississippi Valley. *J. Geol.*, 59: 333–356.
FLINT, R. F., 1941. Glacial geology. In: *Geology, 1888–1938. Geol. Soc. Am., 50th Anniv. Vol.*, pp.19–41.
FOOKES, P. G. and BEST, R., 1969. Consolidation characteristics of some Late Pleistocene periglacial metastable soils of east Kent. *Quart. J. Eng. Geol.*, 2: 103–128.
GAGE, M., 1965. Some characteristics of Pleistocene cold climates in New Zealand. *Trans. Roy. Soc. New Zealand (Geol.)*, 3: 11–21.
GELLERT, J. F., 1962. Das Lössproblem in China. *Petermanns Geograph. Mitt.*, 106: 81–94.
GERASIMOV, I. P., 1964. Loess genesis and soil formation. *Rept., INQUA Congress, 6th, Warsaw, 1961*, 4: 463–468.
GUENTHER, E. W., 1961. *Sedimentpetrographische Untersuchungen von Lössen (part 1).* Bohlau, Cologne, 91 pp.
HOBBS, W. H., 1931. Loess, pebble bands, and boulders from glacial outwash of the Greenland glacier. *J. Geol.*, 39: 381–385.
HOBBS, W. H., 1943a. The glacial anticyclones and the European continental glacier. *Am. J. Sci.*, 241: 333–336.
HOBBS, W. H., 1943b. The glacial anticyclone and the continental glaciers of North America. *Proc. Am. Phil. Soc.*, 86: 368–402.
HOLMES, C. D., 1944. Origin of loess—A criticism. *Am. J. Sci.*, 242: 442–446.
KELLER, W. D., 1970. Review of Ollier (1969). *Geotimes*, 15(1): 31.
KEILHACK, K., 1920. Das Rätsel der Lössbildung. *Z. Deut. Geol. Ges.*, 72: 146–161.
KRINITZSKY, E. L. and TURNBULL, W. J., 1967. Loess deposits of Mississippi. *Geol. Soc. Am., Spec. Papers*, 94: 64 pp.
KUENEN, PH. H., 1969. Origin of quartz silt. *J. Sediment. Petrol.*, 39: 1631–1633.
LEIGHTON, M. M. and WILLMAN, H. B., 1950. Loess formations of the Mississippi Valley. *J. Geol.*, 59: 323–332.
LOZEK, V., 1965. Das Problem der Lössbildung und die Lössmollusken. *Eiszeitalter Gegenwart*, 16: 61–75, (English summary).
LUGN, A. L., 1962. The origin and sources of loess. *Univ. Nebraska Studies*, 26: 105 pp.

MATALUCCI, R. V., SHELTON, J. W. and ABDEL-HARDY, M., 1969. Grain orientation in Vicksburg loess. *J. Sediment. Petrol.*, 39: 969–979.

MCKEAGUE, J. A. and CLINE, M. G., 1963. Silica in soils. *Advan. Agron.*, 15: 339–396.

OLLIER, C. D., 1969. *Weathering*. Oliver and Boyd, Edinburgh, 304 pp.

OROMBELLI, G., 1970. I depositi loessici di Copreno (Milano). *Boll. Soc. Geol. Ital.*, 89: 529–546.

PENCK, A., 1909. Die Morphologie der Wüsten. *Geograph. Z.*, 15: 545–558.

PENCK, A., 1931. Zentral-Asien. *Z. Ges. Erdkunde, Berlin (für 1931)*, p.1–13.

PITCHER, W. S., SHEARMAN, D. J. and PUGH, D. C., 1954. The loess of Pegwell Bay, Kent, and its associated frost soils. *Geol. Mag.*, 91: 308–314.

RAESIDE, J. D., 1964. Loess deposits of the South Island, New Zealand, and soils formed on them. *New Zealand J. Geol. Geophys.*, 7: 811–838.

RATHJENS, C., 1928. Löss in Tripolitanien. *Z. Ges. Erdkunde, Berlin (für 1928)*, p.211–228.

RAY, L. L., 1967. An interpretation of profiles of weathering of the Peorian loess of western Kentucky. *U.S., Geol. Surv., Profess. Paper*, 575-D: D221–D227.

RUSSELL, R. J., 1944a. Lower Mississippi Valley loess. *Bull. Geol. Soc. Am.*, 55: 1–40.

RUSSELL, R. J., 1944b. Origin of loess—A reply. *Am. J. Sci.*, 242: 447–450.

SMALLEY, I. J., 1966. The properties of glacial loess and the formation of loess deposits. *J. Sediment. Petrol.*, 36: 669–676.

SMALLEY, I. J., 1968. The loess deposits and Neolithic culture of northern China. *Man*, 3: 224–241.

SMALLEY, I. J., 1970a. Calcium carbonate encrustations on quartz grains in loess from the Karlsruhe region. *Naturwissenschaften*, 57: 87.

SMALLEY, I. J., 1970b. Cohesion of soil particles and the intrinsic resistance of simple soil systems to wind erosion. *J. Soil Sci.*, 21: 154–161.

SMALLEY, I. J. and CABRERA, J. G., 1970. The shape and surface texture of loess particles. *Geol. Soc. Am. Bull.*, 81: 1591–1595.

SMALLEY, I. J. and PERRY, N. H., 1969. The Keilhack approach to the problem of loess formation. *Proc. Leeds Phil. Lit. Soc., Sci. Sect.*, 10: 31–43.

SMALLEY, I. J. and VITA-FINZI, C., 1968. The formation of fine particles in sandy deserts and the nature of "desert" loess. *J. Sediment. Petrol.*, 38: 766–774.

SMITH, H. T. U., 1968. Nebraska dunes compared with those of North Africa and other regions. In: C. B. SCHULTZ and J. C. FRYE (Editors), *Loess and Related Eolian Deposits of the World*. Proc. INQUA Congr., 7th, Univ. Nebraska Press, Lincoln, 12: 29–47.

SMITH, R. S. and NORTON, E. A., 1935. Parent materials, subsoil permeability and surface character of Illinois soils. In: *Parent Material of Illinois Soils*. Univ. Illinois, Urbana, pp.1–4.

TERUGGI, M. E., 1957. The nature and origin of Argentine loess. *J. Sediment. Petrol.*, 27: 322–332.

THOMASSON, A. J., 1965. Review of Berg (1964). *Geograph. J.*, 131: 414–415.

VISHER, S. S., 1922. The time of glacial loess accumulation in its relation to the climatic implications of the great loess deposits; did they chiefly accumulate during glacial retreat? *J. Geol.*, 30: 472–479.

VON RICHTHOFEN, F., 1877. *China: Ergebnisse eigener Reisen und darauf gegründeter Studien (vol. 1)*. Dietrich Reimer, Berlin, 440 pp.

WASCHER, H. L., HUMBERT, R. P. and CADY, J. G., 1948. Loess in the southern Mississippi Valley: identification and distribution of loess sheets. *Soil Sci. Soc. Am. Proc.*, 12: 389–399.

WEIR, A. H., CATT, J. A. and MADGETT, P. A., 1971. Postglacial soil formation in the loess of Pegwell Bay, Kent (England). *Geoderma*, 5: 131–149.

WILD, L. J., 1919. Note on the mechanical composition of the so-called loess at Timaru. *Trans. New Zealand Inst.*, 51: 286–289.

WOLDSTEDT, P., 1967. The Quaternary of Germany. In: K. RANKAMA (Editor), *The Geologic Systems, The Quaternary*. Interscience, New York, N.Y., 2: 239–300.

ZEUNER, F. E., 1949. Frost soils on Mount Kenya, and the relation of frost soils to aeolian deposits. *J. Soil Sci.*, 1: 20–30.

(Received November 6, 1970)
(Resubmitted April 23, 1971)

46

Reprinted from *Trans. N.Y. Acad. Sci.*, Ser. 2, **34**(6), 534–542 (1972)

THE INTERACTION OF GREAT RIVERS
AND LARGE DEPOSITS OF PRIMARY LOESS*†

Ian James Smalley

Department of Civil Engineering
University of Leeds
Leeds LS2 9JT, England

INTRODUCTION

It is widely known that loess was deposited by eolian action, and other depositional agencies thus tend to be totally discounted. But in the history of the formation of the world's loess deposits it is apparent that fluvial action has played a considerable part. For example, no one is likely to dispute that the great inland loess
deposits in the Kansu and Shensi districts of North China were emplaced by
eolian action, but it is equally true that the enormous amount of loess material
that makes up the North China plain and links Shantung peninsula to the mainland was deposited by the Yellow River, flowing along its various courses. Fluvial
action has also had an effect on more accessible deposits. Although the Great
Plains loess is an eolian deposit (or, rather, a whole collection of eolian deposits)
the Mississippi loess would not exist without the contribution made by fluvial
transportation.

The problem of defining loess must be discussed briefly. There are two distinct
ways of defining loess; either the definition can be nothing more than an elaborate
description of the physical nature of the deposit or it can contain some statement
of origin. This latter type of definition seems more desirable, in that it serves to
distinguish loess deposits from other deposits of silt-sized material. The following
definition has been proposed,[1] "Loess is a clastic deposit which consists predominantly of quartz particles 20–50 microns in diameter and which occurs as wind-laid
sheets," and will be used as a basis for the suggestions in this paper. The transportation of the material by eolian action, at some time in the history of the formation of the deposit, is seen as the critical event, but, of course, other factors
must be considered. This paper is essentially an attempt to show the important,
and often neglected, effect that fluvial transportation has had on the formation of
the deposits of loess that exist at present. The loess systems studied all have one
thing in common; in all cases the primary eolian deposit has interacted in some
way with a large river of considerable sediment-transporting capacity. The effect
of this interaction forms the subject of the paper. Six rivers and the deposits they
affect will be considered: the Yellow River, so called because of the great bulk of
loess material it transports, the Mississippi system, the Rhine, the Danube, and, on
a slightly smaller and more speculative scale, the Po and the Rhone.

THE YELLOW RIVER

The Hwang Ho (Hwang-yellow, Ho-river) gained its name of "China's Sor-
row" because of its tendency to burst out from its course and wander over the

---

* Submitted February 4, 1972; accepted March 22, 1972.
† Supported in part by the Royal Society.

countryside, destroying crops and people. It was this great instability, of course, that allowed it to deposit the North China plain, using as material the loess from the great inland deposit. An attempt to show, in a very simple stylised diagrammatic way, the stages in the formation of the North China plain is given in FIGURE 1. Arguable material appears very early in this scheme, since the eolian material is shown coming essentially from the south rather than from the deserts to the north and west. The case for the Chinese loess as glacial rather than desert material has been presented elsewhere[1, 2] and will not be discussed at length here, but it will be assumed that glacial grinding is the only effective way of producing loess material in sufficient quantities to form the great deposits of China, North America, and Europe.[3, 4]

The eolian deposit forms (1a), the river erodes the deposit and carries material along its course (1b, c), the course of the river changes and the secondary deposit becomes more widespread (1d), and the process continues (1e). Thus, the glacially

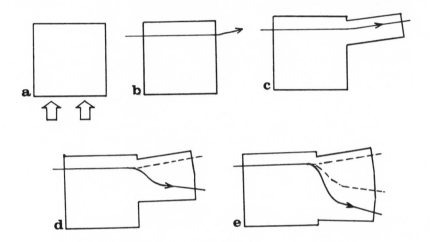

FIGURE 1. Stages in the formation of the North China plain. a— eolian deposition of primary loess; b— river interaction; c— material transported to floodplains; d— secondary deposit spread by change of river course; e— the process continued.

produced material forms two distinctive major sediments, the inland primary loess and the secondary redeposited loess. Ting[5] has been able to show how the continuous formation of the secondary loess deposit connected Shantung Island to the mainland and transformed it into Shantung peninsula. The movement of the Yellow River course has gone on well into historical times, and thus the formation of the North China plain can be seen as a process stretching back from the present day to the Pleisocene events involving the formation of the primary deposit.

There are some interesting prehistoric implications in the formation of the North China plain. The initial settlement in North China appears to have been closely associated with the primary loess, and the spread of the Neolithic culture into Shantung occurred under a simple geomorphological control; the Yellow River had to supply the land before it could be settled.[2]

Thus, when the North China loess is considered, three important sedimentological variables must be studied; the formation of the predominant quartz material, the formation of the primary deposit, and the formation of the secondary deposit. The critical actions appear to be glacial grinding, eolian transportation, and fluvial transportation.

THE MISSISSIPPI SYSTEM

The North American loess scène possesses certain distinct similarities to the North Chinese system but also certain critical differences. Matthes[6] has pointed out the great differences in the sediment carrying characteristics of the Mississippi and the Yellow Rivers. The Mississippi transports 550–600 parts per million of sediment, whereas the Yellow River carries something on the order of 600,000 ppm. The Mississippi collects water from a 1,244,000-sq.-mile watershed, equal to 41% of the area of the (1951) United States, and transports every year some 400,000,000 tons of sediment, 90% fine material in suspension, Most of the suspended material comes from the Missouri branch, and much of it is derived from the primary inland loess deposits. The water flow in the Yellow River is less, and so is the absolute amount of sediment transported, Kingsmill[7] has estimated that about 75,000,000 tons per year are carried.

The major difference in the sediment-transporting characteristics is simply that the Yellow River has a much more direct effect on the loess deposits in its vicinity than does the Mississippi. The main channel of the Yellow River flows through very thick loess deposits, and these occupy a substantial proportion of the drainage basin. The Mississippi drainage basin is so vast that the loess deposits are insignificant in relative terms. The major difference in channel type is that in North America the main channels of the major river were essentially confined and did not have the freedom of movement that circumstances allowed the Yellow River.

FIGURE 2 is an attempt to represent the important stages in the formation of a

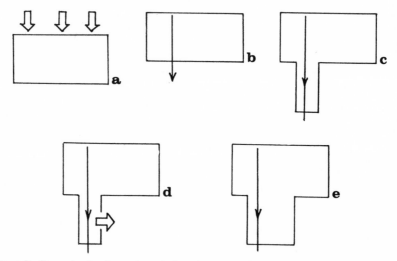

FIGURE 2. Stages in the formation of the Mississippi loess. a—eolian deposition of primary loess; b— river interaction; c— material transported to floodplains; d— eolian transportation; e— second-cycle deposit formed.

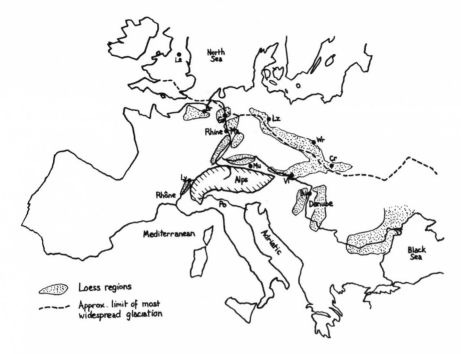

FIGURE 3. Sketch map showing selected aspects of the European loess system. Towns as follows: Br—Brussels, Bu—Budapest, Cr—Cracow, Co—Cologne, Le—Leeds, Ly—Lyons, Lz—Leipzig, Ma—Mainz, Mu—Munich, Vi—Vienna, Wr—Wrocław.

North American loess deposit, using the same graphic terms as in FIGURE 1. Glacial grinding can be invoked with more confidence to provide the bulk of the loess material; this comes from the north by eolian transportation (2a) and forms the inland deposits. It must be emphasized that each glacial advance will tend to provide more material, so sequences are repeated. Some of the loess material deposited in the drainage system of the Mississippi-Missouri rivers tends to be carried to the main channels and eventually into the major Mississippi channel and transported south (2b, c). The loess material is deposited on floodplains (2c) and eventually transported away from the river by eolian action (2d); eventually forming a new eolian deposit (2e).

The loess is not carried far from the river,[8] and thus the deposits are easily associated with their source—or their apparent source; it is evident that much of the material in the Mississippi loess deposits was formed by glacial action in the north and was once part of the primary deposit. Thus, perhaps the Mississippi loess should be thought of as a second-cycle loess just to distinguish it from the primary deposit that sired it.

If the origin of the particles is thought to be a significant event in the history of a sediment, then the Mississippi loess is hardly being adequately described when it is simply referred to as an eolian deposit. The fluvial stage is at least as important as the final short eolian transportation.

THE RHINE

Woldstedt has compared various characteristics of the Rhine and the Mississippi,[9] and they can be further compared with respect to their interaction with loess deposits. The Rhine valley loess is the archetypal loess—it was here that the study of loess began. The study was initiated by Karl Caesar von Leonhard[10, 11] and boosted by Lyell[12] and has been expanding ever since. The Rhine loess system is actually much more complex than that of the Mississippi because the Rhine is associated with two areas of particle production, whereas the Mississippi is associated with only one.

The major Pleistocene glaciations in Europe were those from the north, the Scandinavian and associated ice sweeping down to the south and covering a large part of northern Europe. But there were also glaciations on a smaller scale in the Alpine region, and each set of glaciers produced some loess material. Deposits of loess associated with the North European glaciers stretch in a great band across western Europe and spread wide in the east, and these have been eroded by many rivers, including the Rhine. FIGURE 3 attempts to show, in a very simple way, the setting of the Rhine and other rivers and the location of the related loess deposits and the limits of the glaciated areas. The major European loess deposits appear to be associated directly with the North European glaciers. The transporting winds blew from the glacier[13] and deposited the loess near the fringe of the glaciated region. From the map it will be seen that the loess might reasonably be expected to stretch along the Rhine past Cologne as far as Coblenz or Mainz but not beyond. And yet loess is present in the Rhine valley as far south as the Kaiserstuhl, near Freiburg, and beyond, to Basel.

There is loess near Munich that must be associated with the Alpine glaciers. Is it possible that most of the Rhine valley loess is composed of Alpine material? It seems that this may indeed be the case. The division of drainage basins between the Rhine and the Danube is complicated in the Alpine region, but it is likely that considerable quantities of loess material were deposited in each basin and subsequently transported by the rivers.

This would mean that the Rhine valley loess and the Mississippi loess are indeed very similar and that Lyell's[14] identification of the loess of the Mississippi basin as a counterpart of that bordering the Rhine was very perceptive.

THE DANUBE

The Danube must have received large amounts of Alpine loess, and it is possible that where the course of the river runs near the fringe of the glaciated regions to the north, North European material may have been received—for example, near Vienna and Budapest. But, on the early part of its route to the Black Sea, the Danube appears as a means of transporting Alpine loess material in a generally easterly direction. The situation with respect to loess distribution is a bit confused east of the Alps, and many maps show large areas of loess at increasing distances from possible glacial sources.

There are large deposits of loess in East Europe; at least available maps of loess distribution[15-17] suggest that the loess deposits become much more widespread towards the east. As far as is known, no one has ever suggested a reason for this; it may be that in the east the continental glaciers traversed much more land and were able to pulverize more suitable rock material. There is a very interesting deposit of loess in Rumania at the mouth of the Danube. The delta is far to the south of the glaciated regions, and it appears that it may be similar to the North China plain and consist largely of redeposited loess material. Scheidig[16] shows it as loess, and it would be pedantry to suggest that it be called anything else, but if the distinction is made, in the interests of scientific accuracy, into primary loess and secondary loess, then the delta material is surely secondary and, judging by distribution maps, will have been derived from both Alpine and North European sources.

Loess in the Danube basin and nearby was of great importance in prehistoric times. According to Clark,[18] the best-known exponents of the extensive, shifting agriculture of Neolithic times were the Danubian peasants who colonized the loess of Central Europe. The Danubians worked their way rapidly over the loess in more than one wave; from Moravia they spread, on the one hand, eastwards into Galicia and across Poland as far as the lower Vistula and, on the other, northwards and westwards, over Germany, following the Oder into Silesia, the Elbe into Saxony, and the Danube into Bavaria, ultimately settling the loess lands of the Main, the Neckar, and the Rhine. Here, with respect to the Danube, the people moved in the opposite direction to the loess material, unlike the Neolithic Chinese, who followed their land down the Yellow River valley.

THE PO AND THE RHONE

As the glaciers radiated out from the Alps during the Pleistocene cold periods, they produced loess material in regions all around the Alpine massif. Thus, it would seem reasonable to expect the Alps to be surrounded by loess deposits, and yet the accepted maps of loess distribution show a marked dearth of loess south of the Alps. If this were the true situation it might throw considerable doubt on the preëminence of glacial grinding as a loess-producing mechanism. Luckily the maps were mistaken, and where there should be loess there is loess. Orombelli[19] has described the loess at Copreno, between Como and Milan, and has indicated that there are quite widespread, well-preserved, but hitherto unappreciated loess deposits in Piedmont and Lombardy.[20]

In this case it appears that the river does not actually flow through the deposit, but they are very close in a totally enclosed valley, and it is inevitable that much loess material was carried towards the Adriatic by the Po. The primary deposit is small—insignificant when compared to the Great Plains or the North China deposits—and thus one might expect a small secondary deposit in the Po delta. A material comparison has not yet been made; when it is it should reveal a link between the North Italian loess deposits and the Po delta material. If the North China plain can feature on maps as loess, perhaps this designation should be extended to the Po delta.

Loess material deposited to the west of the Alps has been affected by the river Rhone, which has carried it in an essentially southwards direction toward the Mediterranean. The deposit near Lyons is well shown on the distribution maps based on Grahmann,[15] and the position of the loess in the river valley is clearly indicated. It seems strange that the loess to the west of the Alps should have been recognized and mapped in time for Grahmann to include it in his 1932 loess map of Europe (which is still the standard), and yet similarly produced deposits in a very similar setting south of the Alps were not recognized and appreciated until almost 40 years later.

DISCUSSION AND CONCLUSIONS

The long years of acrimonious dispute about the origin of loess deposits have lead to a degree of oversimplification, and the total acceptance of the eolian deposition concept has obscured the important role played by fluvial action in the transportation of loess material. This is not to say that eolian transportation should not be recognized as the most important transportational factor in the formation of loess deposits but rather that the true picture of significant geological actions contains some other important events.

In TABLE 1 a series of events is listed that might affect the formation of a loess deposit; more complex versions of this chart have been published elsewhere[4] but this simple version serves to highlight important events. Three types of event appear: formation events (P) related to the formation of the loess material, transportation events (T) and deposition events (D). The P events here relate only to the predominant quartz material; the highly variable carbonate content is not considered. In a typical loess deposit the initial material might be formed by glacial grinding (P1), be transported by the glacier (T1) and be incorporated in a mixed deposit when the glacier melts (D1). This mixed deposit can be sorted by the wind, which picks up fine material and transports it away (T2) for a short distance and deposits it as a highly sorted sediment (D2), which is called loess. Event D2 represents the formation of most of the world's loess deposits, and it is therefore emphasized. But subsequent events may occur giving rise to other interesting and identifiable sediments, and these should not be neglected. Richthofen's triumph[21] in establishing stage T2 as critical should not cause the other possible stages to be ignored. In fact, the loess controversy that occupied so much time and energy so long ago can be seen now, with the benefit of considerable hindsight, to have been concentrated on only one important aspect of the loess story.

The "loess problem" was essentially a discussion about what sort of T stage immediately preceded the actual formation of the loess deposit. No thought was given to other aspects of the problem; in terms of TABLE 1, events subsequent to D2 were neglected, and so were events prior to T2. Of the subsequent events it can be seen that they lead to certain important deposits that should be considered as distinct from the major primary D2 deposits. Stage D3 gives rise to the great North China plain and stage D4 to the Mississippi loess. The D3 deposits are conveniently called secondary and thought of as composed of redeposited loess material[1]; the naming of the D4 deposit is more difficult. Since eolian action is the key to loess formation, these eolian deposits are closer to true loess than are the North China plain deposits; they are primary loess, but the history of the material can be seen

TABLE 1
EVENTS IN THE FORMATION OF LOESS DEPOSITS

P1	Quartz particle production by glacial grinding
T1	Transportation of rock debris by glacier
D1	Formation of mixed deposit by outwash and after glacier melts
T2	Silt-sized particles lifted by wind and transported by eolian action
D2	Loess deposit formed: the primary loess
T3	Erosion of primary loess and fluvial transportation of particles
D3	Floodplain deposition: Secondary deposits
T4	Eolian transportation from floodplains
D4	Loess deposit formed: the second-cycle loess

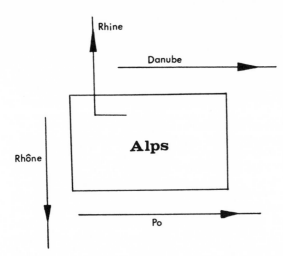

FIGURE 4. A simple view of the Alpine setting with respect to loess-interacting rivers.

as a long and complex one; primary (second-cycle) deposits might be a suitable term—borrowing from the idea of a second cycle sandstone.

These post-D2 events occur because fluvial transportation of the loess material occurs, and the impressive post-D2 deposits are caused because great rivers interact with large deposits of primary loess. Of the six rivers considered, perhaps only four are "great" rivers, but the others, the Po and the Rhone, are of great interest from the loess-interaction point of view. The deposits they interact with could not be described as large deposits of primary loess, but they are significant and important. If a good comprehensive picture of the formation of the world's loess deposits is to be gained, it will be necessary for each once glaciated area to be studied carefully to identify critical factors and events. In many ways, the classic loess areas that have been studied in great detail do not yield the necessary information because the picture is too complex. The Alpine deposits offer a fairly simple system in which the importance of the related events can be assessed, and, in the Po valley, unappreciated and hitherto unknown deposits exist that can be related to predictions and provide new data as an alternative to reprocessing the old.

FIGURE 4 shows the Alpine situation as considered in this paper. The Alps should be surrounded by loess material produced during the Pleistocene glaciations. The four rivers shown should interact with this material and possibly this could lead to secondary deposits. Potentially the most interesting of these rivers is the Po because its region, south of the Alps, is free of all influence of the North European glaciers. As has been pointed out above, the Rhine and the Danube probably both interact with North European material, and so, just possibly, does the Rhone. The North Italian loess is totally Alpine and provides an opportunity for the study of a relatively small, simple loess system.

SUMMARY

In pursuit of the more systematic, worldwide, study of loess deposits it is necessary to consider the contribution made by fluvial transportation in the history of

369

the deposits. The most dramatic illustrations of fluvial transportation occur when a great river interacts with a substantial deposit of primary loess. At least two very famous deposits of loess material have resulted from this sort of interaction; the North China plain and the Mississippi loess. Although the key role played by eolian transportation in the formation of loess deposits must continue to be recognized, it is important that other factors such as the production of the actual material and the fluvial contribution be recognized and studied. Three main groups of deposits might be recognized: primary, secondary and second-cycle primary, with most of the world's deposits falling in the primary class but with significant examples occurring in the others.

Acknowledgments

This paper represents a preliminary part of the Alpine Loess Project. Help from Dr. D. H. Krinsley of Queens College, CUNY, Dr. C. Vita-Finzi, of University College, London, and Dr. G. Orombelli, of the University of Milan, is gratefully acknowledged.

References

1. SMALLEY, I. J. & C. VITA-FINZI. 1968. The formation of fine particles in sandy deserts and the nature of 'desert' loess. J. Sediment. Petrol. **38:** 766–774.
2. SMALLEY, I. J. 1968. The loess deposits and Neolithic culture of northern China. Man **3:** 224–241.
3. SMALLEY, I. J. 1966. The properties of glacial loess and the formation of loess deposits. J. Sediment. Petrol. **36:** 669–676.
4. SMALLEY, I. J. 1971. 'In-situ' theories of loess formation and the significance of the calcium carbonate content of loess. Earth Science Rev. **7:** 67–85.
5. TING, W. S. 1965. The geomorphology of the North China Plain and the history of the early Chinese. Bull. Inst. Ethnol. Acad. Sinica (Taiwan) **20:** 155–162 (in Chinese).
6. MATTHES, G. H. 1951. Paradoxes of the Mississippi. Scientific American **184** (4): 19–23.
7. KINGSMILL, T. W. 1906. The hydraulics of great rivers flowing through alluvial plains. The Shanghai Society of Engineers and Architects 1905–1906, 41p.
8. WAGGONER, P. E. & C. BINGHAM. 1961. Depth of loess and distance from source. Soil Science **92:** 396–401.
9. WOLDSTEDT, P. 1960. Mississippi und Rhein, Ein geologischer Vergleich. Eiszeitalter u. Gegenwart **11:** 31–38.
10. LEONHARD, K. C. VON. 1824. Charakteristik der Felsarten: 3 vols. Joseph Engelmann Verlag. Heidelberg, Germany.
11. KIRCHHEIMER, F. 1969. Heidelberg und der Löss. Zeits. Verein. Freunde Studentschaft Univ. Heidelberg. **46:** 1–7.
12. LYELL, C. 1834. Observations on the loamy deposit called 'loess' of the basin of the Rhine. New Edinburgh Philos. J. **17:** 110–122.
13. HOBBS, W. H. 1943. The glacial anticyclones and the European continental glacier. Amer. J. Science **241:** 333–336.
14. LYELL, C. 1847. On the delta and alluvial deposits of the Mississippi River, and other points in the geology of North America, observed in the years 1845, 1846. Amer. J. Science **3** (ser. 2): 34–39, 267–269.
15. GRAHMANN, R. 1932. Der Löss in Europa. Mitt. Ges. f. Erdk. Leipzig. for 1930–31: 5–24.
16. SCHEIDIG, A. 1934. Der Löss und seine geotechnischen Eigenschaften. Theodor Steinkopff Verlag, Dresden and Leipzig, Germany.
17. POSER, H. 1948. Aolische Ablagerungen und Klima des Spätglazials in Mittel- und Westeuropa. Naturwiss. **35:** 269–276.
18. CLARK, J. G. D. 1952. Prehistoric Europe, The Economic Basis: 95. Methuen. London, England.
19. OROMBELLI, G. 1970. I depositi loessici di Copreno (Milano). Boll. Soc. Geol. Ital. **89:** 529–546.
20. OROMBELLI, G. 1972. Personal communication.
21. RICHTHOFEN, F. VON. 1882. On the mode of origin of the loess. Geol. Mag. **9:** 293–305.

Editor's Comments
on Papers 47 and 48

47 CEGLA
Loess Sedimentation in Poland (summary)

48 CEGLA
Influence of Capillary Ground Moisture on Eolian Accumulation of Loess

Cegla has published detailed studies of loess sedimentation in Poland and his most important contribution to the fundamental study of this process has been in the field of the actual deposition process. He points out (Paper 47) that many of the features of loess once considered typical may not, in fact, be all that characteristic, e.g., lack of structure, vertical fissility, color, and incidence of calcium carbonate concretions. He has stated:

> The presently observed great differentiation of carbonate contents in loess deposits is the result of a great environmental differentiation in which the loesses originated. It is also the result of strongly differentiated processes acting postsedimentally. Thus the calcium carbonate contents index should not be used as a feature of "typical" loess. However the carbonates should be recognized as an important index of sedimentation and diagenesis conditions.

Polish loess investigators have recognized the value of experimental investigations in geological research. In Paper 47 Cegla has reported some interesting experimental observations on loess deposition, but the most significant results so far obtained have related to "periglacial ground structures." Many of these structures are observed in loess and represent an interesting aspect of the study of loess sedimentation (see Butrym et al., 1964, for an excellent review). Cegla has also published preliminary electron microscope studies of loess particles (Cegla et al., 1971); these provide an interesting comparison with the earlier microscopic observations of Butrym (1960).

REFERENCES

Butrym, J. 1960. A study on the morphology of the quartz grains in the consistence of loess. Ann. Mar. Cur. Sklod. Univ. B15, 23–29.

——, Cegla, J., Dzulynski, S., and Nakonieczny, S. 1964. New interpretation of "periglacial structures." Folia Quaternaria No. 17, 34p.

Cegla, J., Buckley, T., and Smalley, I. J. 1971. Microtextures of particles from some European loess deposits. Sedimentology 17, 129–134.

47

Reprinted from *Acta Univ. Wratislav Stud. Geogr.*, **17**(168), 53–71 (1972)

LOESS SEDIMENTATION IN POLAND

Jerzy Cegla

Summary

The paper presents the result obtained in experimental studies on eolian loess sedimentation, and on field materials collected in loess areas of Poland. Further, basing on field-collected data, an attempt was made to render the conditions of sedimentation of Polish loess. The starting point for considering these problems was the conception of the role played by moist ground in the process of eolian loess sedimentation (J. Cegła 1969).

In literature concerning loess problems the term "typical" loess very often can be met. But more and more often the question arises, what by this term should be understood. Many of the "typical" loess features are doubted of. They are: lack of structure, vertical fossily, colour, incidence of calcium carbonate concretion (A. Jahn 1950, 1956, J. Dylik 1954, J. Cegła 1964, 1965). Recent studies show that the same concerns also the remaining features, mainly the calcium carbonate contents and the granulometric composition.

The calcium carbonate contents as one of the features of the "typical" loess, underwent a considerable devaluation. Carbonates are very mobile minerals and can get into the loess dust in various ways and in different periods of time, so in the moment of loess-blanket formation as during their diagenesis (A. G. Czernjachowskij 1966, V. Ambroż 1947, I. J. Smalley, N. H. Perry 1969, L. L. Ray 1967). The presently observed great differentiation of carbonate contents in loess deposits (Table 1) is the result of a great environmental differentiation in which the loesses originated. It is also the result of strongly differentiated processes acting postsedimentally. Thus, the calcium carbonate contents index should not be used as a feature of "typical" loess. However, the carbonates should be recognized as an important index of sedimentation and diagenesis conditions.

The granulometric composition of loess is its most important feature. In these deposits the grain of a fraction of 0,05 - 0,02 mm dominates. Fractions over 0,05 mm and under 0,02 mm appear in a small percentage. Such a fraction system appears in huge territories of Europe (A. Cailleux 1954, P. Krivan 1955, J. Jersak 1965, J. Pelisek 1969, H. Maruszczak 1969, W. K. Łukaszew 1970) of Central Asia (N. I. Kriger, M. R. Moskalew, S. G. Bekerman 1961, A. G. Czernjachowskij 1966), in North American loesses (A. Swineford, J. C. Frye 1951, A. L. Lugn 1962, H. B. Willman, J. C. Frye 1970) and in China (Czżan Czun-hu 1959), in India (F. E. Zeuner 1953), in South America (M. E. Teruggi 1957), in New Zeeland (J. D. Reaside 1964). After A. Malicki (1967) the contents of the 0,05 - 0,02 fraction in "typical" loesses is 50 - 60%. However, the contents of this fraction should be enclosed in wider limits. This

[*Editor's Note:* All figures and plates have been omitted owing to limitations of space.]

is indicated even by the results of the granulometric composition analysis for some of the loesses in Poland (Table I and II). I propose to regard 30 - 60% as the characteristic percentage of the 0,05 - 0,02 mm fraction. The mean grain size (Md) of Polish loess in 0,041 - 0,042 mm (Table II, Fig. 1 and 2). In dependence of the granulometric composition most of the physical properties of loess are formed, between which water permeability is very important (J. Cegła, M. Harasimiuk 1968, 1969).

Accepting various sources of dust material for Polands loesses and a strong differentiation of the processes of transport, deposition and the diagenesis of dust, a notion including strongly differentiated products of their acting should be introduced, and the term "typical" loess, which conveys nothing, should be relinquished. This proper term is loess formation. The composition of loess formation in case of Poland includes loesses of eolian origin, facially differentiated, as well as many types of lithological dust sediments.

EXPERIMENTAL STUDIES

The studies were carried out in the Laboratory of the Wrocław University, Institute of Geography.

Accepting in the experiment the eolic kind of transportation, the possibility of loess forming from dust transported through long distances in high zones of the atmosphere was excluded (W. Zinkiewicz 1949, J. Wojtanowicz, A. Zinkiewicz 1966, 1969). The transportation in conditions of dust storms was accepted. To that kind of dust transportation in many instances a loess-forming role is ascribed (A. A. Lugn 1935, 1962, T. L. Péwé 1955, 1968, B. A. Fiedorowicz 1960, A. Jahn 1961, 1966). The speed of dust cloud translocation during the experiment ranged between 0,5 - 20 m/sec. It guaranteed the possibility of transportation of both the grains of the dust fraction and sand. The experiment was carried out in an open area, and not in a wind-tunnel. The idea was to avoid an artificial limitation and directing of dust clouds by the walls and ceiling of the tunnel. The device used for the experiment consisted of two parts: 1. the deposition area with dry or moisted ground, 2. the wind source and dust material. As dust, loess from a profile in Baborów was used. Its mean grain size was 0,040 mm. The ground in the deposition area was formed by limestone. The ground moisture came from capillary water in the limestone.

During the first experiment three wind-speeds and two kinds of ground — dry and moist — were used. The time of accumulation at every speed was 30 minutes (Fig. 3). At the speed of 0,5 m/sec (Fig. 3 a), dense clouds of loess dust were translocated over two limestone plateaus, one of them dry (A), and the other of a moist surface (B). After 30 minutes a 7 mm strong layer of loess powder on the dry limestone was sedimented. On the moist surface the loess layer was 5 mm strong and the dust was bound as a result of water saturation. Parallel to the loess sedimentation on the moist surface the capillary menisci were rising, keeping in result the "topographical" surface of the growing loess layer constantly moist, catching further dust grains. The increase of wind-speed to 7 m/sec gave different effects (Fig. 3b). The dust clouds became flat. On the dry surface only small amounts of loess were sedimented, filling small depressions in limestone. In conditions of moist ground the loess layers achieved a thickness of 3 mm. At the speed of 15 m/sec (Fig. 3 c) on the dry surface loess was not sedimented; not a single grain was stopped. On the moist surface the loess layer achieved a thickness of 2 mm.

The experiments carried out have shown, that loess dust at various wind-speeds is always retained by moist surfaces, in contradiction to cases of deposition on dry surfaces.

With the presented here problems a difficult question of increments of loess dust thickness is joined. In the experiment a thickness of several millimeters during 30 minutes was obtained, besides, parallel to the increase in wind-speed the increments were smaller. Many authors report yearly loess increments, resulting from eolian sedimentation, in thickness of millimeters, or even fragments of them (Table 3). It is difficult to substantiate in which way such thin layers of dust could outlast the activity of degradation processes during intermittences in sedimentation. This problem remains unexplained even by the unfounded view, that the fixation of dust layers of millimeter thickness is caused by vegetation. It seems that loess-generating pleistocenic dust storms were more frequent and transported larger amounts of material, because of a more full detection of deflation areas. In result the increase in thickness of loess covers progressed faster than it commonly is considered. It seems also, that the increase in thickness obtained experimentally was a little too fast, as it follows from the way, the experiment was carried out.

The next experiment was carried out on an irregularly moistured ground (Fig. 4). Such places of the limestone surface were moistured, which covered the gaps in the impermeable layer. As the dust layer increased the process developed in two directions. Vertically — small water-saturated loess mounds were formed. The dry loess, loose, formed a very thin coat between the mounds (Fig. 4 *b, c, d*). At the same time horizontally an increase of the capillary moistured area progressed (Fig. 4, Phot. 10, 11). The lines of the capillary rise after transgressing the impermeable level, expanded fanlike. Mounds of loess dust increasing in height extended at the same time their bases. After a time moistured fragments associated each other, capturing the drifted loess dust with a growing surface. This way loess mounds were obtained of more or less the same height, but differing in shape of dune forms. The results obtained in this experiment may serve as a contribution for explaining the so called "Loess islands" (J. Cegla 1969).

While carrying out this group of experiments, the formation of fractures resulting from liquid infiltration was noted. The mechanism of their originating has been described previously (J. Cegła et al. 1967). The development of the fracture structures, initiated at the moment when the layer of drifted loess reached the thickness of several millimeters. Once formed fractures did not close during the whole time of dust deposition (Phot. 14, 15). A partial closing of the fractures took place after drying out of the sedimented loess, and at repeated moisturing. This statement is of importance for the disscussion on the vertical cracking of loess (B. Willis 1907, W. Röpke 1928, B. B. Pałynow 1934, R. Russel 1944, W. G. Bondarczuk 1946).

The second series of experiments was carried out under analogous conditions as in case of dust deposition, with the intention to study the loess resistance on deflation.

The experiment was started after a layer of dust of 10 and 7 mm thickness on dry and moist ground was sedimented. The first applied wind-speed was 7 m/sec. The dust material, so dry, powdered, as well as moist, has not been disturbed. Then the wind speed was gradually increased, and only at a speed of 15 m/sec the deflating of dust laying on dry ground has started (Fig. 6 A *c*). Further rapid changes of wind-speed were introduced, which ranged from 7 to 12 m/sec. At one of the blasts loess on dry ground started to deflate, whereas at a constant wind-speed the deflation began not before 15 m/sec. Loess on moist ground did not react on strong wind blasts.

In the next experiment a very rapid removal of dry dust was observed (Fig. 6 A B e and 7). In a wind current of speed of 9 m/sec sand was transported. At the moment when the first grains hit the dry, not consolidated by moisture, layer of dust, the process of deflating had begun, and in result the 10 mm thick layer of loess was destroyed in 10 minutes time. Diametrically different was the reaction on deposition of moist-consolidated sediments. The grains of sand transported by wind sticked to the moist dust surface, building up a layer of 1 mm thickness in a 10 minute time. Adding another cyclus of dust sedimentation, a sandy layer settled between dusts has been obtained (Fig. 8). In the light of the so carried out studies it has to be most emphatically stated, that the existence of a sequence dust—sand—dust, in conditions of dry ground where processes of eolian accumulation occur, is impossible. A dust layer with sand deposits on, not to be deflated in the moment of deposition, has to be fixed by water. This conclusion is of great importance while interpreting sandy bands in loesses, which blanket the uplands and elevated topographical elements.

In one of the experiments a slope of the surface, on which the eolian deposition of the transported material took place, was introduced. In the discussed case (Fig 9), two speeds of the wind current were applied: 0,5 m/sec for loess dust transportation, and 20 m/sec for sand transportation. Well known from the previously carried out experiments situation of sand layer between loess has been obtained. During the experiment on the sloped surface (10° angle) intensive processes deforming the produced layers, already during deposition, have been observed. In this experiment the process of solifluction, undergoing parallelly to eolian sedimentation, was initiated. The term solifluction is used here according to J. G. Anderson's (1906) definition. The floating of water-saturated sand sediments was initiated by unsettling the stability on the experimental slope, by increasing the weight of the sand layer as material has been blown on. Loess deposits underwent also deformation. It did not reveal itself in their structure because of lack of lithological differentiation of the loess used in this experiment. In 1956 A. Jahn has pointed out to the simultaneousness of the eolian sedimentation process and the activity of solifluction.

During the experiments a marked influence of the morphological edge on dust sedimentation was noted. The results obtained in the laboratory (Fig. 10) match with A. Dobrowolski's (1923) statements. While the wind transporting dust blows toward the edge, the current lines ascending it condense on its upper bend (Fig. 10 Ia). The condensation of current lines means an increase of wind flow, and so in that place the wind-speed is the highest. Deposition of dust does not take place, but even if there was dust deposited in other conditions, it will be soon totaly blown away (Fig. 10 IIa). This phenomenon takes place even if there is a soft wind only. However, the sedimentation of dust takes place beneath the edge, as well as in a certain distance behind the edge. In case of a reversed direction of transportation (upper level—edge—lower level) the extending of current lines takes place over the lower level and begins over the edge. Sedimentation takes place on the upper and lower level, and the edge shows a marked dust superstructure (Fig. 10 IIb). The presented problem is of importance in the discussion on the direction of loess transportation in some of the areas of Poland, for instance in case of loesses of the Sandomierz Upland (H. Maruszczak 1963a, 1964, 1967, 1968, S. Z. Różycki 1967).

To summarize, the carried out experiments have shown, that moist ground on which dust is deposited, plays an important role in the process of loess sedimentation. Moisture from the ground, raising in a capillary way into the deposited dust, brings it to a standstill and consolidates it. At the huge dynamics of the geo-morphological degradation processes (especially deflation) the dry, not consolidated by water loess can not be preserved in the place of deposition. "Maintaining" of loess since the time of its deposition is possible only if it géts consolidated by water as early as during the deposition of dust, in that particular case by capillary water.

LOESS STRUCTURE

The genetical differentiation of structures in loess deposits, in spite of appara-nces, is not very high. The following structures are distinctly distinguishable: 1. solifluction and other slope processes, 2. deformational structures resulting from reversed density stratification, 3. wedge structures, 4. gley structures' in loess, and 5. ground ice structures.

1. Structure of solifluction and other slope-processes

This group includes most of the sedimentation structures appearing in loesses. In Polish loesses most common are structures of strated solifluction (types of soli-fluction structures as by A. Jahn 1970). Their development is strongly differentiated, according to the kind of material which underwent the process, to the slope of the surface and the amount of water. Structures of strated solifluction occur as single, somewhat deformated by movement, streaks (Phot. 16), as series of such streaks (Fig. 11 c), or series of strata (Phot. 17). The second type of solifluction structures is the tongue-like solifluction (Phot. 18). This type of structures often forms wide-spread horizons (Phot. 19 s).

In loesses of the slope facies usually lamination occurs. The thickness of the laminae is considerably differentiated, from fractions of one millimeter to 19 milli-meters. The laminae form the loess strata (Phot. 21). A differentiation in colour is also marked. Lamination in loesses is a result of rain-wash processes.

2. Deformational structures resulting from reversed density stratification systems

The mechanism of formation of these structures has been exhaustively discussed in earlier publications (J. M. Anketell et al. 1969, J. Cegla, S. Dłużyński 1970). Defi-nite conditions must exist if unstable systems shall be formed. The most important ones are: 1. a fast progressing sedimentation of rough and heavy material on a small and lighter one, alternately, 2. occurrence of liquefactional deposits, 3. intensive saturation with water of deposits. In periglacial environment it is a periodical satu-ration, where permafrost plays an important role as an impermeable layer (J. Bu-trym et al. 1964). During the time of sedimentation the deformational structures of loess originated in non-mobile and mobile reversed density stratification systems. In the first case diapir structures are dominating, which were formed syndepositio-nally, as well as postdepositionally. In mobile systems a considerable deformation of diapirs as a whole, or partially is often observed (Fig. 14, Phot. 22). In the Kietrz brick-field the author succeeded in uncovering a deformated diapir (Fig. 14 e), which appeared to be a transverse crest in relation to the slightly sloped level of the sur-face. In this case the diapir structure was built by loess (compare — point 6, level "D" — J. Cegla 1971), which intruded into the overlaying level of the gleizated loess. However they are genetically homogeneous structures. Their apparent varia-bility results from the stopping of the deformation processes in a definite stage of their development (J. Cegla, S. Dłużyński 1970).

3. Wedge structures

Only a part of the wedges is bound with cold climate, and can be used as a climatic index, under the stipulation that this sort of conclusions will be drawn with great care (S. Johansson 1959), as there is the possibility that fissures may be formed in result of various processes (J. Sekyra 1960, A. Jahn 1970, J. C. Dionne 1970), which are very similar to ice wedges. Much more complicated is the problem of occurrence of wedges in sediments of the loess formation. Not denying the fact that ice wedges were formed in loesses, it remains to be explained how wedges overlasted the degradation of ice. It is incomprehensible, how it was possible, that loess being not much resistant against liquefaction and erosion processes, during degradation of the ice-vein and before the mineral filling the wedge structure was preserved. To solve this problem further field- and experimental studies are needed, but already now attention should be called to dessication fissures and the role they play in this question. Recently this process is of growing importance in periglacial investigations (A. Washburn 1969). Many from between the known to me wedge structures developed in fissures of various origin. In my opinion a great many of the ice wedges are simply dissication fractures. As such I regard, between others, the wedges cutting the fossil soil in loesses in Hrubieszów ("Feliks" brick-field). The wedges in the plane form a network of polygons of various diameter (Phot. 23), are filled with silt and loess with a considerable admixture of organic material. It should be mentioned here, that the desiccation fissures can be of strongly differentiated size; from small to large ones, of a depth of several meters and with an opening in the upper part reaching 1,5 m. Their sizes depend on the thickness of the desiccation layer.

4. Gley structures in loess

The process of gleization, its geochemical and physical characterization, the meaning of gley horizons for stratygraphy, these are problems widely discussed in pedological, geological and geomorphological literature (I. P. Gierasimow 1959, J. C. Frye, H. B. Willman 1963, R. V. Ruhe 1965, 1969, J. Fink 1966, H. Rohdenburg 1968, G. Bartels, H. Rohdenburg 1968, A. A. Wieliczko 1969, G. Haase et al. 1969). In Poland gleizated loesses are rather common (J. Siuta 1960, J. E. Mojski 1965, J. Jersak 1965, T. Tyrcha-Czyż 1968, J. Cegła 1971). The development of gleization processes in loesses leads to changes in their texture, what in result gives a characteristic structure. The filtration abilities of the loess are disappearing (J. Siuta 1961), and cleyly minerals are formed (C. Bloomfield 1956). Loess changes its colour. On a spotted, gray-blue or greenish background various rusty dyed irony precipitates appear (Fig. 15). The irony precipitates (Phot. 26 and 27) are a very important indicator of acting of geochemical processes; they characterize the water relations in the formed sediments. They also enable a more full reproduction of the sedimentation conditions. The gleizated layers underwent deformation processes. They were translocated by solifluction. They also underwent deformation in systems resulting from reversed density stratification.

5. Ground ice structures

Ground ice structures are short streaks, showing sometimes the shape of strongly flattened lenses. They are of a somewhat darker colour and show usually the, so called, "gneiss structure". Apart from undisturbed streaks also deformed, of wavy course and with many difractions, are mat (Fig. 16, Phot. 28). The formation of ground ice structures is the result of ground freezing in winter, or of appearance of permafrost. In the freezing ground, at the frozen-ground limit, ice lenses are formed, which increase in size as a result of water rise from the beneath seated sediments (R. Rückli 1950). In the ice melting period the loess gets strongly saturated with water, and liquefacting and dislocation of the floating sediments can take place. It seems probable that in such conditions a considerable part of the structures has been destroyed. The presence of ground ice structures in loesses points to the fact, that during formation and disappearing of ice lenses large amounts of water in the ground exist. Carefull investigations (A. Dücker 1939, R. Rückli 1950) have shown, that process of ground ice formation increases thrice the ground moisture.

The above discussed sedimentation structures point unmistikably toward the existence of a strong water saturation of loess during their formation. As exceptions some of the wedge structures can be regarded.

RECONSTRUCTION OF SEDIMENTATION PROCESSES OF LOESSES IN POLAND

The sedimentation of particular deposits should be regarded as a strict connection of four elements: 1. the source of material, 2. kind of transportation, 3. deposition, and 4. redeposition of deposits. The material for Polish loesses was supplied by various systems, beginning from bedrock weathering waste, through alluvial deposits, and ending with glacial deposits. As the most loess generating, the eolian transportation in conditions of dust storm is regarded. To analyse the deposition and redeposition in Poland, the main loess formation profiles were put together (Fig. 17). Loess dust was deposited on a geologically strongly differentiated ground (limestone, glacial-, fluvioglacial-, and alluvial deposits). According to the hypothesis that eolian transported dust is stopped mainly by moist ground, in the particular profiles sections were designated where humidification was permanent or periodical. The moist levels, from which capillary rise mainly took place, were the gleizated levels in loesses, horizons with developed deformation structures, as well as the levels of solifluction loess.

The gley levels are the most important levels for the sedimentation of Polish loesses. They explicitly designate the environmental features of the initiating phase in the last loess sedimentation cyclus. Thus, areas of loess deposition shall be recognized as areas of moist ground, and in some regions even boggy. N. G. Hörner (1936) stresses the importance of atmospheric and ground water moisture for loess accumulation in Central Asia. The moist areas of loess accumulation are also mentioned by W. H. Twenhofel (1932).

Over gley horizons loesses are seated, which show a rich inventary of solifluction, rain wash and ground ice structures. For their formation also considerable amounts of moisture were necessary. From these levels the capillary water rise into the deposited dust also took place.

Structures resulting from ground ice are of special importance for the interpretation of sedimentation processes. The phenomenon of water rise to the frozen ground limit was just this process which enabled the moisturing of loess, even if it was out of range of capillary rise.

The statement, that in the studied loess profiles no clear sedimentation intermissions occur is of importance. The erosion surfaces are not marked, and levels of gliated loess transgress gradually into the overseated loess. This allows to accept the continuity of loess sedimentation on one hand, and the synchronism of acting of the particular processes on the other. As the rate of sedimentation of Polish loess is concerned, I am inclined to accept that the sedimentation is a result of dynamically differentiated dust storms, which in relatively short periods of time deposited huge amounts of dust. They were very intensive acts of sedimentation, including all the mentioned above set of processes.

I would also like to stress, that the environment of loess sedimentation in Poland was strongly differentiated. In western Poland, as compared with eastern areas, a more moist climate was dominating. The winds transporting dust were blowing in various directions and were of changing dynamics. The material deposited in various areas came from different sources. The differentiation of the morphological surface caused the development of syndepositional slope processes and deformation processes of various intensity. In some of the developmental stages the genetical processes extincted, hence, various sorting of sediments. Because of redeposition admixtures of local bedrock are found; also "non-typical" thicknesses are observed.

In such strongly differentiated environment of sedimentation the cause of various development of the loess formation deposits in Poland ought to be suspected.

BIBLIOGRAPHY

A l l e n, V. T., 1962, *Gumbotil, gley and accretion-gley*, Journ. Geol., 70.

A m b r o ż, V., 1947, *Sprase Pahorkatin*, Stat. Geol. Ust. C. Sl. Rep.

A n d e r s s o n, I. G., 1906, *Solifluction, a component of subaerial denudation*, Journ. Geol., 14.

A n k e t e l l, J. M., C e g ł a, J., D ż u ł y ń s k i, S., 1969, *Unconformable surfaces formed in the absence of current erosion*, Geologica Romana VIII.

A n k e t e l l, J. M., C e g ł a, J., D ż u ł y ń s k i, S., 1970, *On the deformational structures in systems with reversed density gradients*, Rocz. Pol. Tow. Geol. XI, z. 1.

A p o l l o w, B. A., 1927, *Wlijanije eołowoj akkumulacii na obmielienije siewiernoj czasti Kaspijskogo moria*, Izv. Centr. Gidromiet. Biuro, 7.

B a r a n o w, P. F., 1913, *Eołowyje nanosy i poczwy na razwalinach Olwii*, Poczwowiedienije 4.

B a r t e l s, G., R o h d e n b u r g, H., 1968, *Fossile Böden und Eiskeilhorizonte in der Ziegeleigrube Breinum (Niedersächsisches Bergland) und ihre Auswertung für die Reliefentwicklung im Jungquartär*, Göttinger Bodenkundliche Berichte 6.

B e r g, L. S., 1932a, *Less kak produkt wywietriwanija i poczwoobrazowanija*, Trudy II Mieżdunar. Konf. AICPJ.

B e r g, L. S., 1932b, *The origin of loess*, Greenlands Beitr. zu Geogr., 35.

B e s k o w, G., 1935, *Tjälbildningen och Tjällyftningen*, Sveriges Geologiska Undersökning, Stockholm.

B l o o m f i e l d, C., 1956, *The experimental production of podzolization*, VI Congrès International de Science du Sol, EV - 3, Paris.

B o n d a r c z u k, W. G., 1946, *O fizyko-gieograficzeskich usłowijach obrazowanija lessa i gumusowych gorizontow juga SSSR.*, Trudy Inst. Gieogr. AN SSSR, 37.

B o r o w i e c, J., N a k o n i e c z n y, S., 1968, *Charakterystyka płytkich utworów lessopodobnych w lewobrzeżnym dorzeczu środkowego Wieprza* (summ.: *A characteristics of the shallow loess-like formation on the left side of the river Wieprz Basin*), Annales UMCS, sectio B, vol. XX.

B o u m a, A. H., 1962, *Sedimentology of some flysch deposits. A graphic approach to facies interpretation*, Elsevier Publishing Company, Amsterdam—New York.

B o u m a, A. H., 1963, *A graphic presentation of the facies model of salt marsh deposits*, Sedimentology, 2.

B u r a c z y ń s k i, J., 1961, *Les vallées de Roztocze Occidental*, Annales UMCS, sectio B, vol. XV.

B u t r y m, J., C e g ł a, J., D ż u ł y ń s k i, S., N a k o n i e c z n y, S., 1964, *New interpretation of „Periglacial structures"*, Folia Quaternaria, 17.

B ü d e l, J., 1951, *Die Klimazonen des Eiszeitalters*, Eiszeitalter und Gegenwart, 1.

C a i l l e u x, A., 1954, *Les loess et limons éoliens de France,* Bull. de Service de la carte géol. de la France, No. 240, vol. 51.

C e g ł a, J., 1959, *Obserwacje nad rozwojem form erozyjnych w obrębie lessowej krawędzi Wyżyny Lubelskiej* (res.: *Observations sur l'évolution des formes d'érosion dans les limites des loess de la bordure du Haut Plateau de Lublin),* Annales UMCS, sectio B, vol. XIII.

C e g ł a, J., 1961a, *A Study of silt formations in the Carpathian basins,* Annales UMCS, sectio B, vol. XV.

C e g ł a, J., 1961b, *On the presence of loess (?) and silt materials in the Carpathians,* Abstracts of Papers, INQUA VIth Congress.

C e g ł a, J., 1961c, *Some analogies between structures inside loessic sediments be formed in pleistocene and in contemporary,* Abstracts of Papers, INQUA VIth Congress.

C e g ł a, J., 1964, *Utwory pyłowe kotlin karpackich i ich stosunek do lessów wyżynnych Polski,* Maszynopis w Bibliotece Instytutu Geograficznego Uniwersytetu Wrocławskiego.

C e g ł a, J., 1965, *Porównanie utworów pyłowych kotlin karpackich z lessami Polski* (summ.: *On the origin of the quaternary silts in the Carpathian Mountains),* Annales UMCS, sectio B, vol. XVIII.

C e g ł a, J., 1969, *Influence of capillary ground moisture on eolian accumulation of loess,* Bull. Acad. Pol. Sc., Sér. sc. geol. et géogr., XVII.

C e g ł a, J., 1971, *Z problematyki lessów i utworów lessopodobnych na Opolszczyźnie,* Studia Geograficzno-Fizyczne z Obszaru Opolszczyzny, II (w druku).

C e g ł a, J., N a k o n i e c z n y, S., 1961, *Izbica — Dependence of loess sedimentation on old relief,* Guide-Book of Excursion E, Lublin Upland, Symposium on loess, INQUA.

C e g ł a, J., H a r a s i m i u k, M., 1968, *Właściwości fizyczne utworów lessowych i lessopodobnych,* Folia Soc. Sci. Lublinensis, sectio D, vol. 7/8.

C e g ł a, J., H a r a s i m í u k, M., 1969, *Niektóre właściwości fizyczne utworów pyłowych kotlin karpackich i lessów wyżynnych* (summ.: *Some physical properties of the silt material of the Carpathian basins and upland loesses),* Annales UMCS, sectio B, vol. XXII.

C e g ł a, J., K l i m e k, K., 1968, *Osady kopalnych zagłębień bezodpływowych jako wskaźnik degradacji moreny dennej w obszarach starszych zlodowaceń* (summ.: *Deposits of closed depressions as indicators of ground moraine degradation in areas of older glaciations),* Przegląd Geograficzny, XL.

C e g ł a, J., D ż u ł y ń s k i, S., 1970, *Układy niestatecznie warstwowane i ich występowanie w środowisku peryglacjalnym* (summ.: *Systems with reversed density gradient and their occurrence in periglacial zones),* Acta Universitatis Wratislaviensis, 124, Studia Geograficzne XIII.

C e g ł a, J., D ż u ł y ń s k i, S., K w i a t k o w s k i, S., 1967, *Fractures resulting from liquid infiltration into dry powdered materials,* Bull. Acad. Pol. Sc., Sér. sc. géol. et géogr., XV.

C e g ł a, J., B u c k l e y, T., S m a l l e y, I., J., 1971, *Microtextures of particles from some European loess deposits, Sedimentology,* 17.

Chałczewa, T. A., Faustowa, M. A., 1970, *O niekotorych razliczijach w minierałogiczeskom sostawie iskopajemych poczw i lossow wierchniego plejstocena centralnoj czasti Russkoj Rawniny*, Biull. Kom. po Izucz. Czetwiert. Pier., 37.

Chepil, W. S., 1955, *Factors that influence clod structure and erodibility of soil by wind: IV. Sand, silt and clay*, Soil Science, 80.

Chepil, W. S., 1957, *Sedimentary characteristics of dust storms: I. Sorting of wind-eroded soil material*, American Journal of Science, 255.

Czeppe, Z., Kozłowski, J. K., Krysowska, M., 1963, *Le gisement paléolithique de loess de Racibórz-Ocice en Haute Silésie*, Folia Quaternaria, 15.

Czerniachowskij, A. G., 1966, *Sowriemiennoje lessoobrazowanije w wysokogornych stiepjach wnutrienniego TjanSzanja*, Sowriemiennyj i czetwierticznyj kontinientalnyj litogieniez, Izdat. „Nauka", Moskwa.

Czżan-Czun-hu, 1959, *O gieniezisie i processie obrazowanija lessow rajona Lundun w Siewiero-Zapadnom Kitaje*, Tr. Kom. po Izucz. Czetw. Pier. AN SSSR, 14.

Dionne, J. C., 1970, *Structures sédimentaires dans du fluvioglaciaire*, Rev. Géogr. Montr., vol. XXIV, no. 3.

Dionne, J. C., 1971, *Contorted structures in unconsolidated quaternary deposits, Lake Saint-Jean and Saguenay regions, Quebec*, Rev. Géogr. Montr., vol. XXV, no. 1.

Dobrowolski, A. B., 1923, *Historia naturalna lodu*.

Dobrzański, B., 1950, *Fizyczne właściwości lessu*, Przegląd Geograficzny, XXII.

Dobrzański, B., 1961, *Kazimierówka — Wierzchowiska, Loesslike covering deposits*, Guide-Book of Excursion E, Lublin Upland, Symposium on Loess, INQUA.

Dobrzański, B., Malicki, A., 1949 *Rzekome loessy i rzekome gleby loessowe w okolicy Leżajska* (summ.: *Pseudoloesses and pseudo-loess soils in the environment of Leżajsk*), Annales UMCS, sectio B, vol. III.

Dobrzański, B., Uziak, S., 1966, *Rozpoznawanie i analiza gleb*, UMCS Lublin.

Dücker, A., 1939, *Neue Erkenntnisse auf dem Gebiete der Frostforschung; Bodenmechanik und neuzeitlicher Strassenbau*, VuR Verlag, Berlin.

Dylik, J., 1952, *Głazy rzeźbione przez wiatr i utwory podobne do lessu w środkowej Polsce* (summ.: *Wind-worn stones and loess-like formations in Middle Poland*), Biull. PIG, 67.

Dylik, J., 1954, *Zagadnienie genezy lessu w Polsce*, Biuletyn Peryglacjalny, 1.

Dylik, J., 1960, *Rhythmically stratified slope waste deposits*, Biuletyn Peryglacjalny, 8.

Dylik, J., 1969, *L'action du vent pendant le dernier âge froid sur le territoire de la Pologne centrale*, Biuletyn Peryglacjalny, 20.

Dżułyński, S., 1963a, *Wskaźniki kierunkowe transportu w osadach fliszowych* (summ.: *Directional structures in flysch*), Studia Geologica Polonica, XII.

Dżułyński, S., 1963b, *Polygonal structures in experiments and their bearing upon some periglacial phenomena*, Bull. Acad. Pol. Sc., Sér. sc. géol. et géogr., XI. nr 3.

Dżułyński, S., Radomski, A., 1955 *Pochodzenie śladów wleczenia na tle teorii prądów zawiesinowych* (summ.: *Origin of groove casts in the light turbidity current hypothesis*), Acta Geol. Polon., V.

E d e l m a n, C. H., 1951, *La topographie nivéo-éolienne des sables de couverture des environs de Didam (Pays-Bas)*, Sediment et Quaternaire France.

F i e d o r o w i c z, B. A., 1960, *Woprosy proischożdienija lessa w swiazi s usłowijami jego rasprostranienija w Jewrazii*, Trudy Inst. Gieogr., 80.

F i n k, J., 1966, *Zamietki k woprosu o lessie*, Sowriemiennyj i Ozietwierticznyj Kontinientalnyj Litogieniez, Izdat. „Nauka", Moskwa.

F i n k, J., 1969, *Bemerkungen zu den Exkursion über das Problem: „Löss — Periglazial — Paläolithikum"*, Loess — Périglaciaire — Paléolithique sur le territoire de l'Europe moyenne et orientale. L'édition préliminaire pour le VIII Congrès de l'INQUA — Paris, 1969.

F r y e, J. C., W i l l m a n, H. B., 1963, *Loess stratigraphy, Wisconsinian classification and accretion-gleys in central western Illinois*, Midwestern Sec. Friends of the Pleistocene, 14th Ann. Mtg., Illinois Geol. Survey Guidebook, Ser. 5.

F r y e, J. C., W i l l m a n, H. B., G l a s s, H. D., 1960, *Gumbotil, accretion-gley, and the weathering profile*, Illinois Geol. Survey, Circular 295.

F r y e, J. C., S h a f f e r, P. R., W i l l m a n, H. B., E k b l a w, G. E., 1960, *Accretion-gley and the gumbotil dilemma*, American Journal of Science, 258.

G e r l a c h, T., K o s z a r s k i, L., 1967, *Observations sur la sédimentation éolienne actuelle dans les Carpathes Polonaises*, Association Géologique Carpato-Balkanique, VIIIème Congrès, Belgrade, Septembre 1967, Rapports Sedimentology.

G i e r a s i m o w, I. P., 1959, *Glejewyje psewdopodzoły centralnoj Jewropy i obrazowanije dwuczlennych pokrownych nanosow*, Izw. Akad. Nauk. SSSR., Ser. Geogr., 3.

G i e r a s i m o w, I. P., 1962, *Lessoobrazowanije i poczwoobrazowanije*, Izw. Akad. Nauk, SSSR, Ser. Geogr., 2.

G i e r a s i m o w, I. P., 1963, *Loess genesis and soil formation*, Report of the VIth International Congress on Quaternary, 4.

G r a b o w s k a - O l s z e w s k a, B., 1963, *Własności fizyczno-mechaniczne utworów lessowych północnej i północno-wschodniej części świętokrzyskiej strefy lessowej na tle ich litologii i stratygrafii oraz warunków występowania* (summ.: *Physicomechanical properties of loess deposits of the Northern and North-Eastern part of the Holy Cross Mts. loess-zone on the background of their lithology, stratygraphy and conditions of occurrence*), Biuletyn Geologiczny, 3.

G u s i e w, A. I., 1958, *Ob iskopajemych „sledach" mierzłoty i „ledianych" klinach w czetwierticznych otłożenijach*, Gieoł. Sborn. Lwowskogo Gieoł. Obszcz., 5.

H a a s e, G., L i e b e r o t h, I., Ruske, R., et al., 1969, *Diffusion et stratigraphie des loess (Allemagne Orientale)*, Loess — Périglaciaire — Paléolithique sur le territoire de l'Europe moyenne et orientale. L'édition préliminaire pour le VIII Congrès de l'INQUA — Paris, 1969.

H a l i c k i, B., S a w i c k i, L., 1934, *Less nowogródzki* (summ.: *Loess of Nowogródek*), Zbiór prac poświęconych E. Romerowi, Lwów.

H a m m e n V a n d e r T., M a a r l e v e l d, G. C., V o g e l, J. C., Z a g v i j n, W. H., 1967, *Stratigraphy, climatic succession and radiocarbon dating of the last glacial in the Netherlands*, Geologien en Mijnbouw.

H e a v e r, A. A., M c G r a t h, L., S m a l l e y, I., J., 1967, *Development of cohesion in granular materials during comminution*, Chemistry and Industry.

H ö r n e r, N. G., 1936, *Geomorphic processes in continental basing of Central Asia*, International Geological Congress. Report of the XVI Session, USA, 1933, 2.

J a h n, A., 1950, *Less, jego pochodzenie i związek z klimatem epoki lodowej* (summ.: *Loess, its origin and connection with the climate of the glacial epoch*), Acta Geologica Polonica, I.

J a h n, A., 1956, *Wyżyna Lubelska, rzeźba i czwartorzęd* (summ.: *Geomorphology and quaternary history of Lublin Plateau*), PAN Prace Geograficzne, 7.

J a h n, A., 1961, *Problemy geograficzne Alaski w świetle podróży naukowej odbytej w 1960 roku* (summ.: *Geographical problems of Alasca in the light of a research journey made in 1960*), Czasopismo Geograficzne, XXXII.

J a h n, A., 1963, *Gleby strukturalne Czarnego Grzbietu i problem utworów pylastych w Karkonoszach* (summ.: *The structural soil of the Czarny Grzbiet and the problem of „loess-like" sediments in the Karkonosze Mts.*), Acta Universitatis Wratislaviensis, 9.

J a h n, A., 1966, *Alaska*, PWN, Warszawa.

J a h n, A., 1968, *Wysoczyzna Głubczycka*, Studia Geograficzno-Fizyczne z Obszaru Opolszczyzny, I.

J a h n, A., 1969a, *Structures périglaciaires dans les loess de la Pologne*, Biuletyn Peryglacjalny, 20.

J a h n, A., 1969b, *Niveo-eoliczne procesy w Sudetach i ich działanie na glebę* (summ.: *The niveo-eolian processes in the Sudetes and their action on the soil*), Problemy Zagospodarowania Ziem Górskich, 5 (18).

J a h n, A., 1970, *Zagadnienia strefy peryglacjalnej*, PWN, Warszawa.

J e r s a k, J., 1965, *Stratygrafia i geneza lessów okolic Kunowa* (rés.: *Stratigraphie et genèse des loess aux environs de Kunów*), Acta Geographica Lodziensia, 20.

J e r s a k, J., 1970, *Główne kierunki wiatrów osadzających less w czasie ostatniego piętra zimnego* (rés.: *Les directions de vents prédominants accumulant les loess au cours du dernier étage froid*), Acta Geographica Lodziensia, 24.

J o h a n s s o n, S., 1959, *True and false icewedges in Southern Sweden*, Geogr. Annaler., XVI.

K a y, G. F., 1916, *Gumbotil, A new term in Pleistocene geology*, Science, 44.

K a y, G. F., P e a r c e, J. N., 1920, *The origin of gumbotil*, Journ. Geology, 28.

K e i l h a c k, K., 1920, *Das Rätsel der Lössbildung*, Z. Deutsch. Geol. Ges., 72.

K ę s i k, A., 1961, *Vallées des terrains loessiques de la partie ouest du Plateau de Nałęczów*, Annales UMCS, sectio B, vol. XV.

K o l a s a, M., 1963, *Geotechniczne własności lessów okolicy Krakowa* (summ.: *Geotechnical properties of loesses from the vicinity of Cracow*), PAN Prace Geologiczne, 18, Kraków.

K o ł o w, S. N., 1930, *Matieriały dla gieołogii i gidrogiełogii oz. Wijlikul Syrdarinskogo okruga Kazachskoj SSR*, Wiestnik Irrigacii, 5.

K o z ł o w s k i, L., 1922, *Starsza epoka kamienna w Polsce (paleolit)*, Prace Kom. Archeol. Pozn. Tow. Przyj. Nauk, I.

K r i g e r, N. I., 1965, *Less, jego swojstwa i swiaz s gieograficzeskoj sriedoj*, Izd. „Nauka", Moskwa.

K r i g e r, N. I., M o s k a l e w, M. R., B e k e r m a n, S. G., 1961, *Ob usłowijach zaleganija i proischożdienija lessa Sriedniej Azii*, Mat. Wsies. Sow. po Izucz. Czetwiert. Pierioda, 3.

K r i v a n, P., 1955, *La division climatique du pléistocene en Europe centrale et le profil de loess de Paks*, A Magyar Allami Földtani Intezet Evkönyve, 43.

K r y n i n e, D. P., J u d d, W. R., 1957, *Principles of engineering geology and geotechnics*, McGraw BC. Inc., New York, Toronto, London.

L a s k a r e v, V., 1951, *O stratigrafii kvartarnich naslaga Vojvodine*, Geoloski Ann. Balkansk. Paluostrva, XIX.

L e i g h t o n, M. M., M c C l i n t o c k, P., 1930, *Weathered zones of the drift sheets of Illinois*, Journ. Geology, 38.

L i n d n e r, L., 1967a, *Wyspa lessowa Borkowic koło Przysuchej* (rés.: *L'île loessique de Borkowice près Przysucha*), Acta Geologica Polonica, XVII.

L i n d n e r, L., 1967b, *Lessy dorzecza Uniejówki* (rés.: *Les loess du bassin de L'Uniejówka*), Acta Geologica Polonica, XVII.

L i n g e n, Van der, G. J., 1969, *The turbidite problem*, New Zealand Journal of Geology and Geophysics, 12.

L o ż e k, V., 1966, *Lessy i lessowidnyje porody Czechosłowakii*, Sowriemiennyj i Czetwierticznyj Kontinientalnyj Litogieniez, Izdat. „Nauka", Moskwa.

L u g n, A. L., 1935, *The Pleistocene geology of Nebraska*, Nebr. Geol. Survey Bull., 10, 2 nd., ser.

L u g n, A. L., 1962, *The origin and sources of loess*, University of Nebraska Studies, New Series, 26.

Ł u k a s z e w, W. K., 1970. *Gieochimija czetwierticznogo litogienieza*, Izdat. „Nauka" i Tiechnika", Mińsk.

Ł u k a s z e w, W., M o j s k i, J. E., 1968, *Badania geochemiczne lessów Wyżyny Lubelskiej* (summ.: *Geochemical examinations of loesses in the Lublin Upland*), Kwartalnik Geologiczny, 12.

M a c o u n, J., 1962, *Stratigrafie sprašových pokryvu na Opavsku* (Zsfg.: *Stratigraphie der Lössdecken im Opava, Troppauer-Gebiet.*) Přirodovědny Časopis Slezský, XXIII.

M a l i c k i, A., 1946, *Kras loessowy*, Annales UMCS, Sectio B, vol. I.

M a l i c k i, A., 1950, *Geneza i rozmieszczenie lessów w środkowej i wschodniej Polsce* (summ.: *The origin and distribution of loess in Central and Eastern Poland*), Annales UMCS, Sectio, B, vol. IV.

M a l i c k i, A., 1961a, *The loess of the Lublin Upland*, Guide-Book Excursion E, INQUA VIth Congr., Poland.

M a l i c k i, A., 1961b, *The loess of the Miechów Upland*, Guide-Book of Excursion, INQUA VIth Congr., Poland.

M a l i c k i, A., 1961c, *The stratigraphic value of the loess profile in Pikulice (near Przemyśl)*, Annales UMCS, Sectio B. vol. XV.

M a l i c k i, A., 1967, *Lessy na obszarze Polski i ich związek z czwartorzędem*, Czwartorzęd Polski, PWN, Warszawa.

M a l i n o w s k i, J., 1959, *Wyniki badań geotechnicznych lessu między Kazimierzem Dolnym a Nałęczowem* (summ.: *Results of geotechnical investigations of loess between Kazimierz Dolny and Nałęczów (Lublin Upland)*, Kwartalnik Geologiczny, 3.

M a l i n o w s k i, J., 1964, *Budowa geologiczna i własności geotechniczne lessów Roztocza i Kotliny Zamojskiej między Szczebrzeszynem i Turobinem* (summ.: *Geological structure and geotechnical properties of loesses in Roztocze and in the Zamość Basin, between Szczebrzeszyn and Turobin*), IG, Prace, XII.

M a r k o v i ć - M a r j a n o v i ć, J., 1968, *Geomorfologia i stratygrafia czwarto-rzędu międzyrzecza Dunaj-Cisa w Jugosławii*. Cz. II (rés.: *Géomorphologie et stratygraphie du Quaternaire de la zone interfluviale Danube-Tisza en Yougoslavie*. II-e Partie), Annales UMCS, Sectio B, vol. XXI.

M a r u s z c z a k, H., 1958, *Charakterystyczne formy rzeźby obszarów lessowych Wyżyny Lubelskiej* (summ.: *Characteristic relief forms of the loess area within the Lublin Upland*), Czasopismo Geograficzne, XXIX.

M a r u s z c z a k, H., 1961, *Puławy-Włostowice. Northern boundary loessy cover. Morphology of the loess edge." The relation between dune sand and loessy cover*, Guide-Book of Excursion E, INQUA VIth Congr., Poland.

M a r u s z c z a k, H., 1963a, *Wind directions during the sedimentation period of the upper loess in the Vistula Basin*, Bull. Acad. Pol. Sc., Sér. sc. géol. et géogr. XI.

M a r u s z c z a k, H., 1963b, *Zróżnicowanie warunków geograficznych rozwoju erozji gleb w obszarach lessowych wschodniej części Europy środkowej*, Wiado-mości Instytutu Melioracji i Użytków Zielonych, III.

M a r u s z c z a k, H., 1963c, *Warunki geologiczno-geomorfologiczne rozwoju erozji gleb w południowej części województwa lubelskiego*, Wiadomości Instytutu Melioracji i Użytków Zielonych, III.

M a r u s z c z a k, H., 1964, *Conditions d'accumulation du loess dans la partie orien-tale de l'Europe Centrale*, Geogr. Polon. 2.

M a r u s z c z a k, H., 1967, *Kierunki wiatrów w okresie akumulacji lessu młodsze-go we wschodniej części Europy środkowej* (summ.: *Wind directions during the accumulation of the younger loess in East-Central Europe*), Rocznik Pol. Tow. Geol. XXXVII, z. 2.

M a r u s z c z a k, H., 1968, *Przebieg zjawisk w strefie peryglacjalnej w okresie ostatniego zlodowacenia w Polsce* (summ.: *The course of phenomena in the periglacial zone during the last glaciation*), IG PAN, Prace Geograficzne, 74.

M a r u s z c z a k, H., 1969, *Une analyse paléogéographique de la répartition du loess polonais et de ses caractères lithologiques directifs*, Biuletyn Perygla-cjalny, 20.

M o j s k i, J., 1965, *Stratygrafia lessów w dorzeczu dolnej Huczwy na Wyżynie Lubelskiej* (summ.: *Loess stratigraphy in the lower Huczwa river in the Lublin Upland*), Z Badań Czwartorzędu w Polsce, 11.

M ü l l e r, J., 1968, *Struktury eolicznych osadów piaszczystych z charakterystycz-nych obszarów wybrzeża Bałtyku i Polski centralnej*, Maszynopis w Katedrze Geologii Uniwersytetu Jagiellońskiego w Krakowie.

M ü l l e r, J., R u d o w s k i, S., 1967, *Zjawiska cementacji współczesnych osadów plażowych południowego Bałtyku* (summ.: *Cementation of recent beach sedi-ments of the Southern Baltic*), Prace Muzeum Ziemi, 11.

N a k o n i e c z n y, S., 1959, *Profil czwartorzędowy w Dębówce a zagadnienie pozio-mów humusowych w lessach* (Zsfg.: *Profil des Quartärs in Dębówka und das Problem der Genesis von Humusschichten in Loess*), Annales UMCS, sectio B, vol. XII.

P e l i s e k, J., 1969, *Caractéristique lithologique des loess*, Loess — Périglaciaire — Paléolithique sur le territoire de l'Europe moyenne et orientale. L'édition préliminaire pour le VIII Congrès de l'INQUA, Paris 1969.

P é w é, T. L., 1955, *Origin of the upland silt near Fairbanks, Alaska*, Geol. Soc. America Bull., 66.

P é w é T. L., 1965a, *Fairbanks area*, Guide-Book for Field Conference F, INQUA VIIth Congr.

P é v é, T. L., 1965b, *Middle Tanana river valley*, Guide-Book for Field Conference F, INQUA VIIth Congr.

P é w é, T. L., 1968, *Loess deposits of Alaska*, Inter. Geological Congr. Report of the Twenty-third Session, Czechoslovakia, 1968.

P é w é, T. L., H o p k i n s, D. M., G i d d i n g s, J. L., 1965, *The Quaternary geology and archeology of Alaska*, The Quaternary of the United States, a review volume for the VII Congress of the INQUA, Princeton Univ. Press.

Periglazialzone Löss und Paläolithikum der Tschechoslowakei, 1969, Brno.

P i e t i e l i n, W. P., 1956, *Uskoriennyj sposob opriedielenija median i kwartilej*, Biull. Mosk. Obszcz. Isp. Prir., otd. gieołog., XXXI.

P o ł y n o w, B. B., 1934, *Kora wywietrywanija*. Cz. 1, Izdat. AN SSSR.

P o ż a r y s k i, W., 1953, *Plejstocen w przełomie Wisły przez wyżyny południowe* (summ.: *The Pleistocene in the Vistula gap across the Southern Uplands*), IG Prace IX.

R a c z k o w s k i, W., 1960, *Less w okolicach Henrykowa na Dolnym Śląsku* (summ.: *Loess in the region of Henryków in Lower Silesia*), Biuletyn Peryglacjalny, 7.

R e a s i d e, J. D., 1964, *Loess deposits of the South Island, New Zealand, and soil formed on them*, New Zealand Journal of Geology and Geophysics, 7.

R a y, L. L., 1967, *An interpretation of profiles of weathering of the Peorian loess of Western Kentucky*, U. S. Geol. Survey Prof. Paper, 575-D.

R e g e r, R. D., P é w é, T. L., H a d l e i g h - W e s t, F., S k a r l a n d, I., 1964, *Geology and archeology of the Yardang Flint Station*, Univ. Alaska Anthrop. Papers, 12.

R o h d e n b u r g, H., 1968, *Jungpleistozäne Hangformung in Mitteleuropa — Beiträge zur Kenntnis, Deutung und Bedeutung ihrer räumlichen und zeitlichen Differenzierung*, Göttinger Bodenkundliche Berichte, 6.

R o h d e n b u r g, H., M e y e r, B., 1966, *Zur Feinstratigraphie und Paläopedologie des Jungpleistozäns nach Untersuchungen an südniedersächsischen und nordhessischen Lössprofilen*, Mitteilungen der Deutschen Bodenkundlichen Gesellschaft, 5.

R o k i c k i, J., 1952, *Warunki występowania utworów pyłowych i lessów na Dolnym Śląsku* (Zsfg.: *Das Vorkommen von Staubbildungen und Loess in Niederschlesien*), Annales UMCS, sectio B, vol. V.

R ö p k e, E., 1928, *Die Struktur des Löss*, Leopoldina, 5.

R ó ż y c k i, S. Z., 1967, *Plejstocen Polski Środkowej*, PWN Warszawa.

R u h e, R. V., 1954, *Relations of the properties of Wisconsin loess to topography in western Iowa*, Amer. Journ. Sci., 252.

R u h e, R. V., 1965, *Quaternary paleopedology*, The Quaternary of the United States, Princeton Univ. Press.

R u h e, R. V., 1969, *Application of pedology to Quaternary research*, Pedology and Quaternary Research, Univ. Alberta Press.

R u s s e l, R. J., 1944, *Lower Mississippi valley loess*, Bull. Geol. Soc. Amer., 55.

R ü c k l i, R., 1950, *Der Frost im Baugrund*, Springer Verlag, Wien.

S a l i s b u r y, J. W., G l a s e r, P. E., S t e i n, B. A., V o n n e g u t, B., 1964, *Adhesive behaviour of silicate powders in ultrahigh vacuum*, Journ, Geophys. Res., 69.

S a w i c k i, L., 1952, *Warunki klimatyczne akumulacji lessu młodszego w świetle wyników badań stratygraficznych stanowiska paleolitycznego lessowego na Zwierzyńcu w Krakowie* (rés.: *Les conditions climatiques de la période de l'accumulation du loess supérieur aux environs de Cracovie*), PIG Biull. 66.

S c h e r f, E., 1935, *Geologische und morphologische Verhältnisse des Pleistozäns und Holozäns der grossen ungarischen Tiefebene*, Jhber. D. Königl. Ung. Geolog. Anstalt.

S c h ö n h a l s, E., 1955, *Kennzahlen für den Feinheitsgrad des Lösses*, Eiszeitalter und Gegenwart, 6.

S i u t a, J., 1960, *O procesach glejowych i wytrąceniach żelazistych w lessach okolic Kazimierza Dolnego* (summ.: *The processes of gley soil formation and ferruginous precipitates in loess in the vicinity of Kazimierz Dolny*), Przegląd Geograficzny, XXXII.

S i u t a, J., 1961, *Wpływ procesu glejowego na kształtowanie się cech morfologicznych i właściwości chemicznych profilu glebowego* (summ.: *Influence of the gleying process on formation of the morphological and chemical properties of soil profiles*), Roczniki Gleboznawcze, X.

S i u t a, J., M o t o w i c k a, T., 1963, *Znaczenie wytrąceń żelazistych w stratygrafii niektórych utworów czwartorzędowych* (summ.: *Importance of ferruginous concretions for the stratigraphy of some sedimentary rocks*), Przegląd Geograficzny, XXXV.

S i u t a, J., M o t o w i c k a - T e r e l a k, T., 1967, *Z badań nad systematyką glebowych wytrąceń żelazistych* (summ.: *Study on the systematics of iron precipitations in soils*), Pamiętnik Puławski — Prace IUNG, 30.

S i u t a, J., M o t o w i c k a - T e r e l a k, T., 1969, *The origin and systematics of ferruginous precipitates in Quaternary formations and in present-day soils*, Biuletyn Peryglacjalny, 18.

S k w o r c o w, J. A., 1965, *O fizyko-gieograficzeskoj obstanowkie lednikowych epoch i obrazowanii Jużno-Russkich lossow*, Naucznyje Trudy, 266, Taszkent.

S m a l l e y, I. J., 1967, *Tensil strength of granular materials*, Nature, vol. 216, no. 5118.

S m a l l e y, I. J., 1970a, *Calcium carbonate encrustation on quartz grains in loess from the Karlsruhe region*, Naturwissenschaften, 57.

S m a l l e y, I. J., 1970b, *Cohesion of soil particles and the intrinsic resistance of simple soil systems to wind erosion*, The Journal of Soil Science, vol. 21, no. 1.

S m a l l e y, I. J., P e r r y, N. H., 1969, *The Keilhack approach to the problem of loess formation*, Proceedings of the Leeds Philosophical and Literary Society, Sci, sec., vol. X, part. III.

S m a l l e y, V., S m a l l e y, I. J., 1964, *Tensile strength of granular materials*, Nature, vol. 202, no. 4928.

S m a l l e y, I. J., T a y l o r, R. L. S., 1970, *Loess — the yellow earth*, Science Journal, vol. 6, no. 2.

S m a l l e y, I. J., H e a v e r, A. A., M c G r a t h, L., 1967, *Variation of cohesion with fineness of mineral powders and development of cohesion during grinding*, Transactions Section C of the Institution of Mining and Metallurgy, 76.

S m i t h, H. T. U., 1966, *Lessowyje otłożenija S. Sz. A. Sowriemiennyj i Czetwierticznyj Kontinientalnyj Litogieniez*, Izdat. „Nauka", Moskwa.

Soil Survey Staff, 1962, *Soil survey manual — identification and nomenclature of soil horizons*, Suppl. to US Dept. Agriculture Handbook 18.

S w i n e f o r d, A., F r e y, J. C., 1951, *Petrography of the Peoria loess in Kansas*, Journ. Geol., 59.

S z c z e p a n k i e w i c z, S., 1968, *Rzeźba doliny Odry na Opolszczyźnie*, Studia Geograficzno-Fizyczne z Obszaru Opolszczyzny, I.

S z c z e p a n k i e w i c z, S., 1970, *Cechy niektórych pokryw późnoczwartorzędowych* (summ.: *Characters of some late Quaternary covers*), Acta Universitatis Wratislaviensis, 124, Studia Geograficzne XIII.

T e r u g g i, M. E., 1957, *The nature and origin of Argentine loess*, Journ. of Sedim. Petrol., 27.

T h o u l e t, J., 1908, *De l'influence du vent dans le remplissage du lit de l'océan*, Compt. Rend. Acad. Sci., Paris vol. 146.

T r i c a r t, J., 1953, *Géomorphologie dynamique de la steppe russe*, Revue de Géomorphologie Dynamique, 4.

T r o w b r i d g e, A. C., 1961, *Accretion-gley and the gumbotil dilemma — discussion*, Amer. J. Sci., 259.

T w e n h o f e l, W. H., 1932, *Treatise on sedimentation*, Dover Publ., Inc., New York.

T y r c h a - C z y ż, T., 1969, *Charakterystyka petrograficzno-chemiczna osadów plejstoceńskich z odkrywki w Białej Prudnickiej* (summ.: *Petrographic-chemical characteristic of the Pleistocene deposits exposed at Biała Prudnicka*), Prace Geologiczno-Mineralogiczne, II.

T y r c h a - C z y ż, T., 1970, *Wpływ czynników fizykochemicznych na powstawanie wytrąceń żelazistych* (summ.: *The influence of some physico-chemical factors upon forming iron precipitation*), Acta Universitatis Wratislaviensis, 124, Studia Geograficzne XIII.

U z i a k, S., 1964, *Zagadnienie typologii niektórych gleb pyłowych Pogórza Karpackiego* (summ.: *Typology of some silt soils of the Carpathian Foothills*), Annales UMCS, sectio B, vol. XVII.

V a š í č e k, M., 1947, *Kritické poznámky o kalciumkarbonáut ve spraších*, Sbornik Československé Společnosti Zeměpisne, 52.

W a r n, F. G., C o x, W. H., 1951, *A sediment study of dust storms in the vicinity of Lubbock, Texas*, Amer. Journ. Sci., 249.

W a s h b u r n, A. L., 1969, *Weathering, frost action, and patterned ground in the Mesters Vig District, Northeast Greenland*, Meddelelser om Gronland, vol. 176,, no. 4.

W i e k l i c z, M. F., 1968, *Stratigrafija lessowoj formacji Ukrainy i sosiednich stran*, Izdat. „Naukowa Dumka", Kijew.

W i e l i c z k o, A. A., 1965, *Woprosy gieochronołogii lessow Jewropy*, Izdat. AN SSSR, ser. gieogr., 4.

W i l l i s, B., 1907, *Quaternary Huang-tu formation in Northwestern China*, Research in China, vol. 1.

W i l l m a n, H. B., F r y e, J. C., 1970, *Pleistocens stratigraphy of Illinois*, Illinois State Geol. Survey, Bull. 94.

W i ł u n, Z., 1969, *Mechanika gruntów i gruntoznawstwo drogowe*, WKŁ, Warszawa.

W o j t a n o w i c z, J., Z i n k i e w i c z, A., 1966, *Występowanie zapylenia eolicznego i opadu pyłu w Polsce*, Folia Soc. Sc. Lubl., sec. D, 5/6, Lublin.

W o j t a n o w i c z, J., Z i n k i e w i c z, A., 1968, *Zapylenie i opad pyłu eolicznego na obszarze województwa lubelskiego w kwietniu 1965 r.* (summ.: *Dust obscuration and eolian dust deposition in the Lublin Voivodeship, April 1965*), Annales UMCS, sectio B, vol. XXI.

Z e u n e r, F. E., 1953, *Das Problem der Pluvialzeiten*, Geol. Rundschau, 41.

Z i e m n i c k i, S., 1951, *Wstępne badania nad erozją lessów Lubelszczyzny*, Annales UMCS, sectio E, vol. VI.

Z i n k i e w i c z, W., 1950, *Perturbacja w przezroczystości atmosfery oraz opad pyłu eolicznego na Lubelszczyźnie w kwietniu 1948 roku* (summ.: *Optic perturbation of the atmosphere and dust deposit on the province of Lublin, April 1948*), Annales UMCS, sectio B. vol. IV.

48

Reprinted from *Bull. Acad. Pol. Sci. Geol. Geog. Ser.*, **17**(1), 25–27 (1969)

Influence of Capillary Ground Moisture on Eolian Accumulation of Loess

by

J. CEGŁA

Presented by M. KSIĄŻKIEWICZ on February 7, 1969

The present paper deals with loess deposits which blanket the uplands and elevated topographical elements. The eolian origin of such loess may be accepted without reservation.

There exists a vast literature on the subject of eolian loess. The questions discussed, however, are mostly those pertaining to the source and transportation of loess particles whereas but few articles have been published on the subject of vertical accumulation of loess. The statement is frequently made that loess particles are trapped by vegetation, for instance, by grass or arctic moss [2], [5], [6]. The dry grassland, however, is not particularly apt to intercept considerable amounts of fine dust. The movements of grass induce turbulence of the air, and fine particles are being swept away and prevented from permanent settling. Assuming that under certain conditions the grassland is buried under loess, one should expect to find beneath it the corresponding soil horizons. In fact, such soil horizons do exist under loess deposits of eastern and southeastern Europe. They are clearly recognizable in the eastern part of the Lublin Upland and in the Roztocze area [8]. However, west of the Bug—Wieprz water-shed [7], the soil horizons tend to disappear, and the loess deposits extending over the southwestern part of Poland are practically devoid of them.

It is known that loess accumulations are not necessarily associated with steppes, though the significance of grass for the preservation of loess is unquestionable.

The rather common occurrence of loess blanketing bare rock surfaces, indicates that some other explanation is needed to show how loess could accumulate on flat rocky surfaces devoid of vegetation. The explanation suggested in this paper is that the loess particles were primarily deposited on surfaces wetted by capillary waters, i.e. they were laid down in places where the capillary fringe rises to the ground--air interface. This possibility has been overlooked by many authors, when discussing the origin of loess deposits. To the author's knowledge, the only reference relevant to this question is the publication by Hörner — 1933 [4] on recent geomorphic processes in the Gobi desert, from which the following quotation is taken: "Fine wind—

blown dust settles inside as well as outside the arid region, but remains only where there is moisture enough in the ground or the atmosphere to bind, directly or through chemical action, the particles that have settled" (l.c., p. 734).

Rocky surfaces wetted by capillary waters act as adhesive tape for dust particles that may settle upon them, and the capillary forces prevent such particles from being swept away. Once a thin coating of dust is formed, further accumulation proceeds in much the same manner. Concomitantly with deposition of loess the capillary meniscus rises so that the ascending loess-air interface contains enough moisture to intercept more dust particles. For the grain-size of loess (0.05—0.02 mm.), the capillary forces are relatively high, and consequently the particles deposited are precluded from further transportation. It shoud be emphasized that loess is particularly suitable for transmitting water by capillarity, and that the height of capillary rise in such sediments is about 20 m. [1]. It is conceivable that under such conditions considerable blankets of loess may form upon bare rock surfaces.

The concept of capillary forces as a decisive factor in loess accumulation explains the occurrence of isolated patches of loess or "loess islands" [5], [7], [9]. In the light of the above remarks these patches may be regarded as determined by the distribution of moisture in the ground. The concept may also help to clarify some of the confusing questions which present themselves in connection with the widely discussed problem of the origin of loess deposits. One of such questions is that raised by Fisk — [3], p. 354, namely: "how loess formations could be laid down on steep eroded slopes as parallel layers unaffected by mass movements". The other is the objection invoked by Russel — 1944 [10], who pointed out that "against eolian origin of loess it may be urged that no actual or hypothetical wind directions could account for its distiribution. It covers slopes leading in all possible directions [...]" (l.c., p. 23).

The introduction of capillary waters as the main factor determining the distribution of loess, renders such objections pointless. This does mean, however, that all deposits described as loess are necessarily of eolian origin.

In view of the above considerations it is suggested that the presence of moisture in the ground is of primary importance for the accumulation of loess on elevated topographical elements. In many instances this only rendered such an accumulation possible. It should be borne in mind, however, that the presence of capillary moisture as postulated here, applies to the areas of accumulation, and not necessarily to the zones of deflation. The hypothesis outlined above does not invalidate the widely accepted view that the climatic conditions attending the accumulation of loess were cold and relatively dry [9]. As shown by Hörner [4], in cold regions even desert ground may contain a considerable amount of moisture.

The problems discussed in this paper are now being investigated experimentally by the present author in the Institute of Geography of the University of Wrocław. The results of these experiments will be presented in a forthcoming publication.

INSTITUTE OF GEOGRAPHY, UNIVERSITY, WROCŁAW
(INSTYTUT GEOGRAFICZNY, UNIWERSYTET, WROCŁAW)

REFERENCES

[1] G. Beskow, *Tjälbildningen och Tjällyftningen,* Sveriges Geologiska Undersökning, Stockholm, 1935.

[2] J. Budel, Eiszeitalter u. Gegenwart, 1 (1951).

[3] H. N. Fisk, Journal of Geol., 59 (1951).

[4] N. G. Hörner, International Geol. Congr., Rep. XVI session, 1933.

[5] J. Jersak, Acta Geogr. Lodz., 20 (1965).

[6] L. Kozłowski, Prace K m. Archeol., Pozn. TPN., 1 (1922).

[7] A. Malicki, *Czwartorzęd Polski* [in Polish], [*The Polish Quaternary*], PWN, Warszawa, 1967.

[8] J. E. Mojski, *Czwartorzęd Polski* [in Polish], [*The Polish Quaternary*], PWN, Warszawa, 1967.

[9] S. Z. Różycki, *Plejstocen Polski Środkowej* [in Polish], [*The Central-Poland Pleistocene*], 1967.

[10] R. J. Russel, Bull. Geol. Soc. of America, 55 (1944).

Editor's Comments
on Papers 49 and 50

49 **RUHE**
Background of Model for Loess-Derived Soils in the Upper Mississippi River Basin

50 **MAROSI**
A Review of the Evolution of Theories on the Origin of Loess

In 1972 a conference, "Soil Development Sequences and Loess Distribution," was held in Illinois. Papers from this conference were published in *Soil Science* (March 1973). Two new research findings formed the theme of the conference; the first concerned the time of deposition of the last major increment in the Peoria loess, and the second the development and application of a chromatographic or energy model to the study of soil development. Fehrenbacher (1973) and Kleiss (1973) reported that the uppermost increment of loess was deposited in a relatively brief interval, not, as had been previously assumed, uniformly during the interval from about 22,000 to 12,500 years B.P. This upper-level material is the parent for many present loess-derived soils, and it is suggested that the idea must be revised that the Mollic Albaqualfs (Planosols) in thin loess are more strongly developed than are Typic Haplaquolls and Argiudolls (Humic Glei and Brunizems) in thick loess because of much greater weathering in the thin loess area during loess deposition.

Paper 49 is from the Illinois conference and provides background for studies on loess material in the Upper Mississippi River basin. Ruhe's views and observations on loess are set out in much more detail in his book *Quaternary Landscapes in Iowa*, along with some useful quotations from Mark Twain.

Paper 50 presents a Roumanian view of the "loess problem" and provides a comparison to the currently predominant North American views on loess which is useful because, unlike so many Russian authors who seem blissfully unaware that loess exists outside Russia, Marosi is well read in Western literature. However, this review does illustrate the

395

persistence of *in-situ*-type theories in the literature of eastern Europe. Class 2 of group V of the Marosi groupings includes the works of Berg and R. J. Russell and is a very large group. It includes Lysenko, and a recent Russian view on loess can be illustrated by a quotation from his book *Loessial Rocks of the European USSR:*

> The group of loessial rocks, or rocks of loessial formation, comprises (according to present-day usage) loess (proper, or typical) and various loesslike rocks. Loess usually means a loose material, light yellow (straw yellow) in color, with porosity 40–50 percent, nonstratified, macroporous (i.e., including fine branching channels visible to the naked eye), calcareous, substantially silty, with coarse-silty particles (0.05–0.01 mm), amounting to 30–50 percent or more and predominating over the fine-silty particles (0.01–0.002 mm), with a tendency to form vertical jointings on slopes. Recently, Kriger (1962) has voiced the opinion that this definition should be complemented by indicating that loess has a mantle-like mode of occurrence and does not contain interlayers of pebbles or sand.
>
>   The loesslike rocks do not show all these features of loess. Some of these features may be indistinct or even absent. The transition between loess and loesslike material is gradual. Obruchev (1948) and some other researchers apply the term loess to eolian rocks only. This, however, is incorrect, because aqueous deposits which have undergone a lengthy stage of subaerial existence may exhibit all the features of loess.

For a digest of Obruchev's views see Paper 13, published in English at about the same time as the paper cited above appeared in Russian. The Lysenko passage points to the very significant fact that many European loess researchers recognize a variety of loess types (see Table 1 in the Introduction); this point has been made very clearly by Pécsi (1965, 1967).

The last word rests with Lysenko:

> Because of its widespread distribution, the uncertainty of its gensis, its specific aspect, constitution, composition and properties, as well as its unfavorable features from an engineering-geological standpoint, loessial rock has long been of interest to researchers. It is being studied from various viewpoints by geologists of the Quaternary period, lithologists, mineralogists, engineering geologists, geographers, soil scientists, construction engineers, meliorators [improvers? development engineers? land reclaimers? ed.] and other specialists. Nevertheless, many aspects of the genesis, stratigraphy, composition and properties of loessial rocks are still insufficiently studied.

REFERENCES

Fehrenbacher, J. B. 1973. Loess stratigraphy, distribution, and time of deposition in Illinois. Soil Sci. 115, 176–182.

Kleiss, H. J. 1973. Loess distribution along the Illinois soil-development sequence. Soil Sci. 115, 194–198.

Lysenko, M. P. 1967. Loessial rocks of the European U.S.S.R. English translation published in 1973 by Israel Program for Scientific Translations, Jerusalem, 176p.

Pécsi, M. 1965. Genetic classification of the deposits constituting the loess profiles of Hungary. Acta Geol. Acad. Sci. Hung. 9, 65–84.

——. 1967. Horizontal and vertical distribution of loess in Hungary. Stud. Geomorph. Carpatho-Balcanica 1, 13–20.

49

Reprinted by permission from *Soil Sci.*, 115(3), 250–253 (1973)

BACKGROUND OF MODEL FOR LOESS-DERIVED SOILS IN THE UPPER MISSISSIPPI RIVER BASIN

ROBERT V. RUHE

Indiana University, Water Resources Research Center

Received for publication September 29, 1972

ABSTRACT

The loess province of the Upper Mississippi River Basin has repetitive soil-geomorphic patterns. Loess thins systematically from a major source area, and as it does so, soils formed on the loess have progressively greater development. Many studies have been made, but the explanations of soil formation continue in argument. The background of the studies and the arguments in the literature are indexed by subject matter and may be useful to students.

The loess province of the Upper Mississippi River Basin in midwestern United States provides a unique model for soil-geomorphology studies. The loess generally thins systematically from a major source area, and as it does so, soils formed on the loess have progressively greater development. Particle size usually becomes finer as the loess thins, and depth to more impermeable paleosols also decreases, so that the solum of the ground soil is progressively closer to the substratum barrier. Generally, relief decreases and ridges broaden in the thin-loess areas. These facts form the basis for three explanations for increased soil development along traverses from thick to thin loess:

(1) Finer textured parent material in the thin-loess area permits more intensive soil formation than coarser textured parent material in the thick-loess area regardless of time. This is usually called the parent-material effect.

(2) Shallow depth to more impermeable paleosols on broad ridges provides a wetter weathering environment than in the thick-loess area regardless of time. This is the wetness effect.

(3) Assuming uniform rate at any site, deposition was slower in the thin-loess area than in the thick-loess area. The loess weathered more intensely during deposition in the thin-loess area and provided a head start over soil formation in the thick-loess area. This is the "effective age" idea.

The argument continues. The last view proposed by Smith (1942) in Illinois and Hutton (1947, 1951) in Iowa is highly untenable in each state. Detailed stratigraphic studies with radiocarbon control demonstrate conclusively that loess deposition was not at a uniform rate at any site and that the bulk of the loess in the thin-loess areas is the youngest increment. These relations have been spelled out in a series of writings in order: Ruhe (1969a); Ruhe, Miller, and Vreeken (1971); Kleiss, H. J. (1972), Loess distribution as revealed by mineral variations, Univ. Illinois unpublished Ph.D. dissertation; and see papers by Fehrenbacher and Kleiss in this issue of *Soil Science*.

There is no need to repeat results reported in the above literature citations. Instead, it is desirable to present a background of the literature on the model of loess and loess-derived soils. These references, organized under subject headings, may be useful to students and researchers. Some of the subjects such as properties, stratigraphy, origin, weathering, and soil formation are routine.

Two subjects require emphasis as they are generally avoided in loess-derived soils studies except by a few. One cannot use radiocarbon dating of loess and associated paleosols without becoming involved in the problems of sample contamination. As loess thins, paleosols become closer to the surface and its soil and vegetation. Thus, there is possibility of contamination of organic matter in buried soils by roots of vegetation growing at higher level and by downward percolation of soluble organic compounds such as humic and fulvic acids from the surface soil.

Problems arise in the validity of radiocarbon dates of the organic matter of the buried soils.

One must also recognize that during deposition of the loess and soil formation on it, there were changing patterns of Quaternary climates and vegetations. In the Upper Mississippi River Basin, loess was deposited mainly between 14,000 and 22,000 years ago. Glacier ice was present in parts of the region during that time and until about 11,000 years ago. Coniferous forest was regionally dominant until 9,000 to 10,000 years ago and the prairie did not become supreme until 7,000 to 8,000 years ago. Since then, there may have been a warmer, drier regime before changing to the current patterns. Soils were forming on the loess at some places during all of these changes. How, then, are their effects determined in the dynamics of the soil-genesis systems?

The following background, not quite exhaustive, is presented for your reading pleasure.

REFERENCES

Organized by subject matter. Three or more authors listed under first author only. Analogues of Mississippi Basin model given. Many papers may be cross-indexed.

I. *Properties, stratigraphy, and origin of loess.*

Barnhisel, R. F., et al. 1971. Loess distribution in central and eastern Kentucky. Soil Sci. Soc. Am. Proc. 35: 483–487.

Caldwell, R. E. and J. L. White. 1956. A study of the origin and distribution of loess in southern Indiana. Soil Sci. Soc. Am. Proc. 20: 258–263.

Fehrenbacher, J. B., et al. 1965. Loess distribution in southeastern Illinois and southwestern Indiana. Soil Sci. Soc. Am. Proc. 29: 566–572.

Fisk, H. N. 1951. Loess and Quaternary geology of the lower Mississippi Valley. J. Geology 59: 333–356.

Franzmeier, D. P. 1970. Particle size sorting of proglacial eolian materials. Soil Sci. Soc. Am. Proc. 34: 920–924.

Frazee, C. J., et al. 1970. Loess distribution from a source. Soil Sci. Soc. Am. Proc. 34: 296–301.

Glass, H. D., et al. 1968. Clay mineral composition, a source indicator of Midwest loess. (*In* The Quaternary of Illinois, Univ. Illinois Coll. Agric. Spec. Publ. 14, pp. 35–40.)

Hanna, R. M. and O. W. Bidwell. 1955. The relation of certain loessial soils of northeastern Kansas to the texture of the underlying loess. Soil Sci. Soc. Am. Proc. 19: 354–359.

Hutton, C. E. 1947. Studies of loess-derived soils in southwestern Iowa. Soil Sci. Soc. Am. Proc. 12: 424–431.

Hutton, C. E. 1951. Studies of the chemical and physical characteristics of a chrono-litho-sequence of loess-derived Prairie soils of southwestern Iowa. Soil Sci. Soc. Am. Proc. 15: 318–324.

Krinitzsky, E. L. and W. J. Turnbull. 1967. Loess deposits of Mississippi. Geol. Soc. Am. Spec. Paper 94.

Krumbein, W. C. 1937. Sediments and exponential curves. J. Geology 45: 577–601.

Malalucci, R. V., et al. 1969. Grain orientation in Vicksburg loess. J. Sediment. Petrol. 39: 969–979.

Péwé, T. L. 1951. An observation on wind-blown silt. J. Geology 59: 399–401.

Péwé, T. L. 1955. Origin of upland silt near Fairbanks, Alaska. Geol. Soc. Amer. Bull. 66: 699–724.

Péwé, T. L. 1968. Loess deposits of Alaska. Intern. Geol. Congr. Proc., Prague, 8: 297–309.

Ruhe, R. V. 1954. Relations of the properties of Wisconsin loess to topography in western Iowa. Am. J. Sci. 252: 663–672.

Ruhe, R. V., et al. 1971. Paleosols, loess sedimentation, and soil stratigraphy. (*In* D. H. Yaalon (ed.), Paleopedology—Origin, nature and dating of paleosols, Israel Univ. Press, Jerusalem. 41–60.)

Russell, R. J. 1940. Lower Mississippi Valley loess. Geol. Soc. Am. Bull. 55: 1–40.

Schultz, C. B. and J. C. Frye (eds.). 1968. Loess and related Eolian deposits of the world. Univ. Nebraska Press, Lincoln.

Frye, J. C., et al. Correlation of midwestern loesses with the glacial succession. 3–21.

Reed, E. C. Loess deposition in Nebraska. 23–28.

Lugn, A. L. The origin of loesses and their relations to the Great Plains in North America. 139–182.

Simonson, R. W. and C. E. Hutton. 1954. Distribution curves for loess. Am. J. Sci. 252: 99–105.

Smith, G. D. 1942. Illinois loess—variations in its properties and distribution. Illinois Agr. Exper. Sta. Bull. 490: 139–184.

Springer, M. E. 1948. The composition of the silt fraction as related to the development of soil from loess. Soil Sci. Soc. Am. Proc. 13: 461–467.

Wascher, H. L., et al. 1947. Loess in the southern Mississippi Valley—identification and distribution of the loess sheets. Soil Sci. Soc. Am. Proc. 12: 389–399.

Willman, H. B. and J. C. Frye. 1970. Pleistocene

stratigraphy of Illinois. Illinois Geol. Survey Bull. 64.

II. *Radiocarbon dating and stratigraphy of loess.*

Daniels, R. B. and R. L. Handy. 1959. Suggested new type section for the Loveland loess in western Iowa. J. Geology 67: 114–119.

Daniels, R. B., et al. 1960. Dark-colored bands in the thick loess of western Iowa. J. Geology 68: 450–458.

Frye, J. C., et al. 1962. Stratigraphy and mineralogy of the Wisconsinan loesses of Illinois. Illinois Geol. Survey Circ. 334.

Frye, J. C., et al. 1968. Mineral zonation of Woodfordian loesses of Illinois. Illinois Geol. Survey Circ. 427.

Frye, J. C. and O. S. Fent. 1947. The late Pleistocene loesses of Central Kansas. Kansas Geol. Survey Bull. 70: 29–52.

Frye, J. C. and A. B. Leonard. 1949. Pleistocene stratigraphic sequence in northeastern Kansas. Am. J. Sci. 247: 883–899.

Goldthwait, R. P. 1968. Two loesses in central southwest Ohio. (*In* The Quaternary of Illinois. Univ. Illinois Coll. Agric. Spec. Publ. 14: 41–47.)

Leighton, M. M. 1931. The Peorian loess and the classification of the glacial drift sheets of the Mississippi Valley. J. Geology 39: 45–53.

Leighton, M. M. and H. B. Willman. 1950. Loess formations of the Mississippi Valley. J. Geology 58: 599–623.

Leighton, M. M. and H. B. Willman. 1964. The stratigraphic succession of Wisconsin loesses in the Upper Mississippi River Valley. J. Geology 73: 323–345.

Ruhe, R. V. 1968. Identification of paleosols in loess deposits in the United States. (*In* Loess and related Eolian deposits of the world. Univ. Nebraska Press, Lincoln. 49–65.)

Ruhe, R. V., et al. 1968. Iowan Drift problem, northeastern Iowa. Iowa Geol. Survey Report Inv. 7.

Ruhe, R. V. 1969a. Application of pedology to Quaternary research. (*In* S. Pawluk (ed.). Pedology and Quaternary research. Univ. Alberta Press, Edmonton. 1–23.)

Schultz, C. B. and T. M. Stout. 1945. Pleistocene loess deposits of Nebraska. Am. J. Sci. 243: 231–244.

Thorpe, J., et al. 1951. Some post Pliocene buried soils of central United States. J. Soil Sci. 2: 1–19.

III. *Weathering and soil formation in loess.*

Beavers, A. H., et al. 1963. CaO-ZrO₂ molar ratios as an index of weathering. Soil Sci. Soc. Am. Proc. 27: 408–412.

Bray, R. H. 1937. Chemical and physical changes in soil colloids with advancing development in Illinois soils. Soil Sci. 43: 1–14.

Chapman, S. L. and M. E. Horn. 1968. Parent material uniformity and origin of silty soils in northwest Arkansas based on Zr-Ti contents. Soil Sci. Soc. Am. Proc. 32: 265–271.

Cowie, J. D. 1964. Loess in the Manawatu District, New Zealand. New Zealand J. Geol. & Geophys. 7: 389–396.

Foss, J. E. and R. H. Rust. 1962. Soil development in relation to loessial deposition in southeastern Minnesota. Soil Sci. Soc. Am. Proc. 26: 270–274.

Glenn, R. C., et al. 1960. Chemical weathering of layer silicate clays in loess-derived Tama Silt Loam of southwestern Wisconsin. 8th Nat. Conf. Clays and Clay Min. Proc. 63–83.

Khangarot, A. S., et al. 1971. Composition and weathering of loess-mantled Wisconsin- and Illinoian-age terraces in central Ohio. Soil Sci. Soc. Am. Proc. 35: 621–626.

Krusekopf, H. H. 1947. Gumbotil—Its formation and relation to overlying soils with clay pan subsoils. Soil Sci. Soc. Am. Proc. 12:413–414.

Ray, L. L. 1963. Silt-clay ratios of weathering profiles of Peorian loess along the Ohio Valley. J. Geology 71: 38–47.

Rieger, S. and R. L. Juve. 1961. Soil development in recent loess in the Matanuska Valley, Alaska. Soil Sci. Soc. Am. Proc. 25: 243–248.

Ruhe, R. V. 1969b. Quaternary Landscapes in Iowa. Iowa State Univ. Press, Ames. 28–54, 70–77, 114–127.

Shrader, W. D., et al. 1953. Soil survey of Holt County, Missouri. USDA Soil Survey Report Ser. 1939. No. 21.

Theisen, A. A. and E. G. Knox. 1959. Distribution and characteristics of loessial soil parent material in northwestern Oregon. Soil Sci. Soc. Am. Proc. 23: 385–388.

Ulrich, R. 1950. Some physical changes accompanying Prairie, Wiesenboden, and Planosol soil profile development from Peorian loess in southwestern Iowa. Soil Sci. Soc. Am. Proc. 14: 287–295.

Ulrich, R. 1951. Some chemical changes accompanying profile formation of the nearly level soils developed from Peorian loess in southwestern Iowa. Soil Sci. Soc. Amer. Proc. 15: 324–329.

IV. *Problems in radiocarbon dating soils in loess.*

Ballagh, T. M. and E. C. A. Runge. 1970. Clay-rich horizons over limestone—illuvial or residual? Soil Sci. Soc. Amer. Proc. 34: 534–536.

Campbell, C. A., et al. 1967a. Factors affecting the accuracy of the carbon-dating method in soil humus studies. Soil Sci. 104: 81–85.

Campbell, C. A., et al. 1967b. Applicability of the carbon-dating method of analysis to soil human studies. Soil Sci. 104: 217–224.

Olson, E. A. and W. S. Broecker. 1958. Sample contamination and reliability of radiocarbon dates. New York Acad. Sci. Trans. 20: 593–604.

Paul, E. A. 1969. Characterization and turnover rate of soil humic constituents. (*In* Pawluk, S. (ed.), Pedology and Quaternary research, Univ. Alberta Press, Edmonton, pp. 63–76.)

Peterson, F. J. 1969. Pedohumus—accumulation and diagenesis during the Quaternary. Soil Sci. 107: 470–479.

Polach, H. A. and J. Golson. 1966. Collection of specimens for radiocarbon dating and interpretation of results. Australian Inst. Aboriginal Studies Manual No. 2. Canberra.

Polach, H. A., et al. 1969. ANU radiocarbon data list III. Radiocarbon 11: 245–262.

Ruhe, R. V. 1969b. Quaternary landscapes in Iowa. Iowa State Univ. Press, Ames, pp. 223–231.

Scharpenseel, H. W., et al. 1968. Bonn radiocarbon measurements I. Radiocarbon 10: 8–28.

Yaalon, D. H. (ed.). 1971. Paleopedology—Origin, Nature and Dating of Paleosols. Israel Univ. Press, Jerusalem.

  Geyh, M. A., et al. Problems of dating Pleistocene and Holocene soils by radiometric methods. 63–75.

Scharpenseel, H. W. Radiocarbon dating of soils —problems, troubles, hopes. 77–88.

Polach, H. A. and A. B. Costin. Validity of soil organic matter radiocarbon dating: Buried soils in Snowy Mountains, southeastern Australia as example. 89–108.

V. *Quaternary environments and loess-derived soils*

Broecker, W. S., et al. 1960. Evidence for an abrupt change in climate close to 11,000 years ago. Am. J. Sci. 258: 429–448.

Bryson, R. A. and P. R. Julian. 1962. Proceeding of the conference on the climate of the eleventh and sixteenth centuries. Nat. Cen. Atmos. Res. Tech. Notes 63–1.

Bryson, R. A. and W. M. Wendland. 1967. Tentative climatic patterns for some late glacial and post-glacial episodes in central North America. (*In* Mayer-Oakes, W. J. (ed.), Life, land, and water, Proc. Conf. on Environmental Studies of Glacial Lake Agassiz, Univ. Manitoba Press, pp. 271–298.)

Bryson, R. A., et al. 1969. Radiocarbon isochrones on the disintegration of the Laurentide ice sheet. Arctic & Alpine Res. 1: 1–14.

Davis, M. B. 1963. On the theory of pollen analysis. Am. J. Sci. 261: 897–912.

Davis, M. B. 1969. Palynology and environmental history during the Quaternary period. Am. Scientist 57: 317–332.

Deevey, E. S. and R. F. Flint. 1957. Postglacial hypsithermal interval. Sciences 125: 182–184.

Dort, W. and J. K. Jones, (eds.). 1970. Pleistocene and Recent environments of the central Great Plains. Univ. Kansas Dept. Geology Spec. Publ. 3. Univ. Kansas Press, Lawrence.

  Ruhe, R. V. Soils, paleosols, and environment. 37–52.

Bryson, R. A., et al. The character of late-glacial and post-glacial climatic changes. 53–74.

Wright, H. E. Vegetational history of the Central Plains. 157–172.

Ross, H. H. The ecological history of the Great Plains—evidence from grassland insects. 225–240.

Schultz, C. B. and L. D. Martin. Quaternary mammalian sequence in the central great Plains. 341–353.

Ruhe, R. V. 1969b. Quaternary landscapes in Iowa. Iowa State Univ. Press, Ames. 169–195.

Walker, P. H. 1966. Postglacial environments in relation to landscape and soils on the Cary Drift, Iowa. Iowa State Univ. Agr. Exp. Sta. Res. Bull. 549: 838–875.

Watts, W. A. and H. E. Wright. 1966. Late-Wisconsinan Pollen and seed analysis from the Nebraska sandhills. Ecology 47: 202–210.

Wells, P. V. 1970. Postglacial vegetational history of the Great Plains. Science 167: 1574–82.

Wright, H. E. 1964. Aspects of the early postglacial forest succession in the Great Lakes region. Ecology 45: 439–448.

Wright, H. E. 1968a. The roles of pine and spruce in the forest history of Minnesota and adjacent areas. Ecology 49: 937–955.

Wright, H. E. 1968b. History of the Prairie peninsula. (*In* The Quaternary of Illinois, Univ. Illinois Coll. Agr. Spec. Publ. 14: 78–88.)

50

A REVIEW OF THE EVOLUTION OF THEORIES ON THE ORIGIN OF LOESS

Paul Marosi

Department of Geology, University of Cluj, Romania

This article was translated expressly for this Benchmark volume by Paul Marosi, University of Cluj, Romania, and Ian J. Smalley, The University of Leeds, from "Scurta Privire asupra Evolutiei Teorlilor Loessogenezei," Cluj Univ. Stud. Ser. Geol. Mineral., 15, 61–73 (1970)

Starting from an analytical study of more than 500 papers (read in the original or in summary) about loess deposits from China, Central and Middle Asia, Siberia, the Russian Plain, southeast, central, northwest, and southwest regions of Europe, as well as the loesses of North and South America, we can show the evolution of ideas and hypotheses about the origin and genesis of loess and loess-like deposits, ranging from the beginning of the twentieth century to the present day.

There are known to be more than 20 main theories about this problem, but taking into account several variants, their number may increase to 50.

"Classical" theories on the origin of loess, developed in the last century or in the first and second decades of this century, present loess as a simple and primary sediment, formed by a particular and external geological factor. A peculiarity of each of these theories and related discussion was the claim to universal validity for a unique origin of loess everywhere throughout its total distribution, and also the rejection of other concurrent theories. The main contradiction among these theories was the manner in which they explained the origin of loess deposits: that is, either by subaqueous accumulation or subaerial (continental?, terrestrial) processes.

The most important theories extant by the end of the last centruy may be classified as follows:

I. THEORIES OF SUBAQUEOUS ORIGIN

1. Theory of marine origin: Benningsen-Förder, 1857; Kingsmill, 1869–1870; Prestwich, 1894; and others.

2. Theory of lacustrine origin: Horner, 1837; Zeuschner, 1851; Wolf, 1860–1867; Pumpelly, 1866; Borisiak, 1867; Williamson, 1870; Richthofen, 1872–1886; Jentszch, 1877; Szabo, 1877; and others.

3. Theory of glacial–lacustrine origin: Braun, 1843; Wahnschaffe, 1878–1886; McGee and Call, 1882; and others.

The above theories now have only historical interest.

4. Theory of river (alluvial) origin: Darwin, 1839; Eichwald, 1846; Konrad, 1846;

Levakovski, 1861; Jentzsch, 1877; Muschetov, 1877-1886; Todd, 1879-1898; Werveke, 1924; Skvortzov, 1932-1961; and many other contemporary scientists.

5. The theory which claims that loess is a sediment of rivers fed by Quaternary glaciers: Lyell, 1834; Suess, 1866; Geikie, 1874-1900; Nehring, 1878-1890; and others. From this theory arises:

6. Theory of fluvioglacial origin: Agassiz, 1867; Kropotkin, 1876; Winchell, 1879; Jamieson, 1882; Dokuchaev, 1886-1892; Tanfiliev, 1922; Afanasiev, 1925-1933; Glinka, 1932; Sobolev, 1937; and many others.

Theories 4, 5, and 6 partially keep their value even today, being appreciated by many workers as important ways for the accumulation of "aleuritical" (silty) material, that is, the raw material of loess and loess-like deposits which will give rise to the loess in the future.

II. THEORIES OF SUBAERIAL ORIGIN (TERRESTRIAL ORIGIN, WITH LITTLE OR NO INVOLVEMENT OF WATER)

1. Theory of eolian origin: Erlich, 1848; Virlet d' Aoust, 1857; Richthofen, 1872-1886; Muschetov, 1886; Romanovski, 1878; Pumpelly, 1879; Middendorf, 1882; Obruchev, 1911-1957; and many other workers up to the present day.

Concerning the origin of material transported and deposited by wind, there are the following important variants of eolian theories:

a. Dust transported from the interiors of remote deserts and deposited on arid and semiarid steppes (Quaternary to present day) that are close to the deserts: Richthofen, 1886; Muschetov, 1886; Inkey, 1892; Halavats, 1897; Sibirtzev, 1900-1901; Treitz, 1903; Passarge, 1904; Sevastos, 1908; Obruchev, 1911-1957; Märzbacher, 1913; Loczy, 1916; Horusitzky, 1918; Schmittänner, 1925-1933; Münichsdörfer, 1926; Berkey and Morris, 1927; Keas, 1932; Moskvitin, 1933-1950; Bulla, 1934-1956; Scherf, 1936; Trofimov, 1946-1950; Ambroz, 1947; Flint, 1947-1963; Kriger, 1951-1962; Karlov, 1953; Popov, 1953; Kaveev, 1954; Fedorovich, 1955-1961; Krivan, 1955; Ian-Cijun-Tzian, 1957; and many workers from Richthofen's and Obruchev's schools, including many contemporary scientists.

b. Dust transported from near glaciers and from periglacial arid regions of Quaternary and contemporary glaciations: Tutkovski, 1899; Soergel, 1909; Obruchev, 1911-1948 (the theory of "warm" and "cold" loess); Krokos, 1926; Mirchink, 1927; Moskvitin, 1933-1950; Bratescu, 1933; Dücker, 1937; Flint, 1947-1963; Tavernier, 1948; Jahn, 1950-1961; Lomonovich, 1950-1961; Dmitriev, 1952; Tricart, 1956-1963; Zamorii, 1957-1961; and other workers.

c. Dust transported from local sources, for example from near glaciers and from neighboring periglacial tundra, or from the floodplains of great rivers: Chamberlin, 1897; Leverett, 1899; Murgoci, 1907; Winter, 1937; Smith and Fraser, 1935; Bryan, 1945; Swineford and Frye, 1945-1955; Flint, 1947-1963; Leighton and Willman, 1950; Schönhals, 1953; Ruhe, 1954; Hanna and Bidwell, 1955; Sitler and Baker, 1960; and others.

We must mention that besides the steadfast "orthodox eolianists" many of the cited authors consider that the eolian mechanism is not the exclusive method of loess formation, but that it is only one of several processes important in loess genesis. There are others who espouse with good reason modern theories of loess genesis; they maintain that the accumulated dust ("aleurite") may be derived both from eolian and other processes and they agree that the material must undergo complicated processes of transformation (diagenesis?) to attain the typical appearance of loess and loess-like deposits. Nevertheless, if not the most discussed theory, from the statistical point of view the theory of eolian origin is today the most widely accepted by workers on the genesis of loess. This theory usually appears in all textbooks, but unfortunately frequently in a very simplistic form, being presented as the only explanation with general validity.

III. THEORIES OF LOESS ACCUMULATION BY SLOPE PROCESSES

1. Theory of deluvial* origin: Folger, 1869; Richthofen, 1872–1886 (additional process to eolian accumulation); Smidt, 1873; Lapparent, 1883–1898; Dokuchaev, 1886–1892; Gurov, 1888; Pavlov, 1889–1904 (elaborated a modern form of the theory); Armashevski, 1903; Neustruev, 1910; Mateescu, 1927 (slope processes); Ilin, 1927–1935; Markov, 1948–1961; Kriger, 1951–1965; Vasilovski, 1951–1963; Liteanu, 1952–1966; Kaveev, 1954; Pécsi, 1966; and others.

2. Theory of proluvial* origin: Pavlov, 1903; Mateescu, 1927; Petrov, 1937–1953; Boganik, 1945; Shantzer, 1948; Mavlianov, 1948–1960; Popov, 1950; Vasilkovski, 1951–1952; Liteanu, 1952–1966; Pavlinov, 1959–1961; and others.

3. Theory of colluvial* origin: Russell, 1944; Sokolov, 1946; Markov, 1948–1961; Fisk, 1951; Vasilkovski, 1951–1952; Kriger et al., 1961; and others.

Some of the authors cited (especially those of our century) assume that slope processes take place in loess formation simultaneously with an aquatic process (water has a place in fact in the majority of slope processes too) or with eolian sedimentation of dust.

IV. THEORIES OF POLYGENETICAL ORIGIN

During the elaboration of the classical theories of loess genesis, few workers have considered the participation of multiple geological forces in loess and loess-like de-

Translators note: The terms *deluvial* and *proluvial* are used in the sense of A. P. Pavlov. Deluvium (from the Latin "deluo-," washing down) was coined by Pavlov in 1888 to designate the weathering products of bedrock deposited on slopes by rainwater streams. He thought that East European loess was deluvium. Proluvial loess accumulates on the plains. Pavlov used the term in 1903; he considered that the loess deposits in Turkestan were proluvium.

The term *colluvial* is used by Marosi in essentially the same way as Russell (see Paper 9) used it. Colluvial loess forms as a result of, or during, slope processes such as soil creep.

Berg referred to his own theory of loess formation by weathering and soil formation as an "eluvial" theory (see Paper 8). The "aleurite," used by Marosi, has been translated as silt; the terms are effectively equivalent in a lithological sense, but different defined size ranges may be involved.

posit formation. Germs of such differentiated (multifactor) ideas may be found in Richthofen's theory of loess genesis where he made a clear distinction between continental typical loess *(Landlöss)* and lacustrine secondary loess *(Seelöss)*. Similar ideas are contained in the papers of many important workers of that time (Nehring, 1878-1890; Sokolov, 1889; Halavats, 1897; Willis, 1907; and others).

In spite of the fact that so many theories of loess genesis had arisen, ideas evolved toward a single explanation of the problem; this convergent evolution was actually promoted, in our opinion, by the very great diversity of the early theories. Their authors invoke very many different geological external forces in order to explain tbe peculiarities of loess-like deposits.

In this century, there has been a major development of applied branches of geology (rock mechanics, hydrogeology, engineering geology, sedimentary petrography, lithology, etc). Important advances have also been made in interdiscplinary research into continental sediments of Quaternary age (geomorphological, mineralogical, petrographical, physicochemical researches as well as researches in neotectonics, paleogeography—especially in paleoclimatology—paleofaunal and paleobotanical studies, etc). Furthermore, there has been the interdisciplinary classification (by geologists and pedologists) of continental superficial deposits and the establishment of the main genetical types of these deposits(classification systems developed by Nikolaev, Shantzer, Iakovlev, and others). In all, a massive collection of data has accumulated as a consequence of widespread investigations, which have changed the essential character of discussions about loess genesis.

The above investigations show that acting in the frame of the territorial relief unit there are usually many factors of denudation, transportation, and resedimentation. In neighboring sectors the contrasting factors may lead to the accumulation at a local level of very different continental sediments of widely differing genetic types. The action of the different factors can simultaneously and convergently take place in one and the same sector, very often causing the accumulation of silty polygenetic sediments, such as diluvial–proluvial–eolian deposits, in combination with silt from landslides and solifluction action, all these being deposited on hillslopes; or if the combinations are alluvial–proluvial–eluvial–eolian, they dominate great tracts of the interior plains.

At the same time it has been shown that under particular geographical conditions (especially climatic ones), in a continental polygenetic layer of sediments and a neighboring accumulation of different origin there may be a convergent facies style or even uniform aspect.

Considering all these ideas that attempt to reconstruct the process of loess genesis, it becomes obvious that the accumulation of loess-like material cannot be controlled by a single universal factor for the whole widespread range of those deposits. Accordingly, we must not look for a universal and exclusive cause for loess genesis; it is necessary to determine, for each territorial unit studied, which continental sedimentation factors operated, and consequently which of them had the dominant role in the accumulation of the loess-like silt layer.

In this way, in the second decade of the present century there appeared the theory of polygenetic origin (Berg, 1916-1947) and afterward several other genetic classifications, offering more and more differentiation for typical loesses and secondary loess-like deposits: Berg, 1916-1947; Zhirmundski, 1925; Grahmann, 1932; Rungaldier, 1933; Polynov, 1934-1961; Gerasimov, 1939; Bondarchuk, 1939; Trofimov, 1946-1950; Reingard, 1947; Liteanu, 1952-1966; Cijan-Tzun-Hu, 1954-1960;

Naum and Grumazescu, 1954; Sümeghy, 1954; Fedorovich, 1955–1961; Iakovlev, 1955, Bulla, 1956; Tricart, 1956–1963; Nalivkin, 1956–1963; Molodyh, 1958–1962; Lukashev, 1961; Kriger et al., 1961; Mikfailov, 1963; and many other workers. many other workers.

V. THE THEORIES OF "LOESSIFICATION"

The development of the polygenetic origin theory was, undoubtedly, a very important step in the evolution of ideas about the problem of loess genesis. Nevertheless, this theory was not to explain satisfactorily the characteristics and the petrographic, structural, and lithological peculiarities of loess and loess-like deposits. It could not clarify either the nature of those common peculiarities which characterize them throughout their widespread occurrence or those distinctive properties which give them special features in different climatic subzones of their overall distribution. Also this theory could not explain how continental sediments with very different origin can acquire similar characteristics as well as the common structural and textural properties that uniquely typify loess and loess-like deposits.

The apparent uniformity of loesses was, in the last century and at the beginning of this century, one of the main arguments used by the partisans of the theory of the eolian origin of loess. Contemporary knowledge about loess suggests rather that most distinctive peculiarities of loess cannot be explained either by subaqueous sedimentation or by simple subaerial accumulation of eolian dust or of various alluvial, glacial, proluvial, or diluvial silts. We consider that loess has the following distinctive characteristics: (1) a general distribution over almost all types of topographic relief and overlying all varieties of geological formations; (2) a characteristic color (yellow) with variations; (3) a high uniform porosity and well-marked macroporosity; (4) a high content of secondary carbonates (especially $CaCO_3$) independent of source rock peculiarities; (5) an apparent lack of stratification, or very weak stratification, with a vertical-fissured structure observed when digging; (6) a relatively homogeneous granulometry (predominantly silt) and characteristic variations of the mechanical properties of various loess-like deposits; (7) a compaction capacity under wet conditions, with different degrees of settlement.

A study of present-day sedimentary processes and a comparison of recent continental sediments with different types of loess deposit has led us to the conclusion that the majority of characteristic properties of loesses are secondary peculiarities, which appeared either in the fresh sediments during their accumulation (syngenetic) or afterward (epigenetic). All these secondary processes were described by the term "loessification," by which we understand natural processes with a geographical zonal character. During this process the most various continental and superficial sediments of the loessification zone (very weakly linked or unlinked, predominantly silty sediments) undergo transformations that give them the common aspect of loess or loess-like deposits. The loessification zone coincides with the forest steppes and especially with cold or warm stepps, and therefore with arid or semiarid regions, adjacent to deserts or the peripheral and periglacial zone of great Quaternary glaciations.

The seeds of these ideas are found in the early period of studies of the classical theories of loess genesis. Charpantier (1841) and Foetterle (1853), who studied the loess of the Roumanian Plain, Wood (1882), Roth (1888), Kudriavtzev (1892),

Bogoslovski (1899), and Sibirtzev (1900–1901) have considered that loess is an epigenetic eluvium. Richthofen, Obruchev, and many others from the schools they founded, have spoken of the steppe-like setting and the soil formation from eolian dust; in this loessification process the grassy vegetation of the steppes played an active part. Wood maintained that loess is a diagenetic product, an altered product that evolved under the influence of freezing conditions; and Sibirtzev (1900–1911) maintained that loess is a product of soil formation processes.

Sibirtzey's opinion gained great popularity and many partisans after Berg's (1916–1948) theory made its appearance. He believed loess to be an epigenetic alluvium and specific product of soil formation. This process takes place in climatic conditions peculiar to the loessification zone. Berg was the first to publish a general synthesis of those physicochemical processes which represent the basis of loessification by soil formation. This opinion has gained more scientific ground since Dücker's researches (1937) (he considered loess to be a product of the periglacial zone), after Tavernier's papers (1948), which elaborated the niveo-eolian origin theory, but especially after Polynov's investigations (1934–1961), which introduced into science the concept of "alteration crust." Polynov, together with other partisans of the new direction (Perelman, 1955–1966; Lukashev, 1961; etc), performed valuable services in the development of the new theory. The theory of geochemical zonation of the altered soil crust opened new points of view for the explanation of the detailed mechanism of the loessification process.

Starting from these new discoveries it is apparent that the genesis of loess involves two natural complex processes, independent, simultaneous, or successive, but anyway convergent in the common results of their action:

a. The accumulation of silty sediment (the "raw material"), which can take place in many different ways during continental sediment genesis, being prevalently polygenetic but locally monogenetic.

b. The actual loessification process, that is, a physicochemical process evolving in simultaneously accumulated deposits and after the sedimentation of mineral and primary material.

To explain the physicochemical processes that are in fact the real content of the loessification process, we can mention the following main theories:

1. Protogenesis theory: Dücker, 1937; Gerasimov, 1939; Morozov, 1951; Glazovskaya, 1954; and others.

2. Eluvial-epigenetic theory or epigenetic soil formation: Berg, 1916–1947; Ganssen, 1922; Sokolov, 1923–1932; Neustruev, 1925; Münichsdörfer, 1926; Kölbl, 1931; Brodski and Samsonova, 1933; Gerasimov and Markov, 1939; Sokolovski, 1934; Russell, 1944; Trofimov, 1946–1950; Piashkovski, 1946–1962; Obruchev, 1948; Bolshakov, 1949; Msvenieradze, 1950; Rozanov, 1951–1952; Karlov, 1953; Popov, 1953; Lysenko, 1955–1962; Sokolovski, 1957–1961; Nikitenko, 1958–1964; and many other workers.

3. Theory of lithogenesis by retransformation from soil: Lichkov, 1945–1957.

4. Diagenesis theory: Wood, 1882; Münichsdörfer, 1926; Mavlianov, 1948–1960; Kriger, 1951–1962; Popov, 1953; Iakovlev, 1955; Bulla, 1956; and others, with the following main variants:

a. Theory of the periglacial alteration and breaking up under the influence of repeated freezing: suggested by Wood; then by Dücker, 1937; Markov, 1946–

1961; Kriger, 1951–1962; Vasilovski, 1951–1953; Dylik, 1952–1961; Kadar, 1952–1956; Pierzchalko, 1954; Kirchev, 1961; and others.

   b. Niveol-eolian theory: Tavernier, 1948; Jahn, 1950–1956; Marechal and Maarleveld, 1955; Tricart, 1956–1963; and others.

   c. Loess as a characteristic component of sialitic, carbonatic altered soil crust, produced by hypergenetic alteration: Polynov, 1934; Gerasimov, 1939; Rozanov, 1951–1952; Bulla, 1956; Lukashev, 1961; and others.

In our opinion, according to contemporary knowledge about the loess genesis, this last theory clarifies completely the physicochemical processes of sediment transformation, which take place by various progressive steps in unique processes of loessification, ending up with the appearance of typical loess. Evidently the typical loess stage may not be reached in every case. Depending on climatic conditions, the loessification may halt at the level of some intermediary stage of loessoid formation. Because diagenetic processes are reversible according to the theory of soil crust alteration, we may logically explain the processes of decomposition of loess-like deposits (deloessification). The phenomenon takes place frequently if the respective sediments remain for a very long period in unfavorable conditions (water-saturated soils) for the maintenance of the dynamic, lithological-structural, and geochemical equilibrium, which maintains the loess-like aspect of these rocks.

SELECTED BIBLIOGRAPHY

1. A g a s s i z L.. *Uber den Unsprung des Löss* (Brief an Prof. Geinitz). „Neues Jahrbuch für Min., Geol. und Pal.". 1867.
2. B a r b o u r G. B.. *Das Lössproblem in China.* „Leopoldina", **6,** (Festschrift für J. Walter), Leipzig. 1930.
3. B e n n i n g s e n - F ö r d e r R., *Beitrag zur Niveaubestimmung der drei nordischen Diluvialmeere.* „Z. d. Dtsch. geol. Ges.", **9,** 1857.
4. B e r g L. S.. *Klimat i jizn.* cap. IX, OGIZ, Moscova. 1947.
5. B o l ş a k o v A. F., *O ghenezise lëssa i lëssovidnih otlojenii.* „Pocivovedenie". 1949, nr. 6.
6. B o r d e s F., *Loess des Etats Unis et loess du Bassin de Paris.* „Anthropologie", **53,** nr. 3—4, 1955.
7. B r a u n A.. *Löss,* „Neues Jahrb für Min.. Geogn.. Geol. u. Petref., H. 1. Stuttgart, 1847.
8. B r ă t e s c u C.. *Profile cuaternare în falezele Mării Negre.* „Opere alese", p. 132—162. Bucureşti, 1967.
9. B r y a n K., *Glacial Versus Desert Origin of Loess.* „Am. J. Sci.", 1945, **243,** nr. 5.
10. B u l l a B.. *Der pleistozäne Löss im Karpathenbecken.* „Földtani Közlöny", 1937—1938. Budapest.
11. C a i l l e u x A., *Les loess et limons éoliens de France.* „Bull. de Serv. de la cart. géol. de la France", 1953, nr. 240, **51,** Paris et Liège, 1954.
12. D m i t r i e v N. I., *K voprosu o proishojdenii lëssa USSR.* „Ucen. zap. Harkovsk. Gos. Univ.", nr. 41, Tr. Gheogr. fak. **1,** 1952.
13. D o e g l a s D. J., *Loess. an Eolian Product.* „J. of Sedim. Petrol.", **22,** nr. 1, 1952.
14. D u b o i s G., *Les principaux types de limons en France septentrionale.* „Revue scient.", **69,** nr. 13, 1931.
15. D y l i k J., *Zagadnenie genezy lessu u Polscé.* „Biul. perygl.". nr. 1. 1954.
16. F e d o r o v i c i B. A.. *Voprosi proishojdenia lëssa v sviazi s usloviami ego rasprostranenia v Evrazii.* .,Tr. Inst. Gheogr. AN SSSR". **80,** 1960.
17. F i s k H. N., *Loess and Quaternary Geology of the Lower Mississippi Valley.* ..J. Geol.". **59,** nr. 4, 1951.
18. F l i n t R. F.. *Glacial Geology and the Pleistocene Epoch.* New York, 1947.
19. G a n s s e n R.. *Die Entstehung und Herkumft des Lösses.* „Mitt. aus d. Labor. d. preuss. geol. Landesanstalt", Berlin, 1922.
20. G e r a s i m o v I. P.. *Lëssoobrazovanie i pocivoobrazovanie.* „Izv. AN SSSR, seria gheol.", nr. 1. 1962.
21. G e r a s i m o v I. P., M a r k o v K. K., *Cetverticinaia gheologhia (paleogheografia cetverticinogo perioda).* „Ucipedghiz", Moscova, 1939.
22. G o r e ţ k i i G. I., *O periglaţialnoi formaţii.* „Biull. Kom. p. izuci. cetvert. perioada AN SSR", nr. 22. 1958.

* Cuprinde numai bibliografia de bază.

23. Grahmann, R., *Der Löss in Europa.* „Mitteilungen d. Ges. f. Erdkunde zu Leipzig", 1931.
24. Holmes Ch. D.. *Origin of Loess — a Criticism.* „Am. J. Sci.", 242, nr. 8, 1944.
25. Horner L., *Sur la géologie des environs de Bonn.* „Bull. Soc. Géol. France", 1936—1837, **8**, Paris, 1837.
26. Horusitzky H., *Über den diluvialer. Sumpflöss.* „Földt. Közl." **33**, f. 5—6, Budapest, 1903.
27. Iakovlev S. A., *Metodiceskoe rukovodstvo po izuceniu i gheologhiceskoi semke cetverticinîh otlojenii.* **1**, „Gosgheoltehizdat", Moscova, 1954.
28. Iakovlev S. A., *K voprosu o proishojdenii lëssa.* În cartea: „Voprosî gheologhii Azii", **2**, Ed. AN SSSR, 1955.
29. Ian-Cijun-Țzian, *Lëss Kitaia.* „Priroda", nr. 5, 1957.
30. Ian-Tze, *Ghenezis lëssovîh otlojenii Severnogo Kitaja.* „Tr. Kom. p. izuci. cetvert. perioada" AN SSSR, **14**, 1959.
31. Ilin R. S., *Proishojdenie lëssov v svete ucenia o zonah prirodî smeşciaiuşcihsia v prostranstve i vremeni.* „Pocivovedenie", nr. 1, 1935.
32. Jahn A., *Less, jego pochodzenie i zwiazek z klimatem epoki lodowej.* „Acta geol. Pol.", **1**, nr. 3, 1950.
33. Jentzsch A., *Über Baron v. Richthofens Lösstheorie.* „Verh. geol. Reichsanst", nr. 15, 1877.
34. Jirmundskii A. M., *K voprosu o proishojdenii turkestanskogo lëssa.* „Biull. MOIP otd. gheol.", nr. 3—4, 1925.
35. Kádár L., *A lösz keletkezése és pusztulása.* „Közlemények a debreceni Tud. Egyet. Földr. Intézetéből", nr. 19, **II (IV)** 3—4, Debrecen, 1954.
36. Keilhack K., *Das Rätsel der Lössbildung.* „Zeitschr. deutsch. geol. Gesellsch. Monatsber.", **72**, 1920.
37. Kerekes J., *Die periglacialen Bildungen Ungarns.* „M. All. Földt. Int. Évkönyve", **37**, n. 4, Budapest, 1948.
38. Kes A. S., *Lëssî i lëssovo-krasnoglinistie porodî kak eolovo-pocivennie obrazovania.* „Tr. kom. p. izuci. cetvert. perioada AN SSSR". **19**, 1962.
39. Kingsmill Th. W., *The Probable Origin of Deposits of „Loess" in North China and Eastern Asia.* „Quart. J. Geol. Soc. London". **27**, 1870.
40. Kölbl L., *Studien über den Löss.* „Mitteilungen d. Geogr. Ges. in Wien", 1930.
41. Kriger N. L., *Lëss, ego svoistva i sviaz s gheograficeskoi sredoi.* Ed. Nauka, Moscova, 1965.
42. Kriván P., *La division climatologique du Pléistocène en Europe Centrale et le profil de loess de Paks.* „M. All. Földt. Int. Évkönyve",**XLIII**, 3. Budapest, 1955.
43. Krokos V. I., *Less i fosilni grunti pivdenno-zahidno Ukraini.* „Vistn. silskogospodar. nauki", **III**, vîp. 3—4. Harkov, 1924.
44. Kropotkin P., *Issledovania o lednikovom periode.* „Zap. Russk. Gheogr. Obşc. po obşc. geogr.", 1876.
45. Kukla J., Lożek V., *Loesses and Related Deposits of Czechoslovakia.* „Prace Inst. Geol.", **34**, Czartozed Europy Srodkovej i Wschodniej. cz. 1, Warszawa, 1961.
46. Leighton M. M., Willman H. B. *Loess Formation of the Mississippi Valley.* „J. Geol.", **58**, nr. 6, 1950.
47. Leonard A. B., Frye J. C., *Ecological Conditions Accompanying Loess Deposition in the Great Plains Region of the United States.* „J. Geol.", 62, nr. 4, 1954.
48. Liteanu E., *Karta cetverticinîh otlojenii vnekarpatskoi ciasti Rumânskoi Narodnoi Respubliki.* „Biull. Kom. p. izuci. cetvert. perioda AN SSSR, nr. 23. 1959.
49. Lîsenko M. P., *Lëssovîe porodî evropeiskoi ciasti SSSR.* Ed. Univ. Leningrad. 1967.

50. Lomonovici M. I., *Ghenezis lëssa iugo-vostocinogo Kazahstana na primere Zailiiskogo Alatau.* „Tr. Kom. p. izuci. cetvert., perioda AN SSSR", **13**, 1957.
51. Lukaşev K. I., *Problema lëssov v svete sovremennih predstavlenii.* Ed. AN BSSR, Minsk. 1961.
52. Lukaşev K. I., *Ocerki po gheohimii ghipergheneza.* Ed. AN BSSR. Minsk, 1963.
53. Lyell Ch., *Observations on the Loamy Deposit Called „Loess" of the Bassin of the Rhine.* „Edinburgh New Phil. Journ.", XVII, nr. 33—34.
54. Maarleveld G. C., *Sur les sédiments périglaciaires en Hollande: formes et phénomènes.* „Biull. perygl.", nr. 4, 1956.
55. Malicky A., *Guide-book of Excursion. E. The Lublin Upland. Symposium on Loess.* INQUA, Lodz, 1961.
56. Maréchal R., *L'étude des phénomènes périglaciaires en Belgique.* „Biul. perygl.". nr. 4, 1956.
57. Markov K. K., Lazukov G. I., Nikolaev B. A., *Cetverticinii period.* II, Cap. I, Ed. Univ. Moscova, 1965.
58. *Materiali vsesoiuznogo soveşciania po izuceniu cetverticinogo perioda* (culegere). I—III, Ed. AN SSSR, Moscova, 1961.
59. Mavlianov G. A., *O proishojdenii lëssa i lëssovidnih porod iujnih raionov Srednei Azii.* „Mat. p. cetvert. periodu SSSR", Vîp. 2. Ed. AN SSSR, 1950.
60. Märzbacher G., *Die Frage der Entstehung des Lösses.* „Petr. Geogr. Mitt.", Bd. 59. I—III, 1913.
61. McGee W. I., Call R. E., *On the Loess and Associated Deposits of Des Moins.* „Am. J. Sci.", 24, sept. 1883.
62. Mihályi-Lányi I., *Klassifikaţia vengherskih raznovidnostei lëssa i procih obrazovanii sîpucei piii.* „Acta Geol. Ac. Sci. Hung.", 2, nr. 1—2, Budapest, 1953.
63. Millette I. F .G., Higbee H. W., *Periglacial Loess I, Morphological Properties.* „Am. J .Sci.", 256, nr. 4. 1958.
64. Mojski J. E., *Periglacial Deposits and Structures in the Stratigraphy of the Quaternary in Poland.* „Prace Inst. Geol.", 34, Czwartozęd Europy Srodkovej i Wschodniej, cz. 2, Warszawa, 1961.
65. Morozov S. S., *Klassifikaţia lëssovih porod.* În cartea: „Inj.-gheol. svoistva gornîh porod i metodi-ih izucenia". Ed. AN SSSR, 1962.
66. Murgoci G. M., *Cimpia Romănă şi Balta Dunării.* „Ghidul de excursii III, Congr. Internaţ. petrol", 1907 şi în „Opere alese", Bucureşti. 1957.
67. Münichsdörfer F., *Der Loess als Bodenbildung.* „Rundschau", XVII, 1926.
68. Nalivkin D. V., *Ucenie o faţiah.* II, Ed. AN SSSR, Moscova-Leningrad. 1956.
69. Naum Tr. şi Grumăzescu H., *Problema loessului.* „Probl. de Geografie", I, Bucureşti. 1954.
70. Nehring A., *The Fauna of Central Europe during the Period of the Loess-Rejoinder to Mr. H. H. Howorth.* „Geol. mag.", NS., dec. 2, 10, nr. 2, 1883.
71. Neustruev S. S., *Pocivennaia ghipoteza lëssoobrazovania.* „Priroda", 1925 nr. 1—3.
72. Obrucev V. A., *K voprosu o proishojdenii lëssa (v zaşitu eolovoi ghipotezî).* Izv. Tomsk. tehnol. Inst.", XXIII, 1911, nr. 3.
73. Obrucev V. A., *Loess Types and their Origin.* „Am. J. Sci." 213, nr. 5, 1945.
74. Obrucev V. A., *Lëss kak osobii vid pocivi. ego proishojdenie, tipi i zadaci izucenia.* „Mat. p. cetvert. periodu SSSR" vîp. 2, Ed. AN SSSR. Moscova-Leningrad. 1950.
75. Obrucev V. A., *Kak i kogda voznikla i slojilas eolovaia teoria proishojdenia lëssa.* „Tr. Inst. Gheol. Nauk AN SSSR, seria gheomorf. i cetvert. gheol.", vîp. 1, 1957.
76. Oprea C. V., Contrea A., *Contribuţii la cunoaşterea formării şi răspîndirii loessului în partea de vest a ţării.* „Studii şi cercet. şt.", Ac. R.P.R. Baza Timişoara. Seria agric., 3, nr. 1—2, 1956.

77. P a v l o v A. P., *Stati po gheomorfologhii i prikladnoi gheologhii* (Opere alese, **II,**), Ed. MOIP, Moscova, 1951.
78. P ă v a i - V a j n a F., *Uber den Löss des sibenbürgischen Beckens.*, Jahresber. Ungar. geol. Reichsanst", f. 1909, Budapest, 1912.
79. P e n c k A., *Das Klima der Eiszeit.* Verh. der III, Internac. Quartär-Konferenz, Wien, Sept. 1936, Wien, 1938.
80. P e n n i s t o n J. B., *Note on the origin of loess.* „Popular Astronomy" **39**, nr. 7, 1931.
81. P e n n i s t o n J. B., *Additional Note on the Origin of Loess.* „Popular Astronomy", **51**, nr. 3, **1943**.
82. P e r e l m a n A. I., *Gheohimia landsafta.* „Gheografghiz", Moscova, 1961.
83. P e r e l m a n A. I., *Gheohimia epigheneticeskih proţessov (zona ghipergheneza).* Ed. Vîsşaia şcola, 1961.
84. P i a ş k o v ş k i i B. V., *Lëss kak glubokopocivennoe obrazovanie.* „Pocivovedenie", nr. 11, 1964.
85. P i a ş k o v s k i i B. V., *Problema lëssov v ee istoriceskom razvitii.* „Vopr. gheogr.", **24**, „Gheografghiz"., 1951.
86. P o l î n o v B. B., *Kora vivetrivania. I.* Ed. AN SSSR., Leningrad, 1934.
87. P o p o v V. I., *Faţialnoe razvitie osadkov gornîh sklonov i podgornîh pustinnih ravnin.* „Mat. p. cetvert. periodu SSSR", vîp. 2, Ed. AN SSSR, Moscova-Leningrad, 1950.
88. P o p o v V. V., *Klassifikaţia lëssovîh porod.* „Tr. Inst. gheol. nauk AN SSSR, seria gheomorf. i cetvert. gheol.", vîp. 1, 1957.
89. P o p o v V. V., *Osnovnie prinţipi podhoda k gheneticeskoi klassifikaţii porod lëssovoi formaţii.* „Tr. 2. Uzbekist. ghidro-gheol. soveşci.", 1958.
90. P o s e r H., *Severnaia graniţa lëssa i pozdnelednikovîi klimat.* „Vopr. gheol, cetvert. perioda', Ed. I. Moscova, 1955.
91. P r e s t w i c h J., *On a possible marine origin of the Loess.* „Geol. mag.", ser. 4. **1**, 1894.
92. P r i k l o n s k i i V. A., *Injenerno-gheologhiceskaia klassifikaţia lëssovîh porod.* „Tr. soveşci. p. inj.-gheol. svoistvam gornîh porod i metodam ih izuci.", **2**. Ed. AN SSSR, 1957.
93. P u m p e l l y R., *The relation of secular rockdesintegration to loess, glacial drift and rock basins.* „Am. J. Sci. and Arts.", **17**, New Haven, 1879.
94. Q u i r i n g H., *Herkunft, Aussprahe und Schreibung des Wortes „Löss".* Z. d. Dtsch. geol. Ges.", **88**, H. 3, 1936.
95. R a s k a t o v G. I., *O ghenezise lëssov Predkarpatskoi ravnini.* „Tr. Gheol. Fak. Voronejsk. Gos. Univ", **39**, 1955.
96. R a t h j e n s C., *Löss in Tripolitanien.* „Z. Ges. f. Erdkunde zu Berlin" 1928.
97. R e m y H., *Der löss am unteren Mittel — und Niederrhein.* „Eiszeit u. Geol." **11**, 1960.
98. R i a b c e n k o v A. S., *O proishojdenii lëssa i lëssovidnih porod Russkoi Ravnini v svete mineraloghiceskih dannîh.* „Biull. MOIP. otd. gheol.", **35**, vîp., 2. 1960.
99. R i c h t h o f e n F., *China. I,* Berlin, 1877.
100. R i c h t h o f e n F., *Bemerkungen zur Lössbildung.* „Verh. geol. Reichsanst.", nr. 13, 1878.
101. R i c h t h o f e n F., *On the mode of origin of the loess.* „Geol. Mag.", dec. 2, **9**, nr. 7, 1882.
102. R o m o d a n o v a A. P., *Lëssovîe porodi Pricernomoria.* „Tr. Inst. Gheol. Nauk AN USSR, seria gheomorf. i cetvert. gheol.", Vîp. 1, 1957.
103. R u h i n L. B., *Bazele litologiei.* Ed. Teh. Bucureşti, 1966.
104. R u n g a l d i e r R., *Bemerkungen zur Lössfrage, bezonderns im Ungarn.* „Z. f. Geomorph.", 8, 1933.
105. R u s s e l R. J., *Lower Mississippi Valley loess.* „Bull. Geol. Soc. Am.", **55**, nr. 1, 1944.
106. R u s s e l R. J., *Origin of loess — a reply.* „Am. J. Sci.", **242**, nr. 8, 1944.

107. S a m o d u r o v P. I., *Gheohimiceskaia suscinost lëssoobrazovatelnogo proţessa.* „Tr. Inst. Geol. Nauk AN USSR, seria gheomorf. i cetvert. gheol.", vîp. 1, 1957.
108. S a m s o n P., R ă d u l e s c u C., *Beiträge zur Kenntniss der Chronologie des „Jungeren Lösses" in Dobrudscha (Rumänische Volksrepublik).* „Eiszeit u. Geol.", **10**, 1959.
109. S a w i c k i L., *Warunki kliṃaticzne akumulacji lessu mlodszego w swietle wynikow badan stratigraficznych stanowiska paleolitycznego lessowego na Zwierzyncu w Krakowie.* „Biull. Panstw. Inst. Geol. n. 66, Z. badan czwartorzedu w Polsce", **2**, 1952.
110. S c h e i d i g A., *Der Löss und seine geotechnischen Eigenschaften.* Dresden und Leipzig, 1934.
111. S c h u l t z C. B., S t o u t T. M., *Pleistocene loess deposits of Nebraska.* „Am. J. Sci.". **243**, nr. 5, 1945.
112. S e d l e ţ k i i I. D., *Kolloidno-dispersnie minerali i eolovoe proishojdenie lëssa nijnego Dona.* DAN SSSR, **84**, nr. 5, 1951.
113. S e d l e ţ k i i I. D., A n a n i e v V. P., K u ţ e n k o A. E., *Sostav i proishojdenie lëssa* **Vengrii.** DAN SSSR, **94**, nr. 5, 1954.
114. S e k y r a J., *Periglacial phenomena (CSSR).* „Pracе Inst. geol.", **34**, Czartorzed Europy Srodkovej i Wschodniej. cz. 1, Warszawa, 1961.
115. S e r g e e v E. M., M i n e r v i n A. V., *Suscinost proţessa oblëssovania v podzolistoi zone.* „Vestnik MGU, Gheologhia", nr. 3, 1960.
116. S h i m e k B., *Loess as lithological term.* „Bull. Geol. Soc. Am.", **23**, nr. 4, 1912.
117. S m i t h G. D., *Illinois loess: variations in ist properties and distribution — a pedologic interpretation.* „Bull. Illinois Agr. Exp. Sta.", **490**, 1942.
118. S m i t h H. T. U., *Periglacial frost features and related phenomena in the United States.* „Biul. perygl.", nr. 11 ,1962.
119. S o e r g e l W., *Lösse, Eiszeiten und paläolitische Kulturen. Eine Gliederung und Altersbestimmung der Lösse.* Jena, 1919.
120. S o k o l o v s k i i A. N., *Lëss kak produkt vivetrivania i pocivoobrazovania.* „Pocivovedenie", nr. 9-10, 1943.
121. S o k o l o v s k i i A. N., *Znacenie fiziko-himiceskih svoistv lëssa dlia poznania ego ghenezisa.* „Tr. Inst. Geol. Nauk AN USSR, seria gheomorf. i cetvert. gheol.", vîp. 1, 1957.
122. S o k o l o v s k i i I. L., *O znacenii mineraloghiceskogo analiza lëssovih porod zapadnoi ciasti USSR dlia ustanovlenia sposoba ih obrazovania.* „Cetvert. period", vîp. 13—15. 1961.
123. S o k o l o v s k i i I. L., *Reghionalnîe i gheneticeskie tipî lëssovih porod.* „Cervert. period.", vîp. 13—15, 1961.
124. *Sovremennîi i cetverticinîi kontinentalnîi litoghenez* (culegere de articole sub red. E. V. Şanţer). Articolele lui I. P. G h e r a s i m o v şi colab., A. G. C e r n i a k o v s k i i, A. S. K e s, N. I. K r i g e r, G. I. G o r e ţ k i i, K. I. L u k a ş c v şi colab. (U.R.S.S.), I. F i n k (Austria), M. P é c s i (R.P.U.), H, T. U. S m i t h (S.U.A.) V. L o ž e k (C.S.R.), E. L i t e a n u (R.S.R.), A. M a l i c k i (R.P.P.). Ed. Nauka, Moscova. 1966.
125. S t r a h o v N. M., *Osnovî teorii litogheneza.* **I—III,** Ed. AN SSSR, Moscova. 1960—1962.
126. Ş t e f a n o v i c s P., K l é h G., S z ö c s L., *A paksi löszfal anyagának talajtani vizsgálata.* „Agrokémia és talajtan", **3**, n. 4. 1954.
127. S u e s s E., *Uber den Löss.* „Jahrbuch f. Min. Geol. u, Pal.", 1867.
128. S w i n e f o r d A., F r y e J. C., *Petrographic comparison of some loess samples form Western Europe with Kansas Loess.* „J. of Sedim. Petrol.", **25**, nr. 1, 1955.
129. Ş a n ţ e r E. V., *Gheneticeskie tipî cetverticinih kontinentalnih osadocinih obrazovanii.* „Mat. p. cetvert. periodu SSSR", vîp. 2, Ed. AN SSSR, 1950.
130. Ş a n ţ e r E. V., *Nekotorîe voprosi gheologhii antropoghena ţentralnoi ciasti velikih ravnin S.S.A. (S.U.A.) i sopredelnih raionov.* VII. kongr. mej. nar. Assoţ. p. izuci. Cetvert. perioda — INQUA (SUA, 1965). Naucinîe itogi i materialî. Ed. Nauka, Moscova. 1967.
131. T a v e r n i e r R., *Les formations quaternaires de la Belgique, en raport avec*

l'évolution morphologipue du pays. „Bul. Soc. belg., géol. paleont., hydrol.", **57**, 1937.

132. T c r u g g i M. E., *The nature and origin of argentine loess.* „J. of Sedim. Petrol.", **27**, nr. 3, 1957.

133. T h o r p J., *Significance of loess in classification of soils.* „Am. J. Sci.", **243**, nr. 5. 1945.

134. T o d d J. E., *Richthofens theory of the loess in the light of the deposits of the Missouri.* „Am. J. Sci. and Arts.", ser. 3, **18**, nr. 103—108, 1879.

135. T o d d J. E., *Degradation of loess.* „Proc. Jowa Acad. Sci. for 1897", nr. 5, 1898.

136. T r o f i m o v I. I., *Kontinentalnii litoghenez v pustiniah i smejnih s nimi predgornih zonah.* „Mat. p. cetvert. periodu SSSR", vîp. 2, Ed. AN SSSR, 1950.

137. T r i c a r t J., *Carte des phénomènes périglaciaires quaternaires en France.* „Bull. Serv. Carte Géol. Fr.", 1956.

138. T r i c a r t J., *Géomorphologie des régions froides.* Paris, 1963.

139. T u t k o v s k i i P., *Ob ozernom i subaeralnom lësse iugo-zapadnoi ciasti Lutkogo uezda.* „Ejegod. gheol. i miner. Rossii", **II**, 1897.

140. T u t k o v s k i i P., *K voprosu o sposobe obrazovania lëssa.* „Zemlevedenie", nr. 1—2, 1899.

141. U n g e r K. P., R a u D., *Gliederung und Altersbestellung der Lössablagerungen im Thüringer Becken und dessen Randgebiet.* „Prace Inst. Geol.", **34**, Czwartozed Europy Srodkowej i Wschodniej cz. 1, Warszawa, 1961.

142. V a l o c h K., B o r d e s F., *Loess de Tchecoslovaquie et loess de France du Nord.* „Anthropologie", 1957, **61**, nr. 3—4, Praha, 1958.

143. V i s l o u h I. K., *Lëss. Ego znacenie i proishojdenie.* „Izv. Russk. gheogr. obsc.", vîp. 2, „Disskussia", vîp. 4, 1915.

144. W a h n s c h a f f e F., *Die lössartigen Bildungen am Rande des norddeutschen Flachlandes.* „Zeitschr. deutsch. geol. Gesell.", **XXXVIII**, H. 2, 1896.

145. W e r v e k e L., *Uber die Entstehung der lothringischen Lehme und des mittelrheinischen Lösses.* „Stiz. ber. Heidelberger Akademie Wiss. math. nat. Kl.", Abt. A., nr. 5, 1924.

146. W i l l i s B., *Quaternary Huang-tu formation Northwestern China.* „Research in China", 1907, **1**, part. 1. Descriptive topography and geology by Bailey Willis, Eliot Blackwelder and R. H. Sargent. Chapter X, Washington.

147. W i n c h e l l N. H., *The loess of Minnesota.* „Am. J. Sci. and Arts.", ser. 3, **17**, I—VI, 1879.

148. W i n t e r H., *Zur Frage der Lössbildung.* „Z. d. Dtsch. geol. Ges.", 86, 1934.

149. W o l d s t e d t P., *Das Eiszeitalter.* Stuttgart, 1929.

150. W o l f H., *Uber die Diluvialablagerungen in dem östlichen Theile Galiziens zwischen Rzeszow und Lemberg.* „Verh. geol. Reichsanst.", Januar, 1866.

151. W o l f H., *Geologisch-geographische Skizze der niederungarischen Ebene.* „Jahrb. d. geol. Reichsanst.", **17**, Wien, 1867.

152. W o o d S. V., *On the origin of the loess.* „Geol. mag.", nr. 218, N. S. dec. 2, **9**, nr. 8, 1882.

153. W r i g h t G. Fr., *Origin and distribution of the loess.* „Bull. Geol. Soc. Am.", **32**, nr. 1, 1921.

154. Z e u n e r F., *The Pleistocene Period.* London, 1959.

155. Z e u s c h n e r L., *Uber den Löss in den Bieskiden und im Tatragebirge.* „Jahrb. d. geol. Reichanst.", **2**, nr. 1, Wien, 1851.

AUTHOR CITATION INDEX

SUBJECT INDEX

About the Editor

IAN JAMES SMALLEY is a lecturer in engineering geology and soil mechanics at the University of Leeds. Before going to Leeds in 1968, he was a lecturer at City University, London, from 1959 to 1966 and then spent two years in a full-time research post at University College, London, working on experimental petrology.

Dr. Smalley completed his undergraduate work at Battersea Polytechnic in 1959 and received his Ph.D at City University, London, in 1966. While at City University he began his work on glacial sediments, with emphasis on the formation of loess material. His current major concerns include continued work on loess, particularly in the Danube basin and associated regions; the study of postglacial quickclays of Canada and Scandinavia; and the investigation of till phenomena, especially drumlins and mineralogy. His major teaching commitment is to the Leeds M.Sc. in engineering geology.